ESSENTIALS OF PIEZOELECTRIC ENERGY HARVESTING

ESSENTIALS OF PIEZOELECTRIC ENERGY HARVESTING

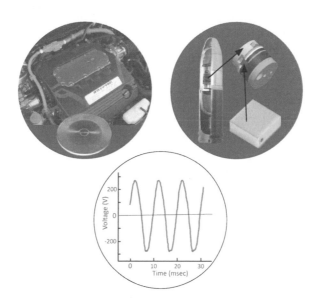

Kenji Uchino

The Pennsylvania State University, USA

World Scientific

NEW JERSEY · LONDON · SINGAPORE · BEIJING · SHANGHAI · HONG KONG · TAIPEI · CHENNAI · TOKYO

Published by

World Scientific Publishing Co. Pte. Ltd.

5 Toh Tuck Link, Singapore 596224

USA office: 27 Warren Street, Suite 401-402, Hackensack, NJ 07601

UK office: 57 Shelton Street, Covent Garden, London WC2H 9HE

Library of Congress Cataloging-in-Publication Data

Names: Uchino, Kenji, 1950–   author.

Title: Essentials of piezoelectric energy harvesting / Kenji Uchino,
  The Pennsylvania State University, USA.

Description: New Jersey : World Scientific, [2021] | Includes bibliographical references and index.

Identifiers: LCCN 2021005228 | ISBN 9789811234637 (hardcover) |
  ISBN 9789811234644 (ebook for institutions) | ISBN 9789811234651 (ebook for individuals)

Subjects: LCSH: Energy harvesting. | Piezoelectric devices. |
  Electric currents--Mathematical models.

Classification: LCC TK2897 .U35 2021 | DDC 621.042--dc23

LC record available at https://lccn.loc.gov/2021005228

British Library Cataloguing-in-Publication Data

A catalogue record for this book is available from the British Library.

For any available supplementary material, please visit
https://www.worldscientific.com/worldscibooks/10.1142/12219#t=suppl

Desk Editors: Balamurugan Rajendran/Steven Patt

Typeset by Stallion Press
Email: enquiries@stallionpress.com

Preface

Recent books on "Renewable Energy" discuss "piezoelectric energy harvesting" occasionally as one of the new energy sources in parallel to "windmills" and "solar cells", the plants of which generate huge electric power in Mega ~ Giga watts. However, the power level of the piezo harvesting device is limited up to 100 W, above which the competitive "electromagnetic generators" have the better privilege in efficiency. Thus, the author believes that the target of "piezo energy harvesting" should not be aligned on the large energy cultivation line, but is to be set on "elimination of batteries", which are classified as hazardous wastes, but recycled by less than 0.5% of 10 s Billion batteries sold per year in the world. Thus, "piezo energy harvesting" is one of the very important technologies in the current "Sustainable Society" from this hazardous waste elimination viewpoint.

The author is known as the "Pioneer of Piezoelectric Actuators", who started to use piezoelectric ceramics to move an object mechanically in the 1970s. When I was working on the vibration control by using piezoelectric actuators in the 1980s, we invented various types of "piezoelectric passive damping mechanisms". That is, the mechanical noise vibration can generate electrical charge on a piezo ceramic via "direct piezoelectric effect". If this electric energy is dissipated, the noise vibration can be reduced significantly. We have developed multiple "damper devices" for precision machinery and automobile engine mats. However, in the 1990s, I realized that just electric energy dissipation was useless. Why not accumulate the converted electric energy into a rechargeable battery? This is the starting point of "piezoelectric energy harvesting devices" historically. Because our patents submitted during the 1980s–1990s already expired, the reader is now free to use the basic idea on "cultivating electric energy from a piezoelectric generated by mechanical vibration".

As one of the pioneers in the piezoelectric energy harvesting, I feel a sort of frustration on 90% of the recent research papers from the following points:

(1) Though the electromechanical coupling factor k is the smallest (i.e., the energy conversion rate from the input mechanical to electric energy is the lowest) among various piezo device configurations, a majority of researchers primarily use the "unimorph" design. Why?

(2) Though the typical noise vibration is in a much lower frequency range, the researchers measure the amplified resonance response (even at a frequency much higher than 1 kHz) and report the "unrealistically" harvested electric energy. Why?

(3) Though the harvested energy is much lower than 1 mW, which is lower than the required electric energy to operate a typical energy harvesting electric circuit with a DC/DC converter (typically around 2–3 mW), the researchers report the result as an energy "harvesting" system. Does this situation mean actually energy "losing"? Why?

(4) Few papers have reported successive energy flow or exact efficiency from the input mechanical energy to the final electric energy in a rechargeable battery via the piezoelectric transducer step by step. Why?

Interestingly, the unanimous answer from these researchers to my question "why" is "because the previous researchers did so"!

My purpose of authoring this textbook is to provide the reader solid theoretical background of piezoelectrics, practical material selection, device design optimization and energy harvesting electric circuits, in order to stop your above "Google Syndrome", and to look forward to the future perspectives in this field. Therefore, I focused on important and fundamental ideas to understand how to design and develop the piezoelectric energy harvesting devices. Many of the studies and example problems cited in this textbook are intentionally from our group's lab notes, by intentionally removing "my" frustrating recent 90% papers, in order to keep the consistency of our development philosophy. Thus, this book is NOT an overall review of this area, which remains for the reader's further search for other scholars' approaches.

Let me introduce the contents. Chapter 1 introduces the overall "Background of Piezoelectric Energy Harvesting". The following two chapters, Chapter 2 "Fundamentals of Piezoelectrics" and Chapter 3 "Principle of Piezoelectric Passive Dampers", summarized the basic prerequisite knowledge on the piezoelectrics for junior researchers, without reading other textbooks. Through the "passive dampers", the reader learns how to dissipate (or cultivate) the electrical energy efficiently, that is, electrical impedance matching. Thus, senior researchers with this knowledge may skip these chapters. Chapter 4 "Mechanical-to-Mechanical Energy Transfer" is the starting chapter of the main content. Not all the mechanical vibration energy will transmit into the piezoelectric devices. You will learn the mechanical impedance matching concept. The "piezoelectric device designing" principle is discussed in Chapters 5 "Mechanical-to-Electrical Energy Transduction". The key to increasing the electromechanical coupling factor in composite structures is demonstrated in this chapter. Chapter 6 treats "Electrical-to-Electrical Energy Transfer", where we provide the principle of designing the electric energy harvesting circuits. Chapter 7 "Case Studies on Energy Flow Analysis" is one of this book's highlights, in which we will discuss the major problems in the current piezoelectric energy harvesting systems. "Hybrid Energy Harvesting Systems" in Chapter 8 introduces hybrid-type electric energy harvesting systems, such as a vibration energy and electromagnetic energy combination. The final Chapter 9 "Conclusions and Future Perspectives" addresses the three development directions: (1) higher electromechanical coupling device utilization, (2) paralleling technology of thousands of MEMS energy harvesting devices and (3) lower energy consumption harvesting-circuit designs, in order to achieve practical piezoelectric energy harvesting systems.

This textbook was written for undergraduate and graduate students, university researchers and industry engineers studying or working in the fields of "piezoelectric energy harvesting systems". This text is designed for self-learning by the reader by himself/herself aided by the availability of the following:

- Chapter Essentials
- Check Points (Quick Answer in this book's Appendix)
- Example Problems ("Solution" is provided successively)
- Chapter Problems ("Hint" available from the Author via e-mail below).

Since this is the first edition, critical review and content/typo corrections on this book are highly appreciated. Send the information directed to Kenji Uchino at 135 Energy and The Environment Laboratory, The Pennsylvania State University, University Park, PA 16802-4800. E-mail: KenjiUchino@PSU.EDU.

For the reader who needs detailed information on ferroelectrics, smart piezoelectric actuators and sensors, *Ferroelectric Devices 2nd Edition* (2010), *Micromechatronics 2nd Edition* (2019) authored by K. Uchino, published by CRC Press, are recommended. Further, *FEM and Micromechatronics with ATILA*

Software published by CRC Press (2008) is a perfect tool for practical device designing, with three losses in simulation. ATILA FEM software code is available from Micromechatronics Inc., State College www.mmech.com.

Even though I am the sole author of this book, it nevertheless includes the contributions of many others. I express my sincere gratitude to my former visiting researchers, associates and graduate students who worked on "piezoelectric energy harvesting" in the Penn State research center: Doctors Kazumasa Ohnishi, Alfredo Vazquez Carazo, Jungho Ryu, Shashank Priya, Hyeoungwoo Kim, Shuxiang Dong, Seyit O. Ural, Gareth J. Knowles, Heath F. Hofmann, and Mr. Amit Batra. I am also indebted to the continuous research fund from the US Office of Naval Research Code 332 during 1991–2021 without intermission through the grants N00014-96-1-1173, 99-1-0754, 08-1-0912, 12-1-1044, 17-1-2088 and 20-1-2309. Finally, my best appreciation goes to my wife, Michiko, who constantly encourages me in my activities.

Kenji Uchino, MS, MBA, Ph.D.
April, 2020 at State College, PA

About the Author

Kenji Uchino, one of the pioneers in piezoelectric actuators and energy harvesting systems, is the Director of International Center for Actuators and Transducers (ICAT), Professor of Electrical Engineering and Materials Science & Engineering, and also currently a Distinguished Honors Faculty at Schreyer Honors College at the Pennsylvania State University. He has been a university professor for 46 years so far, including 18 years in Japanese universities. He has also been a company executive (president or vice president) for 21 years in four companies, most recently the Founder and Senior Vice President of Micromechatronics Inc., a spin-off company from the above ICAT, where he was elaborating to commercialize the ICAT-invented piezo-actuators, transducers and energy harvesters. He was a Government officer for 7 years in both Japan and the US, and recently a "Navy Ambassador to Japan" (2010–2014) for assisting the rescue program for the Big Northern Japan Earthquake. He is currently teaching "Ferroelectric Devices", "Micromechatronics", "FEM Application for Smart Materials" and "Entrepreneurship for Engineers" for the Engineering and Business School graduate students using the textbooks authored by himself, titled the same.

After being awarded his Ph.D. degree from Tokyo Institute of Technology, Japan, Uchino became Research Associate/Assistant Professor in the Physical Electronics Department in 1976 at this university. Then, he joined Sophia University, Japan, as an Associate Professor in Physics in 1985. He was then recruited to Penn State in 1991 under a strong request from the US Navy community. He was also involved with the Space Shuttle Utilizing Committee in NASDA (current JAXA, equivalent to the US NASA), Japan, during 1986–1988, and was the Vice President of NF Electronic Instruments, USA, during 1992–1994. He has his additional Master's degree in Business and Administration from St. Francis University, PA. He has been consulting more than 135 Japanese, US and European industries to commercialize the piezoelectric actuators and electro-optic devices. He was the Chair of Smart Actuator/Sensor Study Committee partly sponsored by the Japanese Government, MITI (1987–2014). He was also the associate editor for *Journal of Materials Technology* (Matrice Technology) and the editorial board member for *Journal of Ferroelectrics* (Gordon & Breach) and *Journal of Electroceramics* (Kluwer Academic), and is currently Editor-in-Chief of *Journal of Insight-Material Science* (PiscoMed Publishing) and Editor-in-Chief of *Journal of Actuators* (MDPI). He also served as an Administrative Committee member for IEEE, Ultrasonics, Ferroelectrics, Frequency Control Society (1998–2000), and Secretary of American Ceramic Society, Electronics Division (2002–2003).

His research interests are in solid state physics — especially dielectrics, ferroelectrics and piezoelectrics, including basic research on materials, device designing and fabrication processes, as well as development of solid state actuators for precision positioners, ultrasonic motors and energy harvesting systems. He has authored 582 papers, 78 books and 33 patents in the piezoelectric actuator and optical device area. He has been a Fellow of the American Ceramic Society from 1997, a Fellow of IEEE since 2012, a Senior Member of National Academy of Inventors since 2019, and is also a recipient of 30 awards, including the Wilhelm R. Buessem Award from the Center for Dielectrics and Piezoelectrics, The Penn State University (2019), Distinguished Lecturer of the IEEE UFFC Society (2018), International Ceramic Award from Global Academy of Ceramics (2016), IEEE-UFFC Ferroelectrics Recognition Award (2013), Inventor Award from Center for Energy Harvesting Materials and Systems, Virginia Tech (2011), Premier Research Award from The Penn State Engineering Alumni Society (2011), the Japanese Society of Applied Electromagnetics and Mechanics Award (2008), the SPIE Smart Product Implementation Award (2007), R&D 100 Award (2007), ASME Adaptive Structures Prize (2005), Outstanding Research Award from Penn State Engineering Society (1996) and Best Paper Award from Japanese Society of Oil/Air Pressure Control (1987).

In addition to his academic carrier, Uchino is an honorary member of KERAMOS (National Professional Ceramic Engineering Fraternity) and has obtained the Best Movie Memorial Award as the director/producer in the Japan Scientific Movie Festival (1989) on several educational video tapes on "Dynamical Optical Observation of Ferroelectric Domains" and "Ceramic Actuators".

Contents

List of Symbols

D	Electric displacement
E	Electric field
P	Dielectric polarization
P_s	Spontaneous polarization
α	Ionic polarizability
γ	Lorentz factor
μ	Dipole moment
ε_o	Dielectric permittivity of free space
ε	Dielectric permittivity
$\varepsilon\ (\varepsilon_r)$	Relative permittivity or dielectric constant (assumed for ferroelectrics: $\varepsilon \approx \chi = \varepsilon - 1$)
κ	Inverse dielectric constant
χ	Dielectric susceptibility
C	Curie–Weiss constant
T_0	Curie–Weiss temperature
T_C	Curie temperature (phase transition temperature)
G	Gibbs free energy
A	Helmholtz free energy
F	Landau free energy density
x	Strain
x_s	Spontaneous strain
X	Stress
s	Elastic compliance
c	Elastic stiffness
v	Sound velocity
d	Piezoelectric charge coefficient
h	Inverse piezoelectric charge coefficient
g	Piezoelectric voltage coefficient
M, Q	Electrostrictive coefficients
k	Electromechanical coupling factor
η	Energy transmission coefficient
Y	Young's modulus
μ	Friction constant
$\tan \delta\ (\tan \delta')$	Extensive (intensive) dielectric loss
$\tan \phi\ (\tan \phi')$	Extensive (intensive) elastic loss
$\tan \theta\ (\tan \theta')$	Extensive (intensive) piezoelectric loss

Prerequisite Knowledge Check

Studying of "Piezoelectric Energy Harvesting" assumes certain basic knowledge. Answer the following questions by yourself prior to referring to the answers on the next page.

Q1 Provide definitions for the *elastic stiffness, c*, and *elastic compliance, s*, using stress (X) – strain (x) equations.

Q2 Sketch a *shear stress* (X_4) by arrows and the corresponding *shear strain* $(x_4)/deformation$ on the square material depicted below.

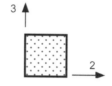

Q3 Describe an equation for the *velocity of sound, v*, in a material with mass density, ρ, and elastic compliance, s.

Q4 Given a rod of length, L, made of a material through which sound travels with a velocity, v, describe an equation for the *fundamental extensional resonance frequency, f_R*.

Q5 When two solid materials are contacted, and moved along the contact plane, friction force is introduced. How do you describe the friction force F in terms of the force N normal to the contact plane and the friction constant μ?

Q6 Provide the capacitance, C, of a capacitor with area, A, and electrode gap, t, filled with a material of *relative permittivity, ε_r*.

Q7 Describe an equation for the *resonance frequency* of the circuit pictured below:

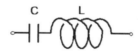

Q8 Given a power supply with an internal impedance, Z_0, what is the optimum circuit impedance, Z_1, required for maximum power transfer?

Q9 Calculate the polarization P of a material with dipole density N (m^{-3}) of the unit cell dipole moment $q \cdot u$ $(\mathrm{C} \cdot \mathrm{m})$.

Q10 Provide the polarization, P, induced in a *piezoelectric* with a piezoelectric strain coefficient, d, when it is subjected to an external stress, X.

Answers [60% or better score is expected.]

Q1 $X = cx$, $x = sX$

[*Note*: c stands for "stiffness" and s stands for "compliance"]

Q2 $x_4 = 2x_{23} = 2\phi$

[*Note*: Radian measure is generally preferred. This shear stress is not equivalent to the diagonal extensional stress]

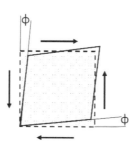

Q3 $v = 1/\sqrt{\rho s}$

Q4 $f = v/2L$

Q5 $F = \mu \cdot N$

Q6 $C = \varepsilon_0 \varepsilon_r (S/t)$

Q7 $f = 1/2\pi\sqrt{LC}$

Q8 $Z_1 = Z_0$ or $Z_1 = Z_0^*$

[*Note*: The current and voltage associated with Z_1 are $V/(Z_0 + Z_1)$ and $[Z_1/(Z_0 + Z_1)]V$, respectively, the product of which yields the power. The maximum power transfer occurs when $Z_0/\sqrt{Z_1} = \sqrt{Z_1}$ when impedance is resistive. When the impedance is complex, $Z_1 = Z_0^*$].

Q9 $P = Nqu$

[*Note*: The unit of the polarization is given by C/m^2, equivalent to the charge density on the surface]

Q10 $P = dX$

[*Note*: This is called the *direct piezoelectric effect*]

Chapter 1

Background of Piezoelectric Energy Harvesting

1.1 Necessity of Piezoelectric Energy Harvesting

The 21st Century is called "The Century of Sustainable Society", which requires serious management of the following:

(1) Power and energy (lack of oil, nuclear power plant, new energy harvesting)
(2) Rare material (rare-earth metal, Lithium)
(3) Food (rice, corn — bio-fuel)
(4) Toxic material

- Restriction (heavy metal, dioxin, Pb)
- Elimination/neutralization (Mercury, Asbestos)
- Replacement material

(5) Environmental pollution
(6) Energy efficiency (piezoelectric device).

Though most of the major countries rely on fossil fuel and nuclear power plants at present, renewable energy development becomes important for complementing the energy deficiency. Energy recovery from wasted or unused power has been the topic of discussion for a long period. Unused power exists in various electromagnetic and mechanical forms such as ambient electromagnetic noise around high-power cables, noise vibrations, water flow, wind, human motion and shock waves. Industrial and academic research units have focused their attention on harvesting energy from vibrations using piezoelectric transducers. Recent books on "Renewable Energy" discuss "piezoelectric energy harvesting" occasionally as one of the new energy sources in parallel to "windmills" and "solar cells", the plants of which generate huge electric power in Mega–Giga watts. However, the power level of the piezo harvesting device is limited up to 100 W, above which the competitive "electromagnetic generators" have better privilege in efficiency.

The author believes therefore that the target of "piezo energy harvesting" should NOT be aligned on the large energy cultivation line, but is to be set on "elimination of batteries", which are classified as "hazardous wastes", but recycled by less than 0.5% of 10 s Billion batteries sold per year in the world. In other words, the research Goal should be related with the above items (4) and (5): elimination of hazardous batteries, rather than mega-power energy harvesting. From this sense, the "piezo energy harvesting" is one of the very important technologies in the current "Sustainable Society" from this hazardous waste elimination viewpoint. The efforts so far put have provided the initial research guidelines and have brought light to the problems and limitations of implementing the piezoelectric transducer.[1] Transferring the present knowledge correctly to the following generation seems to be an important task for one of the pioneers, which is the authoring motivation of this textbook.

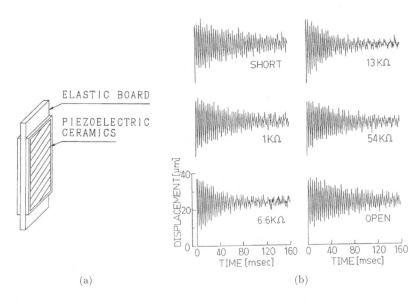

Fig. 1.1: Vibration damping change associated with external resistance change. (a) Bimorph transducer for this measurement. (b) Damped vibration with external resistor.[2]

1.2 From Passive Damping to Energy Harvesting

Historically, the author's group started the research on passive vibration damping using piezoelectric materials in the 1980s. Piezoelectric materials can convert the mechanical stress into electrical energy (i.e., direct piezoelectric effect). Figures 1.1(a) and 1.1(b) show the results for damping vibration generated in a bimorph transducer.[2] An initially deformed piezo device (bimorph) was released under a different resistor shunt condition, and the vibration damping was measured with time lapse. Mechanical energy will convert partially into electrical energy via piezoelectricity, then the induced electrical energy is consumed by the external resistor via Joule heat, so that the original mechanical vibration should be damped significantly. As shown in Fig. 1.1(b), the quickest damping was observed with a $6.6\,\mathrm{k\Omega}$ resistor, which is almost the same value as the electrical impedance of the bimorph (i.e., $R = 1/\omega C$), leading to the quickest damping effectively.[2,3] In addition to the resistive shunt, capacitive, inductive and switch shunts have been successively studied later, aiming at adaptive vibration control after our studies. The details will be discussed in Chapter 3.

However, in the late 1990s, we decided to save and store this generated electrical energy into a rechargeable battery, instead of dissipating it as Joule heat. This research target change is schematically illustrated in Fig. 1.2, which fits with the renewable energy boom in good timing. Note that the maximum energy harvesting condition corresponds to the largest vibration damping: "If you run after two hares (vibration suppression and energy harvesting), you will catch BOTH", rather than "you will catch NEITHER (the original proverb)", which seems to be the best development strategy!

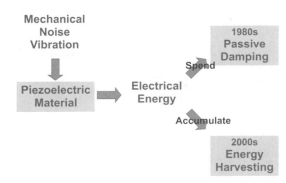

Fig. 1.2: Difference between piezoelectric passive vibration damping and piezoelectric energy harvesting.

1.3 Recent Research Trends

The worldwide annual revenue of piezoelectric devices reached \$22 billion in 2012, and reached \$40 billion at the end of 2017, as shown in Fig. 1.3(a).[4] With respect to the marketing category in Fig. 1.3(b), a

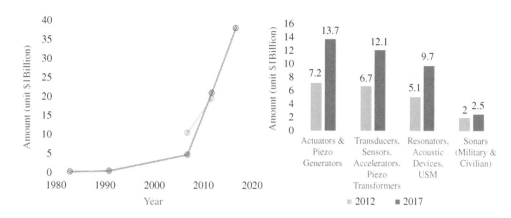

Fig. 1.3: Piezoelectric device market trends.[4]

category of "actuators and piezo generators" increased from $7.2 billion (2012) to $13.7 billion (2017), and in particular, the "ecology and energy harvesting" share increased dramatically from almost zero to 7% of the total revenue during only these several years.

We summarize in this section some research trends primarily after the 2000 s, referencing a book Ref. 5, in order to pick up the problematic issues in the trends to be solved in this textbook. The reader is requested to survey the recent research trends by yourself, since it is not the purpose of this textbook.

1.3.1 *Mechanical Engineers' Approach*

Wallaschek *et al.*[6] and Smithmaitrie[7] reported comprehensive studies on the bimorph transducer design optimization, using a mechanical equivalent system with mass, spring and dashpot in the former, and using a membrane/beam theory in the latter, including various references studied previously. The former indicated that the effective electromechanical coupling factor k can be increased with reducing the thickness ratio (PZT thickness/metal shim thickness), while in the latter paper, the voltage generation in the piezoelectric membrane shows the maximum around the resonance frequency (50 Hz). One of the significant problems of the mechanical engineers is neglect of the piezoelectric ceramics PZT's loss and performance limitation. As we will discuss in Chapter 2, the maximum handling energy level in the current PZTs is only 10–30 W/cm^3 at present. Thus, reducing the PZT thickness in order to increase the electromechanical coupling factor k dramatically reduces the handling energy level (note that the energy conversion amount is given by the product of input energy and the electromechanical coupling factor k^2). We should understand that though we try to increase the handling electric energy level with the mechanical vibration input, additional input energy will convert to just heat generation due to the hysteresis; that is, the PZT becomes a ceramic heater! Knowing the material's limitation is essential prior to extending the theoretical mechanical analysis.

1.3.1.1 *Machinery vibration (resonance usage)*

Beeby and Zhu discussed commercially available energy harvesting products in his paper.[8] Products by Midé Technology (US), Adaptivenergy (US), Arveni (France), Advanced Cerametrics (US) and Prepetuum (UK) were compared. These products are based on relatively large piezo-bimorph/unimorph designs (40–200 cm^3) in order to set the resonance frequency around 50 Hz with a narrow bandwidth less than 10 Hz. Power 1–20 mW is obtained under the acceleration 0.5–2 G. Most of them can be utilized to operate sensors of various machinery health monitoring systems (the machine should be vibrating constantly at 50 Hz) and to transmit the wireless signals of the monitored data. As the reader can imagine, this energy level is not sufficient for general energy storage applications, but suitable for replacing the current battery usage.

Arms *et al.* developed a smart system powered by piezoelectrics in a helicopter.[9] The integrated structural health monitoring and reporting (SHMR) includes strain gauges, accelerometers and load/torque cells. Data were stored in a central location. Wireless sensors were capable of logging by 50,000 sample/s with consuming 9 mA and 3 V DC (27 mW). The energy was supplied by piezoelectric micro-generators capable of harvesting vibration energy from the helicopter gearbox. Micro-generators (similar dimensions of multiple American quarter coins) working under a constant blade rotation condition can generate 37 mW, more than enough to feed the sensors.

1.3.1.2 *Human motion*

NEC-Tokin Corporation, Japan, commercialized two piezoelectric energy harvesting products already in the early 2000 s by utilizing an impact-based mechanism.[10] First, finger-snapping action on a piezo-bimorph generates the electric energy to illuminate LED lamps equipped at the front of a key holder, so that the user can easily find the key hole on the house or car doors. Second, "piezoelectric windmills" were produced and aligned in the northern

Fig. 1.4: Piezoelectric windmill structure (a) for driving an LED traffic light array system (b)[Courtesy by NEC-Tokin].

part of Japan along the major highway roadsides in tunnels, so that the wind generated by running vehicles rotates the windmills, then the rotation of mills generates the steel ball motion as shown in Fig. 1.4(a), which hits the piezo-bimorphs, leading to the electric energy generation for operating an LED traffic light array system which can navigate the vehicle drivers smoothly along the road curb (Fig. 1.4(b)).

Another successful product (million sellers) is "Lightning Switch" commercialized by Face Electronics (PulseSwitch Systems), VA, which is a remote switch for room lights, with using a unimorph ("Thunder") piezoelectric component. In addition to the living convenience, Lightning Switch (Fig. 1.5) can reduce the housing construction cost dramatically, due to a significant reduction of the copper electric wire and the wire aligning labor.[11,12] Harvesting energy from shoes was also reported using the Thunder component by Face Electronics.[12,13]

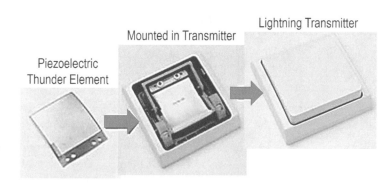

Fig. 1.5: Lightning Switch with piezoelectric Thunder actuator (Courtesy by Face Electronics).

Renaud *et al.*[14] reported a similar impact-based energy harvester, which consists of two piezo-bimorphs and a movable elastic rod. The total volume of the device, 25 cm³, could generate the output power of 60 μW under a motion of 10 Hz with 10 cm linear motion amplitude (which seems to be too quick for human action).

Since the human action is very slow and rather random, the resonance-type devices may not be suitable. Thus, there seem to be two choices for the "wearable" energy harvesting systems: (1) usage of impact-based snap action, or (2) off-resonance, but carefully matching the acoustic/mechanical impedance with soft human tissue and soft piezoelectric devices.

1.3.2 *Electrical Engineers' Approach*

Because the electric energy obtained by the piezoelectric energy harvesting system exhibits high voltage/low current (high impedance), we need to convert the electric impedance to a sufficiently low level ($\sim$50Ω) in order to charge up a rechargeable battery smoothly or drive a typical "ohmic" portable device.

Guyomar *et al.* reviewed studies on "switch" shunt circuits combined with the cyclic piezoelectric energy harvesting systems.[15] They compared initially two cases: the switching device is connected in parallel with the piezoelectric element (parallel synchronized switch harvesting on inductor, parallel SSHI) or in series between the active material and harvesting stage (series SSHI). The discussion was held only on a monochromatic excitation, which narrows the application of this technique.

As you will learn in Chapter 6, because of the randomness of the mechanical noise frequency, our strategy is basically to accumulate the original wide-frequency electric energy in a capacitor as DC charge, then the electric impedance is to be changed using a DC/DC converter in order to facilitate to charge a battery.

1.3.3 *MEMS Engineers' Approach*

Piezoelectric energy harvesting has been ignited significantly in the MEMS or NEMS area, with a new terminology "nanoharvesting". Lopes and Kholkin's paper[16] is a good review to be referred. Designs for the piezoelectric micro-harvesting are currently limited only to the unimorph types: (1) one-end clamp cantilever, and (2) both-end clamp membrane. Because of the combination of this low electromechanical coupling k design and very thin (low volume) piezoelectric film, it is difficult to expect high power or efficiency theoretically. $1-10\,\mu$W harvesting energy is reported at a resonance frequency higher than $1\,$kHz, which is 3 orders of magnitude smaller than a practically usable level ($10\,$mW).

A scientifically interesting topic is the usage of nano piezoelectric fiber (i.e., single crystal). Chang *et al.* reported direct-write, piezoelectric polymeric nanogenerators based on organic non-fibers with higher energy conversion efficiency.[17] These nanofibers are made of poly-vinylidene di-fluoride (PVDF) with high flexibility, minimizing resistance to external mechanical motion in low frequency. PVDF exhibits reasonable piezoelectric and soft mechanical properties, suitable for the medical applications for realizing mechanical impedance matching with soft human tissues. Chang *et al.* utilized electro-spinning process with strong electric field ($> 10^7$V/m) and stretching forces from the naturally aligned dipoles in the nanofiber crystal such that the nonpolar R phase is transformed into polar β phase, determining the polarity of the electro-spun nanofiber. They reported the energy conversion rate of 12% as average, which corresponds to the electromechanical factor $k = 35\%$, which is 3 times higher than the commercial PVDF films.

Another nanowire application is found in ZnO by Wang *et al.*[18] High energy can be produced by making an array of ZnO flexible nanowires where each individual wire can produce electricity. Further improvement was achieved by using a zigzag configuration on the top electrode surface. Though they reported the energy harvesting density $2.7\,$mW/cm^3, 5 times higher density than the PZT cantilever types, the most serious practical problem is how to sum up the harvesting electric output from each nanofiber in phase without canceling each other. The major problem in the current thin film or nanofiber devices exists in the actual power level (just μW or even nano-Watts), instead of the energy volume density, mW/cm^3.

1.3.4 *Military Application — Programmable Air-Burst Munition*

Until the 1960 s, the development of weapons of mass destruction (WMD) was the primary focus, including nuclear bombs and chemical weapons. However, based on the global trend for "Jus in Bello (Justice in War)", environment-friendly "green" weapons became the mainstream in the 21st century, that is, minimal destructive weapons with a pin-point target such as laser guns and rail guns. In this direction, programmable air-bust munition (PABM) was developed from 2003, during the US revenge war against Afghanistan.

After the World Trade Center was attacked by Al Qaeda on September 11, 2001, the US military started the "revenge" war against Afghanistan. The US troops initially destroyed many buildings by bombs, which dramatically increased the war cost for restructuring new buildings. Thus, the US Army changed the war strategy: without collapsing the building, just to kill the Al Qaeda soldiers inside (i.e., pin-point target) by using a sort of micro missile. A micro missile passes the window by making just a small hole in a window glass and explodes in the air 3–5 m (programmable!) inside the window, so that the building structure damage become minimized. For this purpose, each bullet needs to install a microprocessor chip, which navigates the bullet to a certain programmed point. ATK Integrated Weapon Systems, AZ, started to produce button-battery operated programmable air-burst munition (PABM) first.[19] Though they worked beautifully for the initial a couple of months, due to severe weather conditions (incredibly hot in the daytime in Afghanistan), most of the batteries wasted out in three months. No soldier is willing to open a dangerous bullet to exchange a battery!

Under this circumstance, the Army Research Office contacted Micromechatronics Inc., State College, PA (ICAT/PSU spin-off company), to ask for the development of a compact electric energy source to be embedded in the PABM bullet. The initial shooting impact can be converted to electricity via a piezoelectric device, which should fulfill energy to be spent in the micro-processor for 3–5 s during maneuvering the bullet in the 2-km distance. Micromechatronics Inc. developed an energy source for a 25-mm-caliber "Programmable Ammunition". Instead of a battery, a multilayer PZT piezo-actuator is used for generating electric energy under shot mechanical impact to activate the operational amplifiers which ignite the burst according to the command program (Fig. 1.6). This may be the best example of the piezoelectric energy harvesting device for demonstrating "only-the-one" product, not just a replacement of the "battery", and one of the highest revenue products with piezo energy harvesting in these 15 years. The development key is the multilayer usage for realizing high electromechanical coupling factor k and high capacitance, low electrical impedance.

Fig. 1.6: Programmable air-burst munition (PABM) developed by Micromechatronics (25-mm caliber).

According to the current research trends introduced above, since many of which are not satisfactory enough with the author's development philosophy, Uchino decided to introduce his personal development principles in this textbook.

1.4 Uchino's Frustration on the Current Studies

As one of the pioneers in the piezoelectric energy harvesting, the author feels a sort of frustration on 90% of the recent research papers from the industrial application viewpoint, though they are appreciated for their enormous academic and scientific contributions:

(1) Though the electromechanical coupling factor k is the smallest (i.e., the energy conversion rate from the input mechanical to electric energy is the lowest) among various device configurations, the majority of researchers primarily use the "unimorph" design. Why?

(2) Though the typical noise vibration is in a much lower frequency range, the researchers measure the amplified resonance response (even at a frequency higher than 1 kHz) and report the unrealistically harvested electric energy. Why?

(3) Though the harvested energy is lower than 1 mW, which is lower than the required electric energy to operate a typical energy harvesting electric circuit with a DC/DC converter (typically around 2–3 mW), the researchers report the result as an energy "harvesting" system. Does this situation mean actually energy "losing"? Why?

(4) Few papers have reported complete energy flow or exact efficiency from the input mechanical noise energy to the final electric energy in a rechargeable battery via the piezoelectric transducer step by step. Why?

Interestingly, the unanimous answer from these researchers to my questions "why" is "because the previous researchers did so"! This is the main motivation to my authoring this textbook. Since some of the current published papers seem to include strategy-misleading contents as described above, the author tried to select only the necessary papers and fundamental knowledge in this "Essentials of Piezoelectric Energy Harvesting". If the reader would like to check the recent publication trends in general, you are requested to refer to other books or review papers.

1.5 Structure of this Textbook

This textbook is structured as follows (see Fig. 1.7). As you have read, Chapter 1 introduced the overall "Background of Piezoelectric Energy Harvesting". The following two chapters, Chapter 2 "Fundamentals of Piezoelectrics" and Chapter 3 "Principle of Piezoelectric Passive Dampers", summarized most of the basic prerequisite knowledge on the piezoelectrics for junior researchers, without reading other textbooks. Through the "passive dampers", the reader learns how to dissipate (or cultivate) the electrical energy efficiently, that is, electrical impedance matching. Thus, senior researchers with this knowledge may skip these two chapters. Chapter 4 "Mechanical-to-Mechanical Energy Transfer" is the starting chapter of the main content. Not all the mechanical vibration energy will transmit into the piezoelectric devices. You will learn the mechanical impedance matching concept, which is lacking in electrical and materials' engineers, in particular. "Piezoelectric device designing" principle is discussed in Chapters 5 "Mechanical-to-Electrical Energy Transduction". You will learn why the "bimorph or unimorph" designs are not recommended. The key to increasing the electromechanical coupling factor is demonstrated in this chapter. Chapter 6 treats

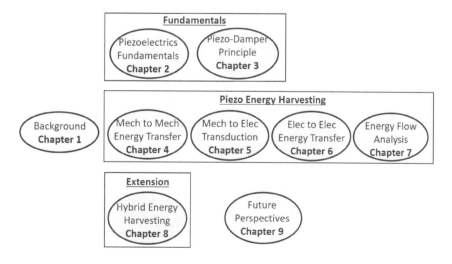

Fig. 1.7: Textbook construction.

"Electrical-to-Electrical Energy Transfer", where we provide the principle of designing the electric energy harvesting circuits, which is lacking in mechanical and materials' engineers. Chapter 7 "Case Studies on Energy Flow Analysis" is one of this book's highlights, in which we will discuss the major problems in the current piezoelectric energy harvesting systems. "Hybrid Energy Harvesting Systems" in Chapter 8 introduces hybrid-type electric energy harvesting systems, such as vibration energy and electromagnetic energy combination. The final Chapter 9 "Conclusions and Future Perspectives" addresses the three development directions: (1) higher electromechanical coupling device utilization, (2) paralleling technology of thousands of MEMS micro-energy harvesting devices and (3) lower energy consumption harvesting-circuit designs, in order to achieve practical piezoelectric energy harvesting systems.

For the reader who needs detailed information on ferroelectrics, smart piezoelectric actuators and sensors, "Ferroelectric Devices 2nd Edition (2010)" and "Micromechatronics 2nd Edition (2019)" authored by K. Uchino, published by CRC Press, are recommended. Further, "FEM and Micromechatronics with ATILA Software" published by CRC Press (2008) is a perfect tool for practical device designing, with three losses in simulation. ATILA FEM software code is available from Micromechatronics Inc., State College www.mmech.com.

Chapter Essentials

1. "Piezoelectric energy harvesting" is an essential technology in the 21st century "Sustainable Society" from the hazardous waste (i.e., battery) elimination viewpoint.

2. Piezo energy harvesting research historically originated from the "piezoelectric damper", and the development strategy is basically the same.

3. The energy density of piezoelectric materials is still limited to 30–40 W/cm^3. Additional input electrical or mechanical energy is converted to heat generation via the material's losses, not to the output electric energy. This material's constraint will restrict the device miniaturization for generating the suitable power level to practical applications, in particular in thin film or nanofiber utilization.

4. Uchino's frustration on the recent research trends on piezo energy harvesting:

   (a) Though the electromechanical coupling factor k is the smallest (i.e., the energy conversion rate from the input mechanical to electric energy is the lowest) among various device configurations, a majority of researchers primarily use the "unimorph" design. Why?

   (b) Though the typical noise vibration is in a much lower frequency range, the researchers measure the amplified resonance response (even at a frequency higher than 1 kHz) and report the unrealistically harvested electric energy. Why?

   (c) Though the harvested energy is lower than 1 mW, which is lower than the required electric energy to operate a typical energy harvesting electric circuit with a DC/DC converter (typically around 2–3 mW), the researchers report the result as an energy "harvesting" system. Does this situation mean actually energy "losing"? Why?

   (d) Few papers have reported complete energy flow or exact efficiency from the input mechanical noise energy to the final electric energy in a rechargeable battery via the piezoelectric transducer step by step. Why?

Check Point

1. (T/F) Since the piezo-MEMS device generates significantly high electric energy density, as long as the piezo-film can endure a high voltage/electric field, we can develop useful energy harvesting systems. True or False?

2. (T/F) In order to develop compact energy harvesting devices from human motion, we had better develop the piezo component which can generate the high output electric energy at its resonance frequency ($\sim 1\,\text{kHz}$). True or False?

3. (T/F) The unimorph piezoelectric structure is most popularly used, because it exhibits the highest energy harvesting rate among various piezo component designs. True or False?

4. Calculate the electrical impedance $1/j\omega C$ of a piezoelectric component with capacitance $1\,\text{nF}$ at the off-resonance frequency $100\,\text{Hz}$, which is larger than the internal impedance of rechargeable batteries ($\sim 50\,\Omega$).

5. (T/F) The target of "piezo energy harvesting" is aligned on the large energy cultivation line, in parallel to "windmills" and "solar cells", the plants of which generate huge electric power in Mega–Giga watts. True or False?

Chapter Problems

1.1 (a) Search the minimum required energy level of the following components/devices to be operated in practice:

- Typical MOSFET
- Blue-tooth transmission device
- DC/DC converter
- Heart pacemaker
- Blood soaking syringe

Hint:

Minimum 1–10 mW power is required for practical applications.

(b) Knowing the PZT material's power density limitation ($30\,\text{W/cm}^3$), consider to develop PZT thin film MEMS pumps, which require to generate $10\,\text{mW}$ for blood sugar rate detection. Calculate the required thin film areas for the two choices of thickness, $1\,\mu\text{m}$ and $50\,\mu\text{m}$.

Hint:

You can understand that less than $1\,\mu\text{m}$ thin films are impractical from the actuator application viewpoint, though popularly commercialized for sensor applications.

1.2 Search the worldwide battery consumption at present, and how these toxic products are disposed or recycled without contaminating the global environment.

References

1. K. Uchino, *Ferroelectric Devices*, 2nd edn. Boca Raton, FL: CRC Press (2010).
2. K. Uchino and T. Ishii, *J. Japan. Ceram. Soc.* 96(8), 863 (1988).
3. K. Uchino and K. Ohnishi, Shock Preventing Apparatus, US Patent No. 4,883,248.
4. Revenue estimated by Uchino from various resources such as iRAP (Innovative Research & Products, https://www.researchandmarkets.com/s/innovative-research-and-products/), Market Publishers (https://marketpublishers.com/) and IDTechEx (https://www.idtechex.com/).
5. N. Muensit (ed.), *Energy Harvesting with Piezoelectric and Pyroelectric Materials*, *Materials Science Foundations*, Vol. 72. Stafa-Zuerich, Switzerland: Trans Tech Pub. (2011).
6. J. Wallaschek, M. Neubauer and J. Twiefel, Electromechanical Models for Energy Harvesting Systems. In *Energy Harvesting with Piezoelectric and Pyroelectric Materials*, *Materials Science Foundations* Vol. 72, Chapter 2. Stafa-Zuerich, Switzerland: Trans Tech Pub. (2011).
7. P. Smithmaitrie, Vibration Theory and Design of Piezoelectric Energy Harvesting Structures. In *Energy Harvesting with Piezoelectric and Pyroelectric Materials*, *Materials Science Foundations*, Vol. 72, Chapter 3. Stafa-Zuerich, Switzerland: Trans Tech Pub. (2011).

8. S. P. Beeby and D. Zhu, Energy Harvesting Products and Forecast, In *Energy Harvesting with Piezoelectric and Pyroelectric Materials, Materials Science Foundations*, Vol. 72, Chapter 9. Stafa–Zuerich, Switzerland: Trans Tech Pub. (2011).

9. S. W. Arms, C. P. Townsend, J. H. Galbreth, L. C. David and P. Nam, Synchronized System for Wireless Sensing, RFID, Data Aggregation & Remot Reporting (2009).

10. K. Uchino, Piezoelectric Actuators 2004 — Materials, Design, Drive/Control, Modeling and Applications. In *Proc. 9th Int'l Conf. New Actuators,* A1.0, Bremen, Germany, June 14–16, 2004, pp. 38–48.

11. K. Uchino, Piezoelectric Actuators 2010 — Piezoelectric Devices in the Sustainable Society. In *Proc. 12th Int'l Conf. New Actuators*, A3.0, Bremen, Germany, June 14–16, 2010.

12. http://www.lightningswitch.com/.

13. H. Kim, Y. Tadesse and S. Priya, In *Energy Harvesting Technologies*, S. Priya and D. Inman (eds.). New York: Springer Science & Business Media LLC (2009).

14. M. Renaud, P. Fiorini, R. van Schaijk and C. van Hoof, *Smart Mater. Struct.* 18, 035001 (2009).

15. D. Guyomar, M. Lallart, N. Muensit and C. Lucat, Conversion Enhancement for Energy Harvesting. In *Energy Harvesting with Piezoelectric and Pyroelectric Materials, Materials Science Foundations*, Vol. 72, Chapter 5. Stafa-Zuerich, Switzerland: Trans Tech Pub. (2011).

16. R. P. Lopes and A. Kholkin, Energy Harvesting for Smart Miniaturized Systems. In *Energy Harvesting with Piezoelectric and Pyroelectric Materials, Materials Science Foundations*, Vol. 72, Chapter 6. Stafa-Zuerich, Switzerland: Trans Tech Pub. (2011).

17. C. Chang, V. H. Tran, J. Wang, Y.-K. Fuh and L. Lin, *Nano Lett.*, 10, 726 (2010).

18. Z. L. Wang and J. Song, *Science*, 321, 242 (2006).

19. http://www.atk.com/MediaCenter/mediacenter_videogallery.asp.

Chapter 2

Fundamentals of Piezoelectrics

Certain materials produce electric charges on their surfaces as a consequence of applying mechanical stress. The induced charges are proportional to the mechanical stress. This is called the *direct piezoelectric effect* and was discovered in quartz by Pierre and Jacques Curie in 1880. "Piezo" is derived from the Greek "piezein", which means to squeeze or press. Materials showing this phenomenon also conversely exhibit a geometric strain proportional to an applied electric field. This is the *converse piezoelectric effect*, discovered by Gabriel Lippmann in 1881. This chapter reviews the fundamentals of piezoelectrics: (a) crystal symmetry, (b) microscopic origins of the electric-field-induced strain, (c) piezoelectric materials. (d) piezoelectric constitutive equations, (e) figures of merit in piezoelectrics, (f) resonance and antiresonance, and (g) equivalent circuit analyses of piezoelectric components.

2.1 Crystal Structure and Piezoelectricity

In the so-called *dielectric* materials, the constituent atoms are considered to be ionized to a certain degree and are either positively or negatively charged. In such ionic (and some covalent) crystals, when an electric field is applied, cations are attracted to the cathode and anions to the anode due to electrostatic interaction. The electron clouds also deform, causing electric dipoles. This phenomenon is known as *electric polarization* of the dielectric, and the polarization is expressed quantitatively as the sum of the electric dipoles per unit volume $[C \cdot m/m^3 = C/m^2]$. Compared with air-filled capacitors, dielectric capacitors can store more electric charge due to the

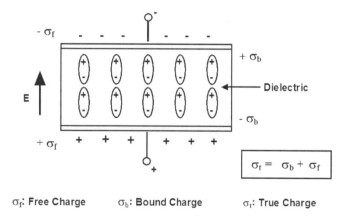

σ_f: **Free Charge** σ_b: **Bound Charge** σ_t: **True Charge**

Fig. 2.1: Charge accumulation in a dielectric capacitor.

dielectric polarization P, as visualized in Fig. 2.1. The physical quantity corresponding to the stored electric charge per unit area is called the *electric displacement D*, and is related to the electric field E by the following expression:

$$D = \varepsilon_0 E + P = \varepsilon \varepsilon_0 E. \tag{2.1}$$

Here. ε_0 is the vacuum permittivity ($= 8.854 \times 10^{-12}$ F/m). ε is the material's *relative permittivity* (also simply called permittivity or *dielectric constant*, and in general is a tensor property). Since the permittivity

11

Table 2.1: Crystallographic classification according to crystal centro-symmetry and polarity.

Polarity	Symmetry	Crystal System											
		Cubic		Hexagonal		Tetragonal		Rhombohedral		Orthorhombic	Monoclinic	Triclinic
Nonpolar (22)	Centro (11)	$m3m$	$m3$	$6/mmm$	$6/m$	$4/mmmm$	$4/m$	$\bar{3}m$	$\bar{3}$	mmm	$2/m$	$\bar{1}$
	Nonecentro (21)	432	23	622	$\bar{6}$	422	$\bar{4}$	32		222	2	
		$\overline{4}3m$		$\bar{6}m2$		$\bar{4}2m$						
Polar (Pyroelectric) (10)				6mm	6	4mm	4	3m	3	mm2	m	1

Note: Inside the bold line are piezoelectrics.

of piezo ceramics such as lead zirconate titanate (PZT) is rather high ($\sim$1000), we approximate $D \approx P$ in this textbook.

Crystals can be classified into 32 *point groups* according to their crystallographic symmetry, and these point groups can be divided into two classes, one with a center of symmetry and the other without, as indicated in Table 2.1. There are 21 point groups which do not have a center of symmetry. In crystals belonging to 20 of these point groups (point group (432) being the sole exception), positive and negative charges are generated on the crystal surfaces when appropriate stresses are applied, which are known as *piezoelectrics*.

Depending on the crystal structure, the centers of the positive and negative charges may not coincide even without the application of an external electric field. Such crystals are said to possess a *spontaneous polarization* (or *pyroelectric*). When the spontaneous polarization of the dielectric can be reversed by an electric field (not exceeding the breakdown limit of the crystal), it is called *ferroelectric*. Two of the most popular piezoelectric materials are quartz (32 symmetry) and PZT (4 mm symmetry), as introduced in Section 2.3; the former does not have spontaneous polarization, while the latter does have.

2.2 Origin of Field-Induced Strain

Solids, especially ceramics (inorganic materials), are relatively hard mechanically, but still expand or contract depending on the change of the external parameters. The *strain* (defined as the *displacement* $\Delta L / initial\, length\, L$) induced by temperature change or stress is known as thermal expansion or elastic deformation, respectively. In insulating materials, the application of an electric field can also cause deformation. This is called *electric field induced strain*. We consider the microscopic origin of this strain induction in this section.

Generally speaking, the word *electrostriction* is used in a wide sense to describe electric-field-induced strain, and hence frequently also implies the "converse piezoelectric effect". However, in solid state theory, the converse piezoelectric effect is defined as a primary electromechanical coupling effect, that is, the strain is proportional to the electric field, while electrostriction is a secondary coupling in which the strain is proportional to the square of the electric field. Thus, strictly speaking, they should be distinguished. However, the piezoelectricity of a ferroelectric, which has a centro-symmetric prototype (high temperature) phase, is considered to originate from the *electrostrictive interaction*, and hence the two effects are related. The above phenomena hold strictly under the assumptions that the object material is a mono-domain single crystal and that its state does not change under the application of an electric field. In a practical piezoelectric ceramic, additional strains accompanied by the reorientation of ferroelectric domains are also important.

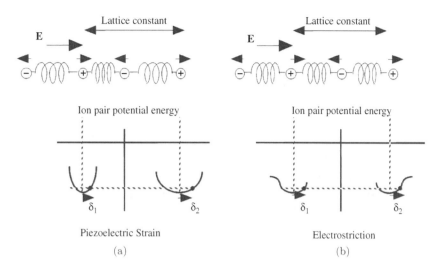

Fig. 2.2: Microscopic explanation of the piezostriction and electrostriction.

Why a strain is induced by an electric field is explained herewith.[1] For simplicity, let us consider an ionic 1D crystal such as a cation–anion chain. Figures 2.2(a) and 2.2(b) show a rigid-ion — spring model of the crystal lattice. The springs represent equivalently the cohesive force resulting from the electrostatic Coulomb energy and the quantum mechanical repulsive energy. Figure 2.2(b) shows the centro-symmetric case, whereas Fig. 2.2(a) shows the more general non-centro-symmetric case. In (b), the springs joining the ions are all the same (symmetric), whereas in (a), the springs joining the ions are different for the longer and shorter ionic distances (asymmetric); in other words, soft and hard springs existing alternately are important. Next, consider the state of the crystal lattice (a) under an applied electric field. The cations are drawn in the direction of the electric field and the anions in the opposite direction, leading to the relative change in the inter-ionic distance. Depending on the direction of the electric field, the soft spring expands or contracts more than the contraction or expansion of the hard spring, causing a strain $x(= (\delta_2 - \delta_1/a)$: unit cell length change rate) in proportion to the electric field E. This is the *converse piezoelectric effect*. When expressed as

$$x = dE, \tag{2.2}$$

the proportionality constant d is called the *piezoelectric constant*.

On the other hand, in Fig. 2.2(b), the amounts of extension and contraction of the spring are nearly the same because of the identical springs, and the distance between the two cations (lattice parameter) remains almost the same; hence, there is no strain, even though the ionic displacements are almost the same as in Fig. 2.2(a). However, more precisely, ions are not connected by such idealized springs (those are called *harmonic springs*, in which force (F) = spring constant (k) × displacement (Δ) holds). In most cases, the springs possess *anharmonicity* ($F = k_1\Delta - k_2\Delta^2$), that is, they are somewhat easy to extend, but hard to contract. Such subtle differences in the displacement cause a change in the lattice parameter, producing a strain which is irrelevant to the direction of the applied electric field ($+E$ or $-E$), and hence is an even-function of the electric field. This is called the *electrostrictive effect*, and can be expressed as

$$x = ME^2, \tag{2.3}$$

where M is the *electrostrictive constant*.

This 1D asymmetric crystal pictured in Fig. 2.2(a) also possesses a spontaneous bias of electrical charge, or a spontaneous dipole moment. The total dipole moment per unit volume is called the *spontaneous polarization*. (In quartz, though there are local dipole moments (asymmetric), the dipole directions are

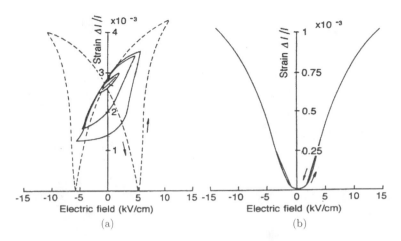

Fig. 2.3: Typical strain curves for (a) piezoelectric lead zirconatetitanate (PZT)-based, and (b) electrostrictive lead magnesium niobate (PMN)-based ceramic.

canceled out in a 3D unit cell (non-pyroelectric), that is, no spontaneous polarization.) When a large reverse bias electric field is applied to a crystal that has a spontaneous polarization in a particular polar direction, a transition "phase" is formed which is another stable crystal state in which the relative positions of the ions are reversed with respect to the horizontal axis. In terms of an untwined single crystal, this is equivalent to rotating the crystal 180° about an axis perpendicular to its polar axis. This transition, referred to as *polarization reversal*, also causes a remarkable change in strain. This particular class of substances is referred to as *ferroelectrics*, as mentioned in Section 2.1. Generally, what is actually observed as a field-induced strain is a complicated combination of the three basic effects: electrostriction, piezostriction and strain associated with the domain reorientation (with some hysteresis).

Figures 2.3(a) and 2.3(b) show longitudinal strain curves (along the electric field direction) for a piezo-electric lead zirconate titanate (PZT)-based and an electrostrictive lead magnesium niobate (PMN)-based ceramic, respectively. An almost linear strain curve in PZT becomes distorted and shows large hysteresis with increasing applied electric field level, which is due to the polarization reorientation.[2] On the other hand, PMN does not exhibit hysteresis under an electric field cycle. However, the strain curve deviates from the quadratic relation (E^2) and saturates at a high electric field level.[3]

We described the converse piezoelectric effect above. Then, what is the normal or *direct piezoelectric effect*? This is the phenomenon whereby charge (polarization P: Coulomb per unit area) is generated by applying an external stress X (force per unit area). Consider the force application on the asymmetric ion chain model in Fig. 2.2(a). It is obvious to imagine the cation–anion relative distance *linear change* because of the hard and soft springs with the applied stress. Note that the same piezoelectric coefficient d is used (as used in Eq. (2.2)), in the relation

$$P = dX \qquad (2.4)$$

2.3 Piezoelectric Materials

This section summarizes the piezoelectric materials: single-crystal materials, piezo ceramics, piezo polymers, composites and piezo films. Table 2.2 shows the material parameters of these piezoelectric materials.[4,5] Quartz with the highest *mechanical quality factor* is used for low loss transducers. PZT family shows high d and k suitable for high-power transducers. "Soft" PZT such as PZT 5H is for off-resonance actuator applications, while "Hard" PZT is for resonance-type ultrasonic motor, transformer and transducer applications. In, particular Sm-doped lead titanates exhibit extremely high mechanical coupling anisotropy (k_t/k_p),

Table 2.2: Piezoelectric properties of representative piezoelectric materials.

Parameter	Quartz	BaTiO$_3$	PZT 4	PZT 5H	(Pb,Sm)TiO$_3$	NKN–Cu	PVDF–TrFE
d_{33} (pC/N)	2.3	190	289	593	65	99	33
g_{33} (10^{-3} Vm/N)	57.8	12.6	26.1	19.7	42	34	380
k_t	0.09	0.38	0.51	0.50	0.50	—	0.30
k_p	—	0.33	0.58	0.65	0.03	0.13	—
$\varepsilon_{33}^{x/\varepsilon_0}$	5	1700	1300	3400	175	331	6
Q_m	>10^5	—	500	65	900	1052	3-10
T_C (°C)	—	120	328	193	355	340	—

suitable for medical transducers. Pb-free (Na.K)NbO$_3$-based ceramics are focused as the next-generation materials, once PZT will be regulated in the future from the ecological and legal regulation. Piezo polymer PVDF has small permittivity, leading to high piezo g constant, in addition to mechanical flexibility, suitable for pressure/stress sensor applications. Piezoelectric parameters in Table 2.2 are explained in Section 2.5.

2.3.1 *Single Crystals*

Although piezoelectric ceramics are widely used for a large number of applications, single-crystal materials retain their utility, being essential for applications such as frequency stabilized oscillators and surface acoustic devices. The most popular single-crystal piezoelectric materials are quartz, lithium niobate (LiNbO$_3$) and lithium tantalate (LiTaO$_3$). The single crystals are anisotropic, exhibiting different material properties depending on the cut of the materials and the direction of bulk or surface wave propagation.

Quartz is a well-known piezoelectric material. α-quartz belongs to the trigonal (rhombohedral) crystal system with point group 32 and has a phase transition at 537°C to its β-form which is not piezoelectric. Quartz has a cut with a zero temperature coefficient. For instance, quartz oscillators, operated in the thickness shear mode of the AT-cut, are used extensively for clock sources in computers and smart phones, and frequency stabilized ones in TVs. On the other hand, an ST-cut quartz substrate with X-propagation has a zero temperature coefficient for *surface acoustic wave*, and so is used for SAW devices with high-stabilized frequencies. Another distinguished characteristic of quartz is an extremely high mechanical quality factor $Q_m > 10^5$.

Lithium niobate and lithium tantalate belong to an isomorphous crystal system and are composed of oxygen octahedron. The Curie temperatures of LiNbO$_3$ and LiTaO$_3$ are 1210°C and 660°C, respectively. The crystal symmetry of the ferroelectric phase of these single crystals is 3 m, and the polarization direction is along c-axis. These materials have high electromechanical coupling coefficients for surface acoustic wave. In addition, large single crystals can easily be obtained from their melt using the conventional Czochralski technique. Thus, both materials occupy very important positions in the SAW device application field.

Highly responsive single-crystal relaxor ferroelectrics from solid solution systems with a *morphotropic phase boundary* (MPB) are one of the epoch-making discoveries in piezoelectrics for applications as ultrasonic transducers and electromechanical actuators.[6] Compositions very near the morphotropic phase boundary tend to show the most promise for these applications. Extremely high values for the electromechanical coupling factor ($k_{33} = 92 - 95\%$) and piezoelectric strain coefficient ($d_{33} = 1500$ pC/N) were first reported for single crystals at the *MPB* of the Pb(Zn$_{1/3}$Nb$_{2/3}$)O$_3$ − PbTiO$_3$ (PZN–PT) system in 1981.[6,7] These compositions belong to so-called "perovskite" structure, isomorphous to BaTiO$_3$ described in the next subsection (see Fig. 2.5). Various electromechanical parameters (at room temperature) are plotted as a function of composition in Fig. 2.4. Note how two different values are plotted for the morphotropic phase boundary composition, $0.91Pb(Zn_{1/3}Nb_{2/3})O_3 - 0.09PbTiO_3$. The highest values for the piezoelectric coefficients and electromechanical coupling factors are observed for the rhombohedral composition only when the single crystal is poled along the perovskite [001] direction, not along [111], which is the direction of the spontaneous polarization.

Approximately 10 years after the discovery, Toshiba, Japan[8] and The Penn State University[9] independently reproduced the above findings, and refined data were collected in order to characterize the material for medical acoustic applications. Strains as large as 1.7% can be induced in single crystals from the morphotropic phase boundary (MPB) composition (0.92)PZN–(0.08)PT of this system in the [001] orientation.

The mechanism for the enhanced electromechanical coupling is basically from the large shear coupling through d_{15}, which is generally dominant for perovskite piezoelectrics. The applied electric field should therefore be applied such that its direction is somewhat ($\sim 50°$) canted from the spontaneous polarization direction in order to produce the optimum piezoelectric response. The exceptionally high strain (up to 1.7%) generated in materials with compositions near the *morphotropic phase boundary* is associated additionally with the field-induced phase transition from the rhombohedral to the tetragonal phase. Recent studies are shifting to single crystals at the MPB of the $Pb(Mg_{1/3}Nb_{2/3})O_3 - PbTiO_3(PMN-PT)$ system, which is isomorphic with PZN. Because of the significant cant angle between the spontaneous polarization and electric field directions, complicated multi-domain structures are created in a single crystal inevitably. Accordingly, the so-called "domain engineering" concept came up to further explain the electromechanical coupling enhancement via the domain–domain interaction and domain interface/wall.[10]

2.3.2 *Polycrystalline Materials*

Barium titanate $BaTiO_3$ (BT) is one of the most thoroughly studied and widely used ferroelectric and capacitor materials with a *perovskite*-type crystal structure, after the discovery independently in Japan, US, Russia and Germany during World War II. As shown in Fig. 2.5, BT has a cubic structure above the Curie temperature (130°C), and transforms into a tetragonal symmetry below the Curie temperature. The vector of the spontaneous polarization points in the [001] direction (tetragonal phase) at room temperature, which reorients in the [011] (orthorhombic phase) below 5°C, and further in the [111]

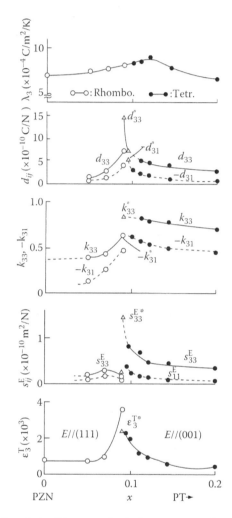

Fig. 2.4: Electromechanical parameters in the $(1 - x)PZN–xPT$ system as a function of PT content, x. All peak at the MPB composition, $x = 0.09$.

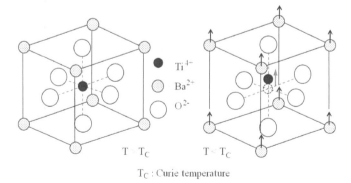

Fig. 2.5: Crystal structure of a perovskite $BaTiO_3$. Left: higher cubic paraelectric, and Right: lower tetragonal symmetry ferroelectric.

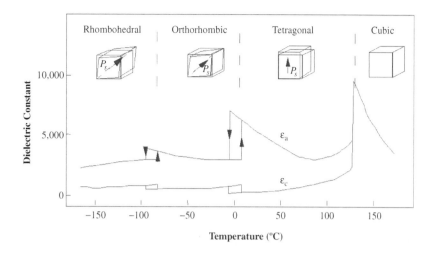

Fig. 2.6: Dielectric constant change with temperature through various phase transitions in BaTiO$_3$.

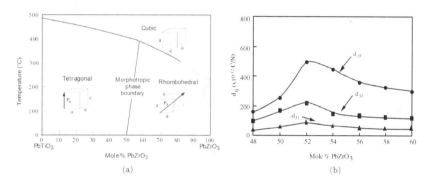

Fig. 2.7: (a) Phase diagram of lead zirconatetitanate (PZT). (b) Dependence of several d constants on composition near the morphotropic phase boundary in the PZT system.

direction (rhombohedral phase) below $-90°$C, successively. Figure 2.6 shows the dielectric constant change with temperature through various phase transitions. The dielectric and piezoelectric properties of ferroelectric ceramic BaTiO$_3$ can be affected by its own stoichiometry, microstructure, and by dopants entering onto the A or B site in solid solution. Modified ceramic BaTiO$_3$ with dopants such as Pb or Ca ions have been developed to stabilize the tetragonal phase over a wider temperature range and are used as piezoelectric materials. Though the initial application was for Langevin-type piezoelectric vibrators, after the discovery of PZTs, the applications to piezoelectric devices faded out until recently when Pb-free materials have the focus.

Lead zirconate titanate (PZT) solid solution systems were discovered in 1954 by Japanese researchers,[11] through systematic studies on BT-isomorphic perovskite materials. Figure 2.7(a) shows the phase diagram of lead zirconate titanate (PZT), where the *Morphotropic Phase Boundary* (MPB) between the tetragonal and rhombohedral phases exists around 52 PZ–48 PT compositions. However, the enormous piezoelectric properties were discovered by Jaffe,[12] Clevite Corporation, and Clevite took the most important PZT patent for transducer applications. Because of this strong basic patent, Japanese ceramic companies were encouraged actually to develop ternary systems to overcome the performance, and more importantly, to escape from the Clevite's patent, that is, PZT + a complex perovskite such as Pb(Mg$_{1/3}$Nb$_{2/3}$)O$_3$(Panasonic), Pb(Zn$_{1/3}$Nb$_{2/3}$)O$_3$(Toshiba), Pb(Mn$_{1/3}$Sb$_{2/3}$)O$_3$, Pb(Co$_{1/3}$Nb$_{2/3}$)O$_3$, Pb(Mn$_{1/3}$Nb$_{2/3}$)O$_3$, Pb(Ni$_{1/3}$Nb$_{2/3}$)O$_3$(NEC), Pb(Sb$_{1/2}$Sn$_{1/2}$)O$_3$, Pb(Co$_{1/2}$W$_{1/2}$)O$_3$, Pb(Mg$_{1/2}$W$_{1/2}$)O$_3$(Du Pont), which have been the basic compositions in recent years.

The crystalline symmetry of this solid-solution system is determined by the Zr content, as shown in Fig. 2.7(a). Pure lead titanate (PT) has a tetragonal ferroelectric phase of perovskite structure. With increasing Zr content, x, the tetragonal distortion decreases, and at $x > 0.52$ the structure changes from the tetragonal *4mm* phase to another ferroelectric phase of rhombohedral *3m* symmetry. The line dividing these two phases is called the MPB. The boundary composition is considered to have both tetragonal and rhombohedral phases coexisting together. Figure 2.7(b) shows the dependence of several piezoelectric d constants on composition near the MPB. The d constants have their highest values near the MPB. This enhancement in piezoelectric effect is attributed to the increased ease of reorientation of the polarization under an applied electric field.

Doping the PZT material with *donor* or *acceptor* ions changes its properties dramatically. Donor doping with ions such as Nb^{5+} or Ta^{5+} provides "soft" PZTs, like PZT-5, because of the facility of domain motion due to the resulting Pb-vacancies. On the other hand, acceptor doping with Fe^{3+} or Sc^{3+} leads to "hard" PZTs, such as PZT-8, because the oxygen vacancies will pin the domain wall motion.

The end member of PZT, lead titanate $PbTiO_3$(PT), has a large crystal distortion: a tetragonal structure at room temperature with its tetragonality c/a = 1.063 with high Curie temperature 490°C. However, densely sintered $PbTiO_3$ ceramics cannot be obtained easily, because they break up into a powder when cooled through the Curie temperature due to the large spontaneous strain. PT ceramics modified by adding small amount of additives exhibit a high piezoelectric anisotropy. Either $(Pb,Sm)TiO_3$[13] or $(Pb,Ca)TiO_3$[14] exhibits an extremely low planar coupling, that is, a large k_t/k_p ratio. Here, k_t and k_p are thickness-extensional and planar electro-mechanical coupling factors, respectively. (Refer to Section 2.5 for the electromechanical coupling factors.) Since these transducers can generate purely longitudinal waves through k_t associated with no transverse waves through k_{31}, clear ultrasonic imaging is expected without "ghost" caused by the transverse wave. $(Pb,Nd)(Ti,Mn,In)O_3$ ceramics with a zero temperature coefficient of surface acoustic wave delay have been developed as superior substrate materials for SAW device applications.[15]

2.3.3 *Relaxor Ferroelectrics*

Relaxor ferroelectrics can be prepared either in polycrystalline form or as single crystals (Section 2.3.1). Lead-based relaxor materials have complex disordered perovskite structures. Though the crystal structure is an isomorph perovskite, different from the previously mentioned normal ferroelectrics such as BT and PZT's, it exhibits a broad phase transition from the paraelectric to ferroelectric state, a strong frequency dependence of the giant dielectric constant (i.e., *dielectric relaxation*) and a weak remnant polarization. This is probably because of the composition distribution nonuniformity of the B-site ions, such as Mg^{2+} and Nb^{5+} in $Pb(Mg_{1/3}Nb_{2/3})O_3$ (PMN).

Ceramics of PMN are easily poled when the electric field is applied near the phase transition temperature, but they are depoled completely when the field is removed as the macrodomain structure reverts to microdomains (with sizes on the order of several 100 Å). This microdomain structure is believed to be the source of the exceptionally large electrostriction exhibited in these materials. The usefulness of the material is thus further enhanced when the transition temperature is adjusted to near room temperature. The longitudinal induced strain at room temperature as a function of applied electric field for $0.9Pb(Mg_{1/3}Nb_{2/3})O_3$–$0.1PbTiO_3$ ceramic is shown in Fig. 2.3(b).[3] Note that the order of magnitude of the electrostrictive strain (10^{-3}) is similar to that induced under unipolar drive in PLZT (7/62/38) through the piezoelectric effect (Fig. 2.3(a)). An attractive feature of this material is the near absence of hysteresis, while the nonlinear strain behavior $(x = ME^2)$ requires a sophisticated drive circuit. Relaxor-type electrostrictive materials are highly suitable for actuator applications, and also exhibit an induced piezoelectric effect under the bias electric field. That is, the electromechanical coupling factor k_t can be changed with the applied DC bias field. As the DC bias field increases, the coupling increases and saturates. Since this behavior is reproducible, these materials can be applied as ultrasonic transducers which are tunable by the bias field.[16] The superior performances in single-crystal relaxor ferroelectrics with the MPB composition have already been discussed in Section 2.3.1.

2.3.4 *PVDF*

Thanks to Kawai's efforts, polyvinylidene difluoride (PVDF or PVF_2) was discovered in 1969.[17] Though the piezoelectric d constant is not as high as piezo ceramics, high piezoelectric g constant due to small permittivity ε is attractive from the sensor and energy harvesting application viewpoint.

The PVDF is a polymer with monomers of CH_2CF_2, isomorphic to ethylene C_2H_4, two H's of which are replaced by fluorine F. Since H and F have positive and negative ionization tendency, the monomer itself has a dipole moment. Crystallization from the melt forms the non-polar α-phase (adjacent monomer dipole moments are canceled out), which can be converted into the polar β-phase by a uniaxial or biaxial drawing operation (monomer dipole moments are aligned in parallel); the resulting dipoles are then reoriented through electric poling (see Fig. 2.8). Large sheets can be manufactured and thermally formed into complex shapes. The copolymerization of vinilydene difluoride with trifluoro-

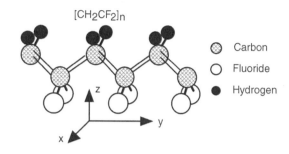

Fig. 2.8: Molecular structure of polyvinylidene difluoride (PVDF).

roethylene (TrFE) results in a random copolymer (PVDF–TrFE) with a stable, polar β-phase. Crystallinity reaches about 50%, and this polymer need not be stretched; it can be poled directly as formed. A thickness-mode coupling coefficient of $k_t = 0.30$ has been reported. Piezoelectric polymers have the following characteristics: (a) small piezoelectric d constants (for actuators) and large g constants (for sensors), due to small permittivity, (b) light weight and soft elasticity, leading to good acoustic impedance matching with water or the human body (~70% water), (c) a low mechanical quality factor Q_M, allowing for a broad resonance band width suitable to sensors. Such piezoelectric polymers are used for directional microphones and ultrasonic hydrophones. Wearable energy harvesting devices are the recent target, which are discussed in Chapter 4.

2.3.5 *Pb-Free Piezo Ceramics*

In 2006, European Union started restrictions on the use of certain hazardous substances (RoHS), which explicitly limits the usage of lead (Pb) in electronic equipment. Basically, we may need to regulate the usage of lead zirconate titanate (PZT), currently most widely used piezoelectric ceramics, in the future. Japanese and European societies may experience the governmental regulation on the PZT usage in these 10 years. Pb (lead)-free piezo-ceramics have started to be developed after 1999. The Pb-free materials include (1) $(K,Na)(Ta,Nb)O_3$ based, (2) $(Bi,Na)TiO_3$ and (3) $BaTiO_3$, which reminds us that "the history will repeat" (i.e., *piezoelectric Renaissance*).[18]

The share of the patents for bismuth compounds (bismuth layered type and $(Bi,Na)TiO_3$ type) exceeds 61%. This is because bismuth compounds are easily fabricated in comparison with other compounds. Honda Electronics, Japan developed Langevin transducers with using the BNT-based ceramics for ultrasonic cleaner applications.[19] Their composition $0.82(Bi_{1/2}Na_{1/2})TiO_3 - 0.15BaTiO_3 - 0.03(Bi_{1/2}Na_{1/2})(Mn_{1/3}Nb_{2/3})O_3$ exhibits $d_{33} = 110 \times 10^{-12} C/N$, which is only 1/3 of that of a hard PZT, but the electromechanical coupling factor $k_t = 0.41$ is larger because of much smaller permittivity ($\varepsilon = 500$)

than that of the PZT. Furthermore, the maximum vibration velocity of a rectangular plate (k_{31} mode) is close to $1\,\mathrm{m/s}$ (rms value), which is higher than that of hard PZTs.

(Na,K)NbO$_3$ (NKN) systems exhibit the highest performance among the present Pb-free materials, because of the morphotropic phase boundary usage. NKN-based ceramics can be utilized for the off-resonance energy harvesting applications, because of high k_{33}. Figure 2.9 shows the current best data reported by Toyota Central Research Lab, where strain curves for oriented and unoriented (K,Na,Li)(Nb,Ta,Sb)O$_3$ ceramics are shown.[20] Note that the maximum strain reaches up to 1500×10^{-6}, which is equivalent to the PZT strain. Drawbacks include their sintering difficulty and the necessity of the sophisticated preparation technique (topo-chemical method for preparing flaky raw powder).

Tungsten-bronze (TB) types are another alternative choice for resonance applications, because of their high Curie temperature and low loss. Taking into account general consumer attitude on disposability of portable equipment, Taiyo Yuden, Japan developed micro ultrasonic motors using non-Pb multilayer piezo actuators.[21] Their composition is based on TB ((Sr,Ca)$_2$NaNb$_5$O$_{15}$) without heavy metal. The basic piezoelectric parameters in TB ($d_{33} = 55 - 80\,\mathrm{pC/N}$, $T_C = 300°\mathrm{C}$) are not very attractive. However, once the c-axis oriented ceramics are prepared, the d_{33} is dramatically enhanced up to $240\,\mathrm{pC/N}$. Further, since the Young's modulus $Y_{33}^E = 140\,\mathrm{GPa}$ is more than twice of that of

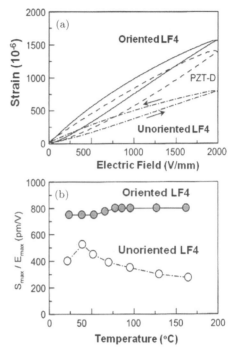

Fig. 2.9: Strain curves for oriented and unoriented (K,Na,Li)(Nb,Ta,Sb)O$_3$ ceramics.[20]

PZT, the higher generative stress is expected, which is suitable for ultrasonic motor applications. Taiyo Yuden developed a sophisticated preparation technology for oriented ceramics with a multilayer configuration, that is, preparation under strong magnetic field, much simpler than the flaky powder preparation.

2.3.6 *Composites*

Piezo composites comprised of a piezoelectric ceramic and a polymer phase are promising materials because of their excellent and readily tailored properties. The geometry for two-phase composites can be classified according to the dimensional connectivity of each phase into 10 structures: 0–0, 0–1, 0–2, 0–3, 1–1, 1–2, 1–3, 2–2, 2–3 and 3–3.[22] The *connectivity* "x–y" in a two-phase composite stands for x-dimensional connection of the primary phase (such as PZT) and y-dimensional connection of the secondary phase (such as polymer). The "0" and "1" con-

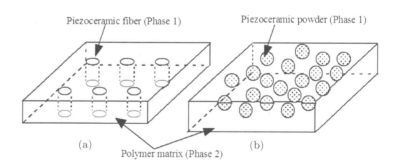

Fig. 2.10: PZT: polymer composites: (a) 1–3 connectivity and (b) 0–3 connectivity.

nection corresponds to "powder" and "needle" shapes, respectively, "2", plane, and "3" is a volumetric connection. The 1–3 composite is illustrated in Fig. 2.10(a). The advantages of this composite are high coupling factors, low acoustic impedance, which matches water or human tissue (70% water), mechanical flexibility, broad bandwidth in combination with a low mechanical quality factor and the possibility of making undiced arrays by structuring the electrodes. The thickness-mode electromechanical coupling of the composite can exceed the k_t (0.40–0.50) of the constituent ceramic, approaching almost the value of the rod-mode electromechanical coupling, k_{33} (0.70–0.80) of that ceramic.[23] Piezoelectric composite materials are especially useful for underwater sonar and medical diagnostic ultrasonic transducer applications.

When needle- or plate-shaped piezoelectric ceramic bodies are arranged and embedded in a polymer matrix, advanced composites can be fabricated, which provide enhanced sensitivity by keeping the actuation function (try Example Problem 2.1 for understanding the principle). These composites exhibit intermediate performances between the solid piezo ceramics and soft piezo polymers. Figure 2.10(a) shows a 1–3 composite device, where PZT rods are arranged in a polymer in a 2D array. The simplest composite from a fabrication viewpoint is a 0–3 connectivity type, which is made by dispersing piezoelectric ceramic powders uniformly in a polymer matrix (Fig. 2.10(b)).

Example Problem 2.1.

A composite consists of two piezoelectric phases, 1 and 2, poled along the 3-axis and arranged in a "parallel" configuration as shown in Fig. 2.11(a). Analogous to the terminology used in electronic circuit analysis, the structures pictured in Figs. 2.11(a) and 2.11(b) are designed as "parallel" and "series" connections, respectively. The volume fraction is $^1V : {}^2V$ $(^1V + {}^2V = 1)$. Assuming that the top and bottom electrodes are rigid enough to prevent surface bending, and that the transverse piezoelectric coupling between phases 1 and 2 is negligibly small in the "parallel" connection, calculate the following physical properties of this composite.

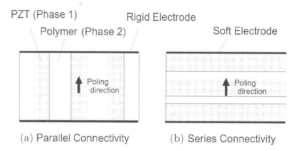

(a) Parallel Connectivity (b) Series Connectivity

Fig. 2.11: Diphasic composites arranged in parallel (a) and Series (b) configurations.

(a) effective dielectric constant ε_3^*,
(b) effective piezoelectric d_{33}^* coefficient,
(c) effective piezoelectric voltage coefficient g_{33}^*.

Use the parameters D_3, E_3, X_3, x_3, s_{33}^E which are the electric displacement, electric field, stress, strain and the elastic compliance along the 3 axis (poling direction), respectively.

Solution:

(a) Since the electrodes are common and E_3 is common to phases 1 and 2,

$$D_3 = {}^1V \, {}^1\varepsilon_3\varepsilon_0 E_3 + {}^2V \, {}^2\varepsilon_3\varepsilon_0 E_3$$
$$= \varepsilon_3{}^*\varepsilon_0 E_3. \tag{P2.1.1}$$

Therefore,

$$\varepsilon_3^* = {}^1V \, {}^1\varepsilon_3 + {}^2V \, {}^2\varepsilon_3. \tag{P2.1.2}$$

(b) If phases 1 and 2 are independently free (no interaction)

$$^1x_3 = {}^1d_{33} \, E_3, \tag{P2.1.3}$$

$$^2x_3 = {}^2d_{33} \, E_3. \tag{P2.1.4}$$

However, when we assume that the top and bottom electrodes are rigid, the strain x_3* must be common to both phases 1 and 2, and the average strain x_3* is given by the following equation from the force balance in two phases:

$$^{1}V(^{1}x_3 - x_3^*)/^{1}s_{33} = {}^{2}V(x_3^* - {}^{2}x_3)/^{2}s_{33}. \tag{P2.1.5}$$

Thus,

$$x_3^* = [(^{1}V\,^{2}s_{33}\,^{1}d_{33} + {}^{2}V\,^{1}s_{33}\,^{2}d_{33})/(^{1}V^{2}s_{33} + {}^{2}V^{1}s_{33})]E_3, \tag{P2.1.6}$$

and consequently, the effective piezoelectric constant is given by

$$d_{33}^* = (^{1}V\,^{2}s_{33}\,^{1}d_{33} + {}^{2}V\,^{1}s_{33}\,^{2}d_{33})/(^{1}V\,^{2}s_{33} + {}^{2}V\,^{1}s_{33}). \tag{P2.1.7}$$

(c) Since $g_{33}^* = d_{33}^*/\varepsilon_0\varepsilon_3*$,

$$g_{33}^* = (^{1}V\,^{2}s_{33}\,^{1}d_{33} + {}^{2}V^{1}s_{33}\,^{2}d_{33})/[(^{1}V^{2}s_{33} + {}^{2}V^{1}s_{33})\varepsilon_0(^{1}V^{1}\varepsilon_3 + {}^{2}\,V^{2}\varepsilon_3)].$$

If we adopt the assumptions $^{1}V << {}^{2}V$, $^{1}\varepsilon_3 >> {}^{2}\varepsilon_3$, we can derive (a) $\varepsilon_3^* \approx {}^{1}V^{1}\varepsilon_3$ (dramatic reduction with smaller ^{1}V); (b) $d_{33}^* \approx^{1} d_{33}$ (keeping the same d constant); $g_{33}^* \approx {}^{1}g_{33}/^{1}V$ (significant enhancement).

The fabrication processes of the 0–3 composites are classified into melting and rolling methods.[24] Figure 2.12 shows a flowchart for the fabrication processes. The powders are mixed with molten polymer in the first method, while the powders are rolled into a polymer using a hot roller in the second method. The original fabrication process of the 1–3 types involves the injection of epoxy resin into an array of PZT fibers assembled with a special rack.[22] After the epoxy is cured, the sample is cut, polished, electroded on

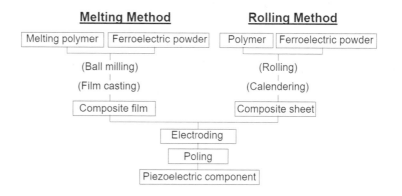

Fig. 2.12: Fabrication process for PZT: polymer composites.

the top and bottom, and finally electrically poled. The die casting technique has also been employed to make rod arrays from a PZT slurry.

A 1–3 composite was applied for developing deformable mirror.[25] The glass mirror is attached to the one surface of the composite on which the common electrode is applied, while the individual PZT rods are addressed from the other side of the device. As each rod in the array can be individually addressed, the surface of the deformable mirror can be shaped to any desired contour. This 1–3 composite design is merely to separate the acting rods. A 0–3 composite comprised of PZT powder in a polymer matrix can be used to make flexible bimorphs. Although a large curvature can be obtained with such a bimorph, excessive heat generation during AC drive is a serious problem for these devices. This is due primarily to the highly thermo-insulating nature of the polymer matrix. The heat generated through dielectric loss of the piezoelectric ceramic is not readily dissipated.

Active fiber composites (AFC) have been developed at MIT, comprised of PZT needles that are arranged in a polymer matrix, pictured in Fig. 2.13.[26] One application for this composite is the helicopter blade vibration control device. Fine PZT needles, which are fabricated by an extrusion technique, are arranged in an epoxy resin, and the composite structure is sandwiched between the interdigital electrodes (Ag/Pd), which are coated on a Kapton substrate. The AFC structures are laminated on the helicopter blade such that the PZT fiber axes are oriented 45 degrees with respect to the blade direction in order to produce a torsional deformation of the blade. A major problem with this design is the limitation in the effective applied electric field strength due to the restricted contact area between the flat electrode surface and the rounded contour of the PZT fibers. One remedy for this problem has been to incorporate conducting particles in the epoxy matrix.

2.3.7 *Thin-Films*

Though the thin-film configuration is a part of composite categories, due to specific constraints such as thickness of films, stress from the substrate and epitaxial growth/film crystal orientation, special interests are obtained from device miniaturization viewpoints (i.e., nanotechnologies).

Both zinc oxide (ZnO) and aluminum nitride (AlN) are simple binary compounds with a Wurtzite-type structure, which can be sputter-deposited as a *c*-axis oriented thin film on a variety of substrates. ZnO has higher piezoelectric coupling and thin films of this material are used in bulk and surface acoustic wave devices, including medical diagnostic transducers. The fabrication of highly oriented (along *c*) ZnO films has been studied and developed extensively. However, the performance of ZnO devices is limited, due to their lower piezoelectric coupling (20–30%) in comparison with PZT.

Micro Electro-Mechanical System (MEMS) is now popular, prepared by so-called "micro-machining process" used conventionally to fabricate silicon devices. Figure 2.14 illustrates the PZT-MEMS micropump for a blood tester developed at the end of the 1990s:[27] The micromachining process used to fabricate this PZT-MEMS micro-pump and a schematic diagram of the structure of a PZT micropump (actual size: 4.5 mm × 4.5 mm × 2 mm) are shown in Figs. 2.14(a) and 2.14(b), respectively.[27] The blood sample and test chemicals entering the system through the

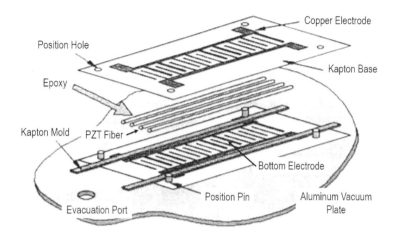

Fig. 2.13: AFC, comprised of oriented PZT fibers and epoxy.[26]

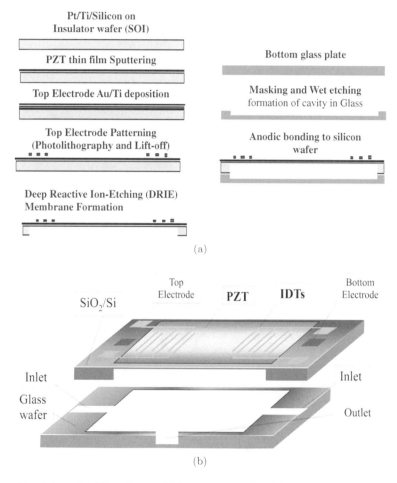

Fig. 2.14: (a) The micromachining process used to fabricate a PZT-MEMS micropump. (b) A schematic diagram of the structure of a PZT micropump. Actual size: 4.5 mm × 4.5 mm × 2 mm.[27]

two inlets, identified in Fig. 2.14, are mixed in the central chamber, and finally are passed through the outlet for analytical instrument. The movement of the liquids through the system occurs through the bulk bending wave of the PZT diaphragm in response to the drive voltage provided by the *interdigital surface electrode*. The key issue for the actuator-type MEMS device is the output mechanical power level, which should be higher than 1 mW. Since the PZT handling power is limited around 30 W/cm³, thin films less than 10 µm are useless in general from the electro-mechanical actuation viewpoint. Thin films less than 10 µm have been commercialized for sensor developments or zero-load optical reflector/mirror control applications.

2.4 Piezoelectric Component Designs

A classification of piezoelectric actuator/transducer structures is presented in Fig. 2.15. Simple disk/plate devices directly use the longitudinally (d_{33}) or transversely (d_{31}) induced strain. The multilayer (ML) types make use of the longitudinal strain (d_{33}) under smaller voltage (higher electric field due to the layer thinness) with higher capacitance. The author invented the ML structure at the end of the 1970s primarily to reduce the driving voltage from 1 kV to 100 V in order to use an inexpensive low-voltage power supply. The ML design does not amplify the displacement, but just increases the total displacement by the lamination. Complex unimorph, bimorph and cymbal devices do not use the induced transversal d_{31} strain directly, but rather a magnified displacement, produced through a spatial magnification. The unimorph (single piezo plate) and bimorph (double piezo plates) are the most commonly used *bending mode* structures. The cymbal exhibits a flex-tensional

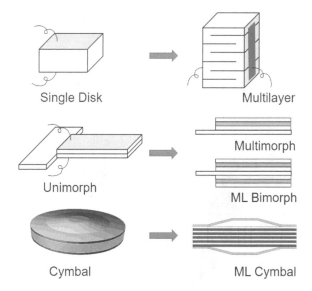

Fig. 2.15: Classification of piezoelectric transducer designs.

up-down displacement from the transversal d_{31} strain. The multilayer type (5 mm × 5 mm × 10 mm size) produces only modest displacements (10 µm); it offers a respectable generative force (1 kN), a quick response speed (10 µs), long lifetime (10^{11} cycles) and a high electromechanical coupling factor k_{33} (70%), while the bimorph type provides large displacements (500 µm), but can only offer a relatively low generative force (1 N), a much slower response speed (1 ms), a shorter lifetime (10^8 cycles) and a rather low electromechanical coupling factor k_{eff} (10%). The cymbal was developed for fulfilling the intermediate performances:[28] displacement 100 µm, force 100 N, response time 100 µs and k_{eff} 30%. In order to reduce the drive voltage, the ML piezo components have primarily been used for all designs recently. The detailed analysis on these three designs is conducted in Chapter 5.

Example Problem 2.2.

Using a PZT-based ceramic with a piezoelectric strain coefficient of $d_{31} = -300$ pC/N, design a shim-less bimorph with a total length of 30 mm (where 5 mm is used for cantilever clamping) which can produce a tip displacement of 40 µm under an applied voltage of 20 V. Then, calculate the response speed of this bimorph. The mass density of the ceramic is $\rho = 7.9$ g/cm³ and its elastic compliance is $s_{11}^E = 16 \times 10^{-12}$ m²/N.

Hint:

Two shim-less bimorph designs are illustrated in Fig. 2.16, where two poled piezo ceramic plates of equal thickness and length are bonded together with either (a) their polarization directions opposing each other or (b) parallel to each other. When the devices are operated under an applied voltage, V, with one end clamped (i.e., *cantilever condition*), the tip displacement, δ, is given by

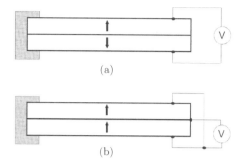

$$\delta = (3/2)d_{31}(l^2/t^2)V, \qquad (P2.2.1a)$$

$$\delta = 3d_{31}(l^2/t^2)V, \qquad (P2.2.1b)$$

where d_{31} is the piezoelectric transversal strain coefficient of the ceramic, t is the combined thickness of the two ceramic plates and l is the length of the bimorph. Equation (P2.2.1a) applies to the antiparallel polarization condition and Eq. (P2.2.1b) to the parallel polarization condition. Notice that the difference between the two cases arises from the difference in the electrode gap. The separation between the electrodes is equal to the combined thickness of the two plates for the antiparallel polarization case (a) and half that

Fig. 2.16: Two types of piezoelectric bimorphs: (a) the antiparallel polarization type and (b) the parallel polarization type.

thickness for the parallel polarization case (b). The merit of type (a) is simpler manufacturing process, leading to cheaper cost (the lead-wire connection from the center electrode in type (b) is tricky). The fundamental resonance frequency in both cases is determined by the combined thickness of the two plates, t, according to the following equation:

$$f_o = 0.161 \left[\frac{t}{l^2 \sqrt{\rho\, s_{11}^E}} \right] \qquad (P2.2.2)$$

where ρ is the mass density of the ceramic, and s_{11}^E is its elastic compliance.[29]

Solution:

When the device is to be operated under low voltages, the parallel polarization type of device pictured in Fig. 2.16(b) is preferred over the antiparallel polarization type pictured in Fig. 2.16(a) because it produces a twice larger displacement under these conditions. Substituting a length of $l = 25$ mm into Eq. (P2.2.1b), we obtain the combined piezoelectric plate thickness, t:

$$t = l\,\sqrt{3(d_{31}\,V/\delta)}$$

$$= 25 \times 10^{-3}\,(m)\,\sqrt{3\left[300 \times 10^{-12}(C/N)\,20\,(V)/40 \times 10^{-6}\,(m)\right]}$$

$$\rightarrow t = 530\,\mu m.$$

The piezo ceramic is sliced into plates $265\,\mu$m in thickness, 30 mm in length and 4–6 mm in width. The plates are electroded, poled and then bonded together in pairs. The width of the bimorph is usually chosen such that $[w/l < 1/5]$ in order to optimize the magnitude of the bending displacement.

The response time is estimated by the resonance period. We can determine the fundamental resonance frequency of the structure from the following equation:

$$f_o = 0.161 \left[\frac{t}{l^2 \sqrt{\rho\, s_{11}^E}} \right]$$

$$= 0.161 \left[\frac{530 \times 10^{-6}\,(\mathrm{m})}{[25 \times 10^{-3}\,(\mathrm{m})]^2 \sqrt{7.9 \times 10^3\,(\mathrm{kg/m^3})16 \times 10^{-12}(\mathrm{m^2/N})}} \right] = 378\,(\mathrm{Hz})$$

$$\rightarrow \quad \text{Response Time} \approx \frac{1}{f_o} = \frac{1}{378\,(\mathrm{s^{-1}})} = 2.6(\mathrm{ms})$$

This bimorph was used for the camera shutter in practice in the 1980s.

Example Problem 2.3.

A unimorph bending actuator can be fabricated by bonding a piezo ceramic plate to a metallic shim.[30] The tip deflection, δ, of the unimorph supported in a cantilever configuration is given by

$$\delta = \frac{d_{31} E \, l^2 \, Y_c \, t_c}{(Y_m \, [t_o^2 - (t_o - t_m)^2] + Y_c \, [(t_o + t_c)^2 - t_o^2])} \tag{P2.3.1}$$

Here, E is the electric field applied to the piezoelectric ceramic; d_{31}, the piezoelectric strain coefficient; l, the length of the unimorph; Y, Young's modulus for the ceramic or the metal; and t is the thickness of each material. The subscripts c and m denote the "ceramic" and the "metal", respectively. The quantity, t_o, is the distance between the *strain-free neutral plane* and the bonding surface, and is defined according to the following:

$$t_o = \frac{t_c \, t_m^2 (3 \, t_c + 4 t_m) Y_m + t_c^4 Y_c}{6 t_c \, t_m \, (t_c + t_m) \, Y_m} \tag{P2.3.2}$$

Assuming $Y_c = Y_m$, calculate the optimum (t_m/t_c) ratio that will maximize the deflection, δ, under the following conditions:

(a) A fixed ceramic thickness (you are a purchaser of a standard PZT plate), t_c, and
(b) a fixed total thickness (you need to keep the resonance frequency), $t_c + t_m$.

Solution:

Setting $Y_c = Y_m$, the Eqs. (P2.3.1) and (P2.3.2) become

$$\delta = \frac{d_{31} \, E \, l^2 \, t_c}{([t_o^2 - (t_o - t_m)^2] + [(t_o + t_c)^2 - t_o^2])}, \tag{P2.3.3}$$

$$t_o = \frac{t_c \, t_m^2 \, (3 t_c + 4 \, t_m) + t_c^4}{6 t_c \, t_m \, (t_c + t_m)} \tag{P2.3.4}$$

Substituting t_o as it is expressed in Eq. (P2.3.4) into Eq. (P2.3.3) yields

$$\delta = \frac{d_{31} \, E \, l^2 \, 3 \, t_m \, t_c}{(t_m + t_c)^3}. \tag{P2.3.5}$$

(a) The function $f(t_m) = (t_m t_c)/(t_m + t_c)^3$ must be maximized for a fixed ceramic thickness, t_c.

$$\frac{d \, f(t_m)}{d \, t_m} = \frac{(t_c - 2 \, t_m) \, t_c}{(t_m + t_c)^4} = 0 \tag{P2.3.6a}$$

The metal plate thickness should be $t_m = t_c/2$ and $t_o = t_c/2$.

(b) Equation (P2.3.6a) becomes under a fixed total thickness, $t_{tot} = t_c + t_m$:

$$\frac{d \, f(t_m)}{d \, t_m} = \frac{(t_{tot} - 2 \, t_m)}{t_{tot}^3} = 0 \tag{P2.3.6b}$$

Thus, it is determined that both the metal and ceramic plate thickness should be $t_m = t_c = t_{tot}/2$ and $t_o = t_{tot}/3$.

2.5 Piezoelectric Constitutive Equations

Now, we describe the necessary formulas for analyzing the piezoelectric effect, that is "piezoelectric constitutive equations" with electromechanical coupling factor k. Section 2.6 handles "piezoelectric figures of merit", in particular, the above k and "energy transmission coefficient" in detail.

2.5.1 *Piezoelectric Constitutive Equation Derivation*

2.5.1.1 *Extensive parameter description*

According to the International Union of Pure, and Applied Chemistry (IUPAC), an *extensive* parameter depends on the volume of the material (e.g., length, charge or entropy S becomes half by cutting the material in half), while an *intensive* parameter is the ratio of two extensive ones, and therefore, is independent on the volume of the material (e.g., force, voltage or temperature T does not change by cutting the material in half).[31] Consequently, stress (X) and electric field (E) are intensive parameters, which are externally controllable, while strain (x) and electric displacement (D) (almost the same as polarization (P) in this textbook) are extensive parameters, which are internally determined in the material. When we consider the free energy in terms of the "extensive" (i.e., material-related) parameters of strain, x, and electric displacement, D, we start from the differential form of the *Helmholtz free energy* designated by F, such that

$$dA = X\,dx + E\,dD - S\,dT \tag{2.5}$$

When we consider the equilibrium status ($dT = 0$) (i.e., *isothermal*), A is supposed to be expressed a

$$A = (1/2)c^D x^2 - h\,xD + (1/2)\kappa_0\kappa^x D^2, \tag{2.6}$$

where (x^2) or (D^2) terms correspond to the "intra-" coupling in between strain or in between electric displacement, respectively, and the product term (xD) is introduced to exhibit the "inter-" coupling between the strain and dielectric displacement. Note that the higher order "intra-" coupling terms (D^4) and/or (D^6) should be integrated to explain the temperature-dependent phase transition such as from para- to ferro-electric phase. We obtain from this energy function the following two linear *piezoelectric constitutive equations*:

$$X = \frac{\partial A}{\partial x} = c^D x - hD, \tag{2.7}$$

$$E = \frac{\partial A}{\partial D} = -hx + \kappa_0\kappa^x D, \tag{2.8}$$

where c^D is the elastic stiffness at constant electric displacement D (open-circuit conditions), h is the inverse piezoelectric charge coefficient, κ^x is the inverse dielectric constant at constant strain x (mechanically clamped conditions) and $\kappa_0 = 1/\varepsilon_0$ [vacuum permittivity $\varepsilon_0 = 8.854 \times 10^{-12}\mathrm{F/m}$].

2.5.1.2 *Intensive parameter description*

To the contrary, when we start with the *Gibbs free energy*, G, given in a general differential form in terms of the "intensive" (i.e., externally controllable) parameters of stress, X, and electric field, E as

$$dG = -x\,dX - D\,dE - S\,dT, \tag{2.9}$$

which in this case may be expressed under the equilibrium status ($dT = 0$) as

$$G = -(1/2)s^E X^2 - dXE - (1/2)\varepsilon_0\varepsilon^X E^2, \tag{2.10}$$

where X and E are the stress and electric field, respectively. Equation (2.10) is the energy expression in terms of the intensive physical parameters X and E. The temperature dependence of the function is associated with the elastic compliance, s^E, the dielectric constant, ε^X, and the piezoelectric charge coefficient, d.

We obtain from the Gibbs energy function the following two *piezoelectric constitutive equations*:

$$x = -\frac{\partial G}{\partial X} = s^E X + dE, \tag{2.11}$$

$$D = -\frac{\partial G}{\partial E} = dX + \varepsilon_0 \varepsilon^X E. \tag{2.12}$$

Note that the Gibbs energy function provides "intensive" physical parameters: E-constant (short-circuit condition) elastic compliance s^E and X-constant (stress-free) permittivity ε^X.

As described in Section 2.2, when an electric field is applied to a piezoelectric material, deformation (ΔL) or strain ($\Delta L/L$) arises. If the applied electric field and the generated stress are not large, the stress X and the dielectric displacement D can be represented by the following linear equations:

$$\begin{cases} x_i = s_{ij}^E X_j + d_{mi} E_m \\ D_m = d_{mi} X_i + \varepsilon_0 \varepsilon_{mk}^X E_k \end{cases} (i, j = 1, 2, \ldots, 6; m, k = 1, 2, 3) \tag{2.13a}$$
$$\tag{2.14a}$$

These are called the matrix-form *piezoelectric constitutive equations*. The strain x is generated in proportion to the applied stress X (Hooke's Law) and to the applied electric field E as the piezoelectric converse coupling effect, while the electric displacement D is generated in proportion to the applied electric field E and the applied stress X as the piezoelectric direct coupling effect. These *linear* piezoelectric constitutive equations do not include nonlinear effects (such as electrostrictor), nor the hysteresis in its original formula construction concept. By introducing the *complex parameters* with loss tangent dissipation factors, small hysteresis influence and the mechanical quality factor can be discussed.[32] Though, precisely speaking, the *dissipation function* must be incorporated with the phenomenological physics in order to discuss the loss and associated heat generation, the engineer sometimes cheats to obtain approximate solutions by neglecting the physical accuracy. As described in Section 2.5.2, the complex parameter analysis provides relatively good approximation to evaluate the losses in piezoelectrics, and the associated heat generation during operation.

The number of independent parameters for the lowest symmetry triclinic crystal is 21 for s_{ij}^E, 18 for d_{mi} and 6 for ε_{mk}^X. The number of independent parameters decreases with increasing crystallographic symmetry. Concerning the popular polycrystalline ceramics (Curie group $C_{\infty v}(\infty mm)$) such as PZT, the poled axis is usually denoted as the z-axis and the ceramic is isotropic with respect to this z-axis. The number of non-zero matrix elements in this case is 10 (s_{11}^E, s_{12}^E, s_{13}^E, s_{33}^E, s_{44}^E, d_{31}, d_{33}, d_{15}, ε_{11}^X, and ε_{33}^X): 10 tensor components, as given below [note $s_{66} = 2(s_{11} - s_{12})$], which are equivalent to $4\,mm$, $6\,mm$ in Table 2.3:

$$\begin{bmatrix} s_{11} & s_{12} & s_{13} & 0 & 0 & 0 \\ s_{12} & s_{11} & s_{13} & 0 & 0 & 0 \\ s_{13} & s_{13} & s_{33} & 0 & 0 & 0 \\ 0 & 0 & 0 & s_{44} & 0 & 0 \\ 0 & 0 & 0 & 0 & s_{44} & 0 \\ 0 & 0 & 0 & 0 & 0 & s_{66} \end{bmatrix}, \quad \begin{bmatrix} 0 & 0 & 0 & 0 & d_{15} & 0 \\ 0 & 0 & 0 & d_{15} & 0 & 0 \\ d_{31} & d_{31} & d_{33} & 0 & 0 & 0 \end{bmatrix},$$

$$\begin{bmatrix} \varepsilon_{11} & 0 & 0 \\ 0 & \varepsilon_{11} & 0 \\ 0 & 0 & \varepsilon_{33} \end{bmatrix}. \tag{2.15}$$

Equations (2.13a) and (2.14a) can be rewritten as matrix product combination:

$$\begin{pmatrix} x_1 \\ x_2 \\ x_3 \\ x_4 \\ x_5 \\ x_6 \end{pmatrix} = \begin{pmatrix} s_{11}^E & s_{21}^E & s_{31}^E & s_{41}^E & s_{51}^E & s_{61}^E \\ s_{12}^E & s_{22}^E & s_{32}^E & s_{42}^E & s_{52}^E & s_{62}^E \\ s_{13}^E & s_{23}^E & s_{33}^E & s_{43}^E & s_{53}^E & s_{63}^E \\ s_{14}^E & s_{24}^E & s_{34}^E & s_{44}^E & s_{54}^E & s_{64}^E \\ s_{15}^E & s_{25}^E & s_{35}^E & s_{45}^E & s_{55}^E & s_{65}^E \\ s_{16}^E & s_{26}^E & s_{36}^E & s_{46}^E & s_{56}^E & s_{66}^E \end{pmatrix} \begin{pmatrix} X_1 \\ X_2 \\ X_3 \\ X_4 \\ X_5 \\ X_6 \end{pmatrix} + \begin{pmatrix} d_{11} & d_{21} & d_{31} \\ d_{12} & d_{22} & d_{32} \\ d_{13} & d_{23} & d_{33} \\ d_{14} & d_{24} & d_{34} \\ d_{15} & d_{25} & d_{35} \\ d_{16} & d_{26} & d_{36} \end{pmatrix} \begin{bmatrix} E_1 \\ E_2 \\ E_3 \end{bmatrix}, \qquad (2.13b)$$

Table 2.3: Piezoelectric strain coefficient (d) matrices.[33]

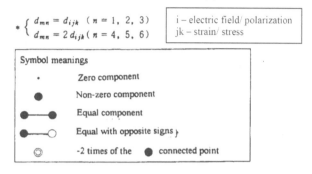

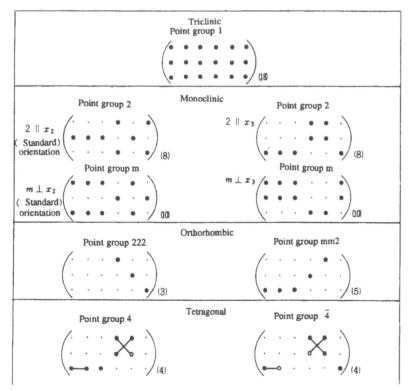

I Centro symmetric point group

Point group $\bar{1}$, $2/m$, mmm, $4/m$, $4/mmm$, $m3$, $m3m$, $\bar{3}$, $\bar{3}m$, $6/m$, $6/mmm$ All components are zero

II Non-centro symmetric point group

(*Continued*)

Table 2.3:   (*Continued*)

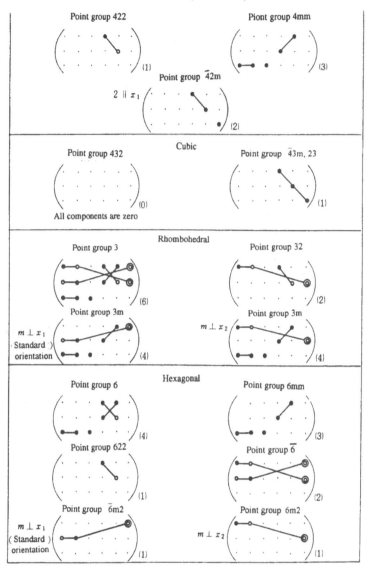

$$\begin{bmatrix} D_1 \\ D_2 \\ D_3 \end{bmatrix} = \begin{bmatrix} d_{11} & d_{12} & d_{13} & d_{14} & d_{15} & d_{16} \\ d_{21} & d_{22} & d_{23} & d_{24} & d_{25} & d_{26} \\ d_{31} & d_{32} & d_{33} & d_{34} & d_{35} & d_{36} \end{bmatrix} \begin{bmatrix} X_1 \\ X_2 \\ X_3 \\ X_4 \\ X_5 \\ X_6 \end{bmatrix} + \varepsilon_0 \begin{bmatrix} \varepsilon_{11}^X & \varepsilon_{12}^X & \varepsilon_{13}^X \\ \varepsilon_{21}^X & \varepsilon_{22}^X & \varepsilon_{23}^X \\ \varepsilon_{31}^X & \varepsilon_{32}^X & \varepsilon_{33}^X \end{bmatrix} \begin{bmatrix} E_1 \\ E_2 \\ E_3 \end{bmatrix}. \tag{2.14b}$$

Table 2.3 summarizes the piezoelectric strain coefficient (*d*) matrices for all 32 crystal point groups.[33]

Example Problem 2.4.

(a) Suppose that a shear stress X_{31} is applied on a crystal cube with a square cross-section such that it is deformed in a rhombus by 1° angle (Fig. 2.17). Shear strain is directly related with the deformed angle of the square in the unit (radian). Calculate the induced strain magnitude $x_5 (= 2x_{31})$.

(b) Calculate the deformed angle in a PZT ceramic with $d_{15} = 500\,\mathrm{pC/N}$ under $E_1 = 1\,\mathrm{kV/mm}$ applied.

Solution:

(a) Since $x_5 = 2x_{31} = \tan\theta \approx \theta$ and $1° = (\pi/180)$ rad, $x_5 = 0.017$. Remember that the angle of the rhombus less than $90°$ is taken as positive strain. (b) When we apply $1\,\mathrm{kV/mm}$ on a PZT ceramic with $d_{15} = 500\,\mathrm{pC/N}$, we can obtain $x_5 = d_{15}E_1 = 5 \times 10^{-4}$ (radian), which corresponds to 0.029 deg or 1.7 min, a very small angle change!

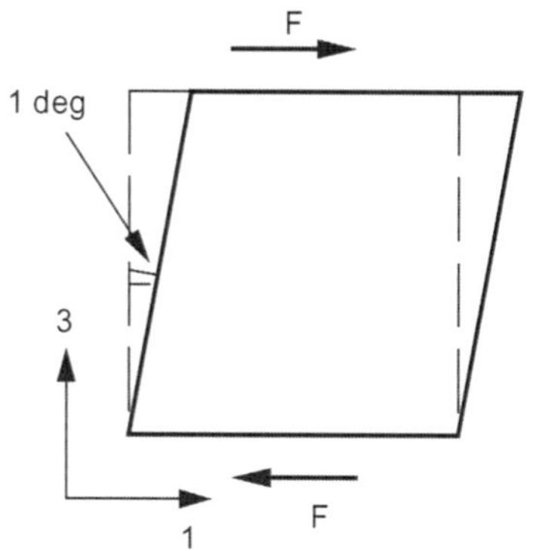

Example Problem 2.5.

Barium titanate (BaTiO$_3$, BT) has tetragonal crystal symmetry (point group 4 mm) at room temperature. The appropriate piezoelectric strain coefficient matrix is therefore of the following form from Table 2.3:

$$d_{ij} = \begin{pmatrix} 0 & 0 & 0 & 0 & d_{15} & 0 \\ 0 & 0 & 0 & d_{15} & 0 & 0 \\ d_{31} & d_{31} & d_{33} & 0 & 0 & 0 \end{pmatrix}$$

(a) Determine the strain induced in the BT, when an electric field is applied along the crystallographic c axis.

(b) Determine the strain induced in the BT, when an electric field is applied along the crystallographic a axis.

Hint:

The matrix equation that applies in this case is

$$\begin{bmatrix} x_1 \\ x_2 \\ x_3 \\ x_4 \\ x_5 \\ x_6 \end{bmatrix} = \begin{pmatrix} 0 & 0 & d_{31} \\ 0 & 0 & d_{31} \\ 0 & 0 & d_{33} \\ 0 & d_{15} & 0 \\ d_{15} & 0 & 0 \\ 0 & 0 & 0 \end{pmatrix} \begin{bmatrix} E_1 \\ E_2 \\ E_3 \end{bmatrix} \tag{P2.5.1}$$

Solution:

We can derive expressions for the induced strains

$$x_1 = x_2 = d_{31}E_3, x_3 = d_{33}E_3, x_4 = d_{15}E_2, x_5 = d_{15}E_1, x_6 = 0 \tag{P2.5.2}$$

so that the following determinations can be made:

(a) When E_3 is applied, elongation in the c direction ($x_3 = d_{33}E_3, d_{33} > 0$) and contraction in the a and b directions ($x_1 = x_2 = d_{31}E_3, d_{31} < 0$) occur (Fig. 2.18(a)). Note that PZT and other oxide piezoelectrics usually exhibit positive d_{33}, that is, extension along the electric field direction, while polymer PVDF shows negative d_{33}.

(b) When E_1 is applied, a shear strain $x_5 (= 2x_{31}) = d_{15}E_1$ is induced. The case where $d_{15} > 0$ and $x_5 > 0$ (less than $90°$) is illustrated in Fig. 2.18(b).

Fig. 2.17: Strain under a shear stress.

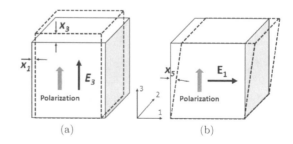

Fig. 2.18: Longitudinal (a) and shear (b) deformation under electric field, E3 and E1, respectively.

Example Problem 2.6.

For a cube-shaped specimen, tensile stress X and compressive stress $-X$ are applied simultaneously along the $(1\ 0\ 1)$ and $(\bar{1}\ 0\ 1)$ axes, respectively (Fig. 2.19). When we take the prime coordinates ($1'$ and $3'$) as illustrated in Fig. 2.19, the stress tensor is represented as

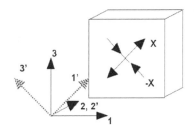

$$\begin{pmatrix} X & 0 & 0 \\ 0 & 0 & 0 \\ 0 & 0 & -X \end{pmatrix} \quad\quad (P2.6.1)$$

Fig. 2.19: Application of a pair of stresses X and –X to a cube of material.

(a) Using the transformation matrix $[A] = \begin{pmatrix} \cos\theta & 0 & \sin\theta \\ 0 & 1 & 0 \\ -\sin\theta & 0 & \cos\theta \end{pmatrix}$

(i.e., $\theta = -45°$ rotation along $2'$ axis in Fig. 2.19), calculate the rotated stress tensor $A \cdot X \cdot A^{-1}$.

(b) Then, verify that the above stress is equivalent to a pure shear stress in the original (non-prime) coordinates.

Solution:

Using $\theta = -45°$, we can obtain the transformed stress representation:

$$A \cdot X \cdot A^{-1} = \begin{pmatrix} 1/\sqrt{2} & 0 & -1/\sqrt{2} \\ 0 & 1 & 0 \\ 1/\sqrt{2} & 0 & 1/\sqrt{2} \end{pmatrix} \begin{pmatrix} X & 0 & 0 \\ 0 & 0 & 0 \\ 0 & 0 & -X \end{pmatrix} \begin{pmatrix} 1/\sqrt{2} & 0 & 1/\sqrt{2} \\ 0 & 1 & 0 \\ -1/\sqrt{2} & 0 & 1/\sqrt{2} \end{pmatrix} = \begin{pmatrix} 0 & 0 & X \\ 0 & 0 & 0 \\ X & 0 & 0 \end{pmatrix} \quad (P2.6.2)$$

The off-diagonal components X_{13} and X_{31} have the same magnitude X, and represent a pure shear stress. Note that a shear stress is equivalent to a combination of extension and contraction stresses. Only an extensional stress applied along a diagonal direction $1'$ may exhibit an apparently similar diagonal distortion (rhombus) of the crystal. However, precisely speaking, without the contraction along the $3'$ direction, this is not exactly equivalent to the pure shear deformation, with a volume expansion. The contraction occurs only from the Poisson's ratio of the extension.

2.5.1.3 *Extensive and intensive parameter correlations*

It is important to consider the conditions under which a piezoelectric material will be operated when characterizing the dielectric constant and elastic compliance. When a constant electric field is applied to a piezoelectric sample as illustrated in Fig. 2.20(a), the total input electric energy (*left*) under mechanically free condition should be equal to a combination of the energies associated with two distinct mechanical conditions that may be applied to the material: (1) stored electric energy under the *mechanically clamped state*, where a constant strain (i.e., *zero strain*) is maintained and the specimen cannot deform, and (2) converted mechanical energy under the *mechanically free state*, in which the material is not constrained and is free to deform. This situation can be expressed by

$$\left(\frac{1}{2}\right)\varepsilon^X\varepsilon_0 E_0^2 = \left(\frac{1}{2}\right)\varepsilon^x\varepsilon_0 E_0^2 + \left(\frac{1}{2s^E}\right)x^2 = \left(\frac{1}{2}\right)\varepsilon^x\varepsilon_0 E_0^2 + \left(\frac{1}{2s^E}\right)(dE_0)^2 \quad (2.16)$$

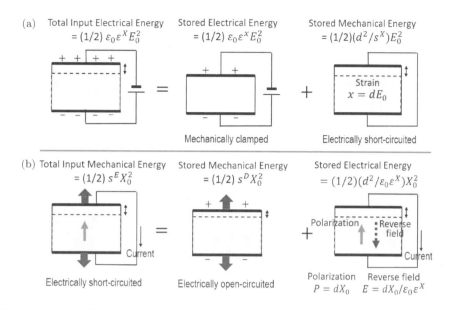

Fig. 2.20: Schematic representation of the response of a piezoelectric material under: (a) constant applied electric field and (b) constant applied stress conditions.

such that

$$\varepsilon^X \varepsilon_0 = \varepsilon^x \varepsilon_0 + \left(\frac{d^2}{s^E} \right), \tag{2.17a}$$

$$\varepsilon^x = \varepsilon^X (1 - k^2) \left[k^2 = \frac{d^2}{\varepsilon^X \varepsilon_0 s^E} \right]. \tag{2.17b}$$

When a constant stress is applied to the piezoelectric as illustrated in Fig. 2.20(b), the total input mechanical energy under short-circuit condition will be a combination of the energies associated with two distinct electrical conditions that may be applied to the material: (1) stored mechanical energy under the *open-circuit state*, where a constant electric displacement is maintained, and (2) converted electric energy under the *short-circuit condition* (i.e., depolarization field), in which the material is subject to a constant electric field. This can be expressed as

$$\left(\frac{1}{2} \right) s^E X_0^2 = \left(\frac{1}{2} \right) s^D X_0^2 + \left(\frac{1}{2} \right) \varepsilon^X \varepsilon_0 E^2 = \left(\frac{1}{2} \right) s^D X_0^2 + \left(\frac{1}{2} \right) \varepsilon^X \varepsilon_0 (d/\varepsilon_0 \varepsilon^X)^2 X_0^2. \tag{2.18}$$

which leads to

$$s^E = s^D + \left(\frac{d^2}{\varepsilon^X \varepsilon_0} \right), \tag{2.19a}$$

$$s^D = s^E (1 - k^2) \left[k^2 = \frac{d^2}{\varepsilon^X \varepsilon_0 s^E} \right]. \tag{2.19b}$$

Hence, we obtain the following equations for the permittivity and elastic compliance under constrained and unconstrained conditions:

$$\varepsilon^x / \varepsilon^X = (1 - k^2), \tag{2.20}$$

$$s^D / s^E = (1 - k^2), \tag{2.21}$$

where

$$k^2 = \frac{d^2}{s^E \varepsilon^X \varepsilon_0}. \tag{2.22}$$

We may also write equations of similar forms for the corresponding reciprocal quantities:

$$\kappa^X / \kappa^x = (1 - k^2), \tag{2.23}$$

$$c^E / c^D = (1 - k^2), \tag{2.24}$$

where, in this context,

$$k^2 = \frac{h^2}{c^D (\kappa_0 \kappa^x)}. \tag{2.25}$$

This new parameter k is also the *electromechanical coupling factor* in the extensive parameter description, and identical to the k in Eq. (2.22). It will be regarded as a real quantity for the cases we examine in this text (see Example Problem 2.7).

There are the following three relationships between the intensive and extensive parameters: permittivity under constant stress ($\varepsilon^X \varepsilon_0$), the elastic compliance under constant electric field (s^E) and the piezoelectric charge coefficient (d) in terms of their corresponding reciprocal quantities: inverse permittivity under constant strain $\kappa(\kappa_0 \kappa^x)$, the elastic stiffness under constant electric displacement (c^D) and the inverse piezoelectric coefficient (h). Note $k^2 = \frac{d^2}{s^E \varepsilon^X \varepsilon_0} = \frac{h^2}{c^D (\kappa_0 \kappa^x)}$.

$$\varepsilon^X \varepsilon_0 = \frac{1}{(\kappa_0 \kappa^x) \left[1 - \frac{h^2}{c^D (\kappa_0 \kappa^x)}\right]} = \frac{1}{(\kappa_0 \kappa^x)(1 - k^2)}, \tag{2.26}$$

$$s^E = \frac{1}{c^D \left[1 - \frac{h^2}{c^D (\kappa_0 \kappa^x)}\right]} = \frac{1}{c^D (1 - k^2)}, \tag{2.27}$$

$$d = \frac{\frac{h^2}{c^D (\kappa_0 \kappa^x)}}{h \left[1 - \frac{h^2}{c^D (\kappa_0 \kappa^x)}\right]} = \frac{k^2}{h(1 - k^2)}. \tag{2.28}$$

Example Problem 2.7.

(1) Verify the relationship

$$\frac{d^2}{s^E \varepsilon^X \varepsilon_0} = \frac{h^2}{c^D (\kappa^x / \kappa_0)}. \tag{P2.7.1}$$

This value is defined as the square of an electromechanical coupling factor (k^2), which should be the same even for different energy description systems (i.e., intensive or extensive description).

(2) Derive the relationships Eqs. (2.26)–(2.28).

Solution.

(1) When Eqs. (2.7) and (2.8) are combined with Eqs. (2.11) and (2.12), we obtain

$$X = c^D (s^E X + dE) - h(dX + \varepsilon_0 \varepsilon^X E), \tag{P2.7.2}$$

$$E = -h(s^E X + dE) + (\kappa_0 \kappa^x)(dX + \varepsilon_0 \varepsilon^X E), \tag{P2.7.3}$$

or upon rearranging

$$(1 - c^D s^E + hd)X + (h \varepsilon_0 \varepsilon^X - c^D d)E = 0, \tag{P2.7.4}$$

$$[h s^E - (\kappa_0 \kappa^x)d]X + [1 - (\kappa_0 \kappa^x)\varepsilon_0 \varepsilon^X + hd]E = 0. \tag{P2.7.5}$$

Combining the latter two equations yields

$$(1 - c^D s^E + hd)[1 - (\kappa_0 \kappa^x)\varepsilon_0 \varepsilon^X + hd] - (h\varepsilon_0 \varepsilon^X - c^D d)[hs^E - (\kappa_0 \kappa^x)d] = 0 \qquad (P2.7.6)$$

which, when simplified, produces the desired relationship

$$\frac{d^2}{s^E \varepsilon^X \varepsilon_0} = \frac{h^2}{c^D(\kappa^x \kappa_0)}. \qquad (P2.7.7)$$

(2) From the two constitutive equations

$$\begin{bmatrix} x \\ D \end{bmatrix} = \begin{bmatrix} s^E & d \\ d & \varepsilon_0 \varepsilon^X \end{bmatrix} \begin{bmatrix} X \\ E \end{bmatrix}, \text{ and } \begin{bmatrix} X \\ E \end{bmatrix} = \begin{bmatrix} c^D & -h \\ -h & \kappa_0 \kappa^x \end{bmatrix} \begin{bmatrix} x \\ D \end{bmatrix}, \qquad (P2.7.8a,b)$$

we obtain

$$\begin{bmatrix} x \\ D \end{bmatrix} = \begin{bmatrix} s^E & d \\ d & \varepsilon_0 \varepsilon^X \end{bmatrix} \begin{bmatrix} c^D & -h \\ -h & \kappa_0 \kappa^x \end{bmatrix} \begin{bmatrix} x \\ D \end{bmatrix}. \qquad (P2.7.9)$$

Thus, the following equation should be satisfied:

$$\begin{bmatrix} s^E & d \\ d & \varepsilon_0 \varepsilon^X \end{bmatrix} \begin{bmatrix} c^D & -h \\ -h & \kappa_0 \kappa^x \end{bmatrix} = \begin{bmatrix} 1 & 0 \\ 0 & 1 \end{bmatrix}. \qquad (P2.7.10)$$

Accordingly, $s^E c^D - dh = 1$, $-s^E h + d\kappa_0 \kappa^x = 0$, $dc^D - \varepsilon_0 \varepsilon^X h = 0$, $-dh + \varepsilon_0 \varepsilon^X \kappa_0 \kappa^x = 1$. Then, we obtain the relationships Eqs. (2.26)–(2.28), by expressing the intensive parameters $\varepsilon_0 \varepsilon^X$, s^E, d in terms of the extensive parameters $\kappa_0 \kappa^x$, c^D, h and k^2.

2.5.2 *Loss Integration in Piezoelectric Constitutive Equations*

Heat generation is one of the significant problems in piezoelectrics for high-power density applications. In this subsection, we review the loss phenomenology in piezoelectrics, including three losses: dielectric, elastic and piezoelectric losses. Heat generation at off-resonance is attributed mainly to intensive dielectric loss $\tan \delta'$, while the heat generation at resonance mainly originates from the intensive elastic loss $\tan \phi'$. The loss effect on the electromechanical resonance is discussed in Section 2.7.

2.5.2.1 *Intensive losses in piezoelectrics*

The terminologies, "*intensive*" and "*extensive*" losses are introduced here, in the relation with "intensive" and "extensive" parameters in the phenomenology by extending the concept introduced in Section 2.5.1. These are not related with the "intrinsic" and "extrinsic" losses which were introduced to explain the loss contribution from the mono-domain single-crystal state and from the others.[34] Our discussion in piezoelectrics is focused primarily on the "extrinsic" losses, in particular, losses originated from domain wall dynamics in the piezo ceramics. However, physical parameters of their performance such as permittivity and elastic compliance still differ in piezoelectric materials, depending on the boundary conditions: mechanically free or clamped, and electrically short-circuit or open-circuit. These are distinguished as "intensive" or "extensive" parameters and their associated losses.

We start from the following two *piezoelectric constitutive equations with losses* (refer to Eqs. (2.11) and (2.12) without losses):

$$x = s^{E*}X + d^*E, \qquad (2.29)$$

$$D = d^*X + \varepsilon^{X*}\varepsilon_0 E, \qquad (2.30)$$

where x is strain, X, stress, D, electric displacement, and E, electric field. Equations (2.29) and (2.30) are expressed with respect to *intensive* (i.e., externally controllable) physical parameters X and E. The elastic compliance s^{E*}, the dielectric constant ε^{X*} and the piezoelectric constant d^* are temperature-dependent in general. Note that the piezoelectric constitutive equations cannot yield a delay-time-related loss, in phenomenology, without taking into account *irreversible thermodynamic equations* or *dissipation functions*. However, the "dissipation functions" are mathematically equivalent to the introduction of "*complex physical constants*" into the phenomenological equations, if the loss is small and can be treated as a perturbation (*dissipation factor tangent* $<< 0.1$). Based on this mathematical principle, therefore, we introduce complex parameters ε^{X*}, s^{E*} and d^*, using $*$, in order to consider the small hysteresis losses in dielectric, elastic and piezoelectric constants:[35]

$$\varepsilon^{X*} = \varepsilon^X (1 - j \tan \delta'), \tag{2.31}$$

$$s^{E*} = s^E (1 - j \tan \phi'), \tag{2.32}$$

$$d^* = d(1 - j \tan \theta'). \tag{2.33}$$

θ' is the phase delay of the strain under an applied electric field, or the phase delay of the electric displacement under an applied stress. Both delay phases should be exactly the same if we introduce the same complex piezoelectric constant d^* into Eqs. (2.19) and (2.30). δ' is the phase delay of the electric displacement to an applied electric field under a constant stress (e.g., zero stress) condition, and ϕ' is the phase delay of the strain to an applied stress under a constant electric field (e.g., short-circuit) condition. The negative sign in front of the loss tangent comes from a general consensus that the output will be slightly delayed after the input. We will consider these phase delays as "intensive" losses.

Figures 2.21(a)–2.21(d) correspond to the model hysteresis curves for practical experiments: D vs. E curve under a stress-free condition, x vs. X under a short-circuit condition, x vs. E under a stress-free condition and D vs. X under a short-circuit condition for measuring current, respectively. Note that these measurements are easily conducted in practice. For example, $D - X$ relation under a short-circuit condition can be obtained from the integration of the measured current by changing the external stress. The average slope of the $D - E$ hysteresis curve in Fig. 2.21(a) corresponds to the permittivity $\varepsilon^X \varepsilon_0$, where the superscript stands for $X =$ constant (occasionally zero). Thus, $\tan \delta'$ is called "intensive" dielectric loss tangent. The situation of s^E is similar; the slope of the $x - X$ relation is the elastic compliance under $E =$ constant condition. It is worth noting that the actual hysteresis curve exhibits rather sharp edges at the maximum and minimum external parameter (electric field or stress) points, though the complex parameter usage in Eqs. (2.31)–(2.33) should generate rounded edges because it is an ellipse shape theoretically. You can easily imagine a sort of limitation of this complex parameter approach from the discrepancy with the experimental result.

Since the areas on the $D - E$ and $x - X$ domains exhibit directly the electrical and mechanical energies, respectively (see Figs. 2.21(a) and 2.21(b)), the stored energies (during a quarter cycle) and hysteresis losses (during a full electric or stress cycle) for pure dielectric and elastic energies can be calculated as

$$U_e = (1/2)\varepsilon^X \varepsilon_0 E_0^2, \tag{2.34}$$

$$w_e = \pi \varepsilon^X \varepsilon_0 E_0^2 \tan \delta', \tag{2.35}$$

and

$$U_m = (1/2)s^E X_0^2, \tag{2.36}$$

$$w_m = \pi s^E X_0^2, \tan \phi'. \tag{2.37}$$

The dissipation factors, $\tan \delta'$ and $\tan \phi'$, can experimentally be obtained by measuring the dotted hysteresis area and the stored energy area, that is, $(1/2\pi)(w_e/U_e)$ and $(1/2\pi)(w_m/U_m)$, respectively. Note that the

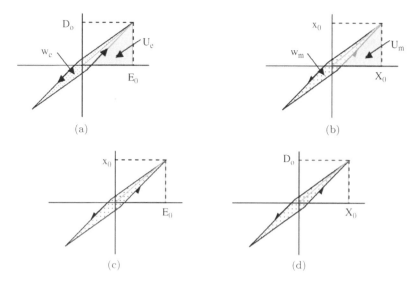

Fig. 2.21: (a) D vs. E (stress-free), (b) x vs. X (short-circuit), (c) x vs. E (stress-free) and (d) D vs. X (short-circuit) curves with hysteresis.

factor (2π) comes from integral per cycle (refer to Example Problem 2.8 on the above derivation). The electro-mechanical hysteresis loss calculations, however, are more complicated, because the areas on the $x - E$ and $P - X$ domains do not directly provide energy. The areas on these domains can be calculated as follows, depending on the measuring ways; when measuring the induced strain under an electric field, the electromechanical conversion energy can be calculated as follows, by converting E to stress X:

$$U_{\text{em}} = \int x dX = \left(\frac{1}{s^E}\right) \int x dx = (d^2/s^E) \int_0^{E_0} E dE = (1/2)(d^2/s^E)E_0^2, \tag{2.38}$$

where $x = d \cdot E$ was used. Then, using Eqs. (2.31) and (2.32), and from the imaginary part, we obtain the loss during a full cycle as

$$w_{\text{em}} = \pi(d^2/s^E)E_0^2 \left(2\tan\theta' - \tan\phi'\right). \tag{2.39}$$

Note that the area ratio in the strain vs. electric field measurement should provide the combination of piezoelectric loss $\tan\theta'$ and elastic loss $\tan\phi'$ (not $\tan\theta'$ directly!).

To the contrary, when we measure the induced charge under stress, the stored energy U_{me} and the hysteresis loss w_{me} during a quarter and a full stress cycle, respectively, are obtained similarly as

$$U_{\text{me}} = \int P dE = (1/2)(d^2/\varepsilon_0\varepsilon^X)X_0^2, \tag{2.40}$$

$$w_{\text{me}} = \pi(d^2/\varepsilon_0\varepsilon^X)X_0^2 \left(2\tan\theta' - \tan\delta'\right). \tag{2.41}$$

Now, the area ratio in the charge vs. stress measurement provides the combination of piezoelectric loss $\tan\theta'$ and dielectric loss $\tan\delta'$. Hence, from the measurements of D vs. E and x vs. X, we obtain $\tan\delta'$ and $\tan\phi'$, respectively, and either the piezoelectric (D vs. X) or converse piezoelectric measurement (x vs. E) provides $\tan\theta'$ through a numerical subtraction. The above equations provide a traditional off-resonance loss measuring technique on piezoelectric actuators, an example of which is introduced in Section 2.7.2.

Example Problem 2.8.

When the observed variation in electric displacement, D, can be represented as if it had a slight phase lag with respect to the applied electric field,

$$E^* = E_0 e^{j\omega t}, \tag{P2.8.1}$$

$$D^* = D_0 e^{j(\omega t - \delta)}, \tag{P2.8.2}$$

If we express the relationship between $D*$ and $E*$ as

$$D^* = \varepsilon^* \varepsilon_0 E^*, \tag{P2.8.3}$$

where the *complex dielectric constant*, ε^*, is

$$\varepsilon^* = \varepsilon' + j\varepsilon'', \tag{P2.8.4}$$

$$\varepsilon''/\varepsilon' = \tan\delta. \tag{P2.8.5}$$

The integrated area inside the hysteresis loop, labeled w_e in Fig. 2.21(a), is equivalent to the energy loss per cycle per unit volume of the dielectric. It is defined for an isotropic dielectric as

$$w_e = -\int DdE = -\int_0^{\frac{2\pi}{\omega}} D\frac{dE}{dt}dt \tag{P2.8.6}$$

(1) Substituting the real parts of the electric field, E^*, and electric displacement, D^*, into Eq. (P2.8.6), we obtain the following equations:

$$w_e = \pi\varepsilon''\varepsilon_0 E_0^2 = \pi\varepsilon'\varepsilon_0 E_0^2 \tan\delta \tag{P2.8.7}$$

(2) Verify an alternative expression for the dissipation factor:

$$\tan\delta = (1/2\pi)(w_e/U_e), \tag{P2.8.8}$$

where U_e, the integrated area so labeled in Fig. 2.21(a), represents the energy stored during a quarter cycle.

Solution:

(1) The integrated area inside the hysteresis loop, labeled w_e in Fig. 2.21(a), is equivalent to the energy loss per cycle per unit volume of the dielectric. It is defined as $w_e = -\int DdE = -\int_0^{\frac{2\pi}{\omega}} D\frac{dE}{dt}dt$. Substituting the real parts of the electric field, E^*, and electric displacement, D^*, into Eq. (P2.8.6) yields

$$w_e = \int_0^{\frac{2\pi}{\omega}} D_0 \cos(\omega t - \delta)[E_0\omega \cdot \sin(\omega t)]dt = D_0 E_0 \cdot \sin(\delta) \int_0^{\frac{2\pi}{\omega}} \omega \cdot \sin^2(\omega t)dt$$

$$= \pi E_0 D_0 \sin(\delta) \tag{P2.8.9}$$

so that

$$w_e = \pi\varepsilon''\varepsilon_0 E_0^2 = \pi\varepsilon'\varepsilon_0 E_0^2 \tan\delta. \tag{P2.8.10}$$

(2) When there is a phase lag, an energy loss (or non-zero w_e) will occur for every cycle of the applied electric field, resulting in the heat generation in the dielectric material. The quantity $\tan\delta$ is referred to as the *dissipation factor*. The electrostatic energy stored during a half cycle of the applied electric field is $2U_e$, where U_e, the integrated area so labeled in Fig. 2.21(a), represents the energy stored during a quarter cycle.

$$2U_e = 2[(1/2)(E_0 D_0 \cos\delta)] = (E_0 D_0)\cos\delta. \tag{P2.8.11}$$

Knowing that $\varepsilon'\varepsilon_0 = (D_0/E_0)\cos\delta$, Eq. (P2.8.11) may be rewritten in the form

$$2U_e = \varepsilon'\varepsilon_0 E_0^2 \qquad (P2.8.12)$$

Then, an alternative expression for the dissipation factor can be obtained:

$$\tan\delta = (1/2\pi)(w_e/U_e). \qquad (P2.8.13)$$

Note that the factor 2π comes from the integration process for one cycle.

2.5.2.2 *Extensive losses in piezoelectrics*

So far, we discussed the "intensive" dielectric, mechanical and piezoelectric losses (with prime notation) in terms of "intensive" parameters X and E. In order to consider "physical" meanings of the losses in the material (e.g., domain dynamics), we will introduce the "extensive" losses[32] in respect to "extensive" parameters x and D. In practice, intensive losses are easily measurable, but extensive losses are not, in the pseudo-DC measurement. That is, it is experimentally difficult to keep completely open circuit without leakage current for a long period (typically maximum 10 minutes or so), or a perfectly clamped condition of the sample. We usually obtain them from the intensive losses by using the *K-matrix* introduced later. The extensive losses are essential when we consider a physical microscopic or semi-macroscopic domain dynamics model. We start again from the piezoelectric constitutive equations with losses in respect to extensive parameters x and D,

$$X = c^{D*}x - h^*D, \qquad (2.42)$$

$$E = -h^*x + \kappa^{x*}\kappa_0 D, \qquad (2.43)$$

where c^{D*} is the elastic stiffness under D = constant condition (i.e., electrically open-circuit), κ^{x*} is the inverse dielectric constant under x = constant condition (i.e., mechanically clamped) and h^* is the inverse of the piezoelectric constant d^* (i.e., inverse tensor). We introduce the *extensive* dielectric, elastic and piezoelectric losses as

$$\kappa^{x*} = \kappa^x(1 + j\tan\delta), \qquad (2.44)$$

$$c^{D*} = c^D(1 + j\tan\phi), \qquad (2.45)$$

$$h^* = h(1 + j\tan\theta). \qquad (2.46)$$

The sign "+" here in front of the imaginary "j" is taken by a general induction principle, that is, "'polarization induced after electric field application" and "strain induced after stress application". However, the sign of the coupling factor (piezoelectric) loss is not very trivial. Regarding the intensive loss $\tan\theta'$, the meaning is simple and may be positive, that is, the time delay of the strain induced by the field or the delay of the polarization induced by the applied stress, while what is the physical meaning of the extensive piezo loss $\tan\theta$? This may be related with the piezoelectric origin. We can consider two piezoelectric origin models: ferroelectricity is primary, coupled with elasticity, or ferroelasticity is primary, coupled with polarization (phase transition order parameter difference). At present, we presume that all loss factors are positive. It is notable that the permittivity under a constant strain (e.g., zero strain or completely clamped) condition, ε^{x*}, and the elastic compliance under a constant electric displacement (e.g., open-circuit) condition, s^{D*} can be provided as an inverse value of κ^x and c^D, respectively, in this simplest 1D expression. Thus, using exactly the same losses in Eqs. (2.31) and (2.32),

$$\varepsilon^{x*} = \varepsilon^x(1 - j\tan\delta), \qquad (2.47)$$

$$s^{D*} = s^D(1 - j\tan\phi), \qquad (2.48)$$

we will consider these phase delays again as "extensive" losses. Care should be taken in the case of a general 3D expression, where this part must be translated as "inverse" *matrix components* of κ^{x*} and c^{D*} tensors.

2.5.2.3 *Correlation between the intensive and extensive losses — [K]-matrix*

Here, we consider the physical property difference between the boundary conditions: E constant and D constant, or X constant and x constant in a simplest 1D model. Referring to Figs. 2.21(a) and 2.21(b) again, we derived the following relations:

$$\varepsilon^x/\varepsilon^X = (1-k^2), \quad s^D/s^E = (1-k^2),$$

$$\kappa^X/\kappa^x = (1-k^2), \quad c^E/c^D = (1-k^2),$$

$$k^2 = \frac{d^2}{s^E \varepsilon^X \varepsilon_0} = \frac{h^2}{c^D(\kappa^x/\varepsilon_0)} \tag{2.49}$$

This k is called the *electromechanical coupling factor*, which is defined as a real number in this textbook. In order to obtain the relationships between the intensive and extensive losses, the following three equations (already derived in Section 2.5.1.3) are essential:

$$\varepsilon^X \varepsilon_0 = \frac{1}{\left(\frac{\kappa^x}{\varepsilon_0}\right)\left[1 - \frac{h^2}{c^D(\kappa^x/\varepsilon_0)}\right]}, \tag{2.50}$$

$$s^E = \frac{1}{c^D\left[1 - \frac{h^2}{c^D(\kappa^x/\varepsilon_0)}\right]}, \tag{2.51}$$

$$d = \frac{\frac{h^2}{c^D(\kappa^x/\varepsilon_0)}}{h\left[1 - \frac{h^2}{c^D(\kappa^x/\varepsilon_0)}\right]}. \tag{2.52}$$

Replacing the parameters in Eqs. (2.50)–(2.52) by the complex parameters in Eqs. (2.31)–(2.33), (2.44)–(2.46), we obtain the relationships between the intensive and extensive losses:

$$\tan\delta' = (1/(1-k^2))[\tan\delta + k^2(\tan\phi - 2\tan\theta)], \tag{2.53}$$

$$\tan\phi' = (1/(1-k^2))[\tan\phi + k^2(\tan\delta - 2\tan\theta)], \tag{2.54}$$

$$\tan\theta' = (1/(1-k^2))[\tan\delta + \tan\phi - (1+k^2)\tan\theta], \tag{2.55}$$

where k is the *electromechanical coupling factor* defined by Eq. (2.49), and here as a real number. It is important that the *intensive* dielectric, elastic and piezoelectric losses (with prime) are mutually correlated with the extensive dielectric, elastic and piezoelectric losses (non-prime) through the electromechanical coupling k^2, and that the denominator $(1-k^2)$ comes basically from the ratios, $\varepsilon^x/\varepsilon^X = (1-k^2)$ and $s^D/s^E = (1-k^2)$, and this real part reflects the dissipation factor when the imaginary part is divided by the real part. Knowing the relationships between the intensive and extensive physical parameters, and the electromechanical coupling factor k, the intensive (prime) and extensive (non-prime) loss factors have the following relationship:[36]

$$\begin{bmatrix}\tan\delta'\\\tan\phi'\\\tan\theta'\end{bmatrix} = [K]\begin{bmatrix}\tan\delta\\\tan\phi\\\tan\theta\end{bmatrix}, \tag{2.56}$$

$$[K] = \frac{1}{1-k^2}\begin{bmatrix}1 & k^2 & -2k^2\\k^2 & 1 & -2k^2\\1 & 1 & -1-k^2\end{bmatrix}, \quad k^2 = \frac{d^2}{s^E(\varepsilon^X\varepsilon_0)} = \frac{h^2}{c^D(\kappa^x\kappa_0)}. \tag{2.57}$$

The matrix $[K]$ is proven to be *"invertible"*, i.e., $K^2 = I$, or $K = K^{-1}$, where I is the identity matrix. Hence, the conversion relationship between the intensive (prime) and extensive (non-prime) exhibits full symmetry. The author emphasizes again that the extensive losses are more important for considering the physical micro/macroscopic models, and can be obtained mathematically from a set of intensive losses, which can be obtained more easily from the experiments (in particular, pseudo-DC measurement).

2.6 Figures of Merit in Piezoelectrics

In order to compare the performance superiority of actuator, transducer materials, we usually use "Figure of Merit (FOM)". There are five types of important FOMs for transducers, in particular, piezoelectric materials: (1) the piezoelectric coefficient, d, g, etc., (2) the electromechanical coupling factor, k, energy transmission coefficient, λ, efficiency, η, (3) the mechanical quality factor Q_m, (4) the acoustic impedance Z, and (5) the maximum vibration velocity $v_{\max}$. Each of these quantities is defined in this section.

2.6.1 *Piezoelectric Constants*

Let us start from the piezoelectric constitutive equations. There are four pairs of description types, depending on the intensive (E, X)/extensive (D, x) parameters:

$$\begin{cases} x = s^E X + dE \\ D = dX + \varepsilon_0 \varepsilon^X E \end{cases}, \tag{2.58}$$

$$\begin{cases} X = c^D x - hD \\ E = -hx + \kappa_0 \kappa^x D \end{cases}, \tag{2.59}$$

$$\begin{cases} x = s^D X + gD \\ E = -gX + \kappa_0 \kappa^x D \end{cases}, \tag{2.60}$$

$$\begin{cases} X = c^E x - eE \\ D = ex + \varepsilon_0 \varepsilon^X E \end{cases}. \tag{2.61}$$

An *intensive* quantity is one whose magnitude is independent of the size of the system, whereas an *extensive* quantity is one whose magnitude is additive for subsystems (IUPAC definition). In practice, intensive E and X are externally controllable parameters, while extensive D and x are internal material's parameters. Accordingly, there are four types of piezoelectric coefficients, d, h, g, and e. The magnitude of the strain, x, induced by an applied electric field, E, is characterized by the *piezoelectric strain coefficient*, d, as

$$x = (d)E. \tag{2.62}$$

This quantity is an important *figure of merit for actuators* (Eq. (2.58a)). The induced electric field, E, is related to the applied stress, X, through the piezoelectric voltage coefficient, g, as

$$E = (g)X. \tag{2.63}$$

This quantity is an important *figure of merit for sensors* (Eq. (2.60b)). Recall that the direct piezoelectric effect is described by $P = (d)X$, where P is the induced polarization (almost equal to D for a large permittivity material). When we combine this expression with Eq. (2.63), we obtain an important relationship between g and d:

$$g = d/\varepsilon_0 \varepsilon^X, \tag{2.64}$$

where ε^X is the dielectric constant/relative permittivity under a free (unclamped) condition. Equation (2.61) is popularly used for analyzing piezoelectric thin films, where the film strain is constrained/clamped by the thick substrate. The electric displacement measured via the short-circuit current by changing the strain via the substrate bending can provide the piezoelectric e constant, which is related as

$$e = d/s^E, \tag{2.65}$$

where s^E is the elastic compliance under a short-circuit condition. Finally, h is basically an inverse component of the d tensor.

2.6.2 *Electromechanical Coupling Factor k and Related Coefficients*

The terms, *electromechanical coupling factor*, *energy transmission coefficient* and *efficiency* are sometimes confused by the junior researchers. All are related to the conversion rate between electrical energy and mechanical energy under the isothermal (T = constant) condition, but their definitions are different.[37]

2.6.2.1 *The electromechanical coupling factor k*

The piezoelectric can transduce the input electric energy to the output mechanical energy, and vice versa. Thus, let us introduce the *electromechanical coupling factor k*, which corresponds to the rate of electromechanical transduction. Five different definitions are introduced in this subsection, which are actually equivalent.

(a) *Mason's Definition*: When we apply the electric field on a piezoelectric material pseudo-statically ($\omega \to 0$), the electro-mechanical coupling factor is provided as follows:

$$k^2 = \text{(Stored mechanical energy/Input electrical energy)} \tag{2.66a}$$

or when we apply the mechanical force on the sample:

$$k^2 = \text{(Stored electrical energy/Input mechanical energy)} \tag{2.66b}$$

This definition was introduced by Mason.[38] Let us calculate Eq. (2.66a), referring to Fig. 2.22(a). Since the input electrical energy is $(1/2)\,\varepsilon_0\varepsilon E^2$ per unit volume and the stored mechanical energy per unit volume under zero external stress is given by $(1/2)\,x^2/s^E = (1/2)(dE)^2/s^E$ (we assume the piezoelectric strain $x = dE$), k^2 can be calculated as

$$k^2 = [(1/2)(dE)^2/s^E]/[(1/2)\varepsilon_0\varepsilon^X E^2] = d^2/\varepsilon_0\varepsilon^X \cdot s^E. \tag{2.67a}$$

On the contrary, regarding Eq. (2.66b), when a stress X is applied to a piezoelectric material (Fig. 2.22(b)), since the input mechanical energy is $(1/2s^E)X^2$ per unit volume and the converted electrical energy per unit volume under short-circuit (or E constant) condition is given by $(1/2\varepsilon_0\varepsilon^X)P^2 = (1/2\varepsilon_0\varepsilon^X)(dX)^2$ (we assume the polarization $P = dX$), k^2 can be calculated as

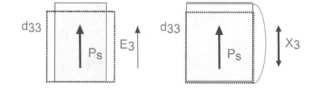

Fig. 2.22: Two ways for calculating the electromechanical coupling factor k under (a) field, and (b) stress.

$$k^2 = [(1/2\varepsilon_0\varepsilon^X)(dX)^2]/[(1/2s^E)X^2] = d^2/\varepsilon_0\varepsilon^X \cdot s^E. \tag{2.67b}$$

Note that the final expressions Eqs. (2.67a) and (2.67b) for the electromechanical coupling factor k are exactly the same as you expect. Though the k is called the *electromechanical coupling factor*, k^2 has the actual physical meaning as the *energy conversion rate*. The k is not merely the material's constant, but mode-dependent.

(b) *Definition in Materials*: The internal energy (per unit volume) U of a piezoelectric is given by summation of the mechanical energy U_M $(= \int x\,dX)$ and the electrical energy U_E $(= D\,dE)$. U is calculated as follows, when linear relation of Eqs. (2.58) is applicable [equivalent to Eq. (2.10)]:

$$U = U_M + U_E = \left[(1/2)\sum_{i,j} s_{ij}{}^E X_j X_i + (1/2)\sum_{m,i} d_{mi} E_m X_i \right] + \left[(1/2)\sum_{m,i} d_{mi} X_i E_m \right.$$

$$\left. + (1/2)\sum_{k,m} \varepsilon_0 \varepsilon_{mk}{}^X E_k E_m \right]$$

$$= U_{MM} + 2U_{ME} + U_{EE} = (1/2)\sum_{i,j} s_{ij}{}^E X_j X_i$$

$$+ 2\cdot(1/2)\sum_{m,i} d_{mi} E_m X_i + (1/2)\sum_{k,m} \varepsilon_0 \varepsilon_{mk}{}^X E_k E_m \tag{2.68}$$

The s and ε terms represent purely mechanical and electrical energies (U_{MM} and U_{EE}), respectively, and the d term denotes the energy transduced from electrical to mechanical energy or vice versa through the piezoelectric effect (U_{ME}). The electromechanical coupling factor k is defined by

$$k^2 = U_{ME}^2 / U_{MM} U_{EE}. \tag{2.69}$$

Using s^E, $\varepsilon_0 \varepsilon^X$ and d notations,

$$k^2 = \frac{\left(\frac{1}{2} dEX\right)^2}{\left(\frac{1}{2} s^E X^2\right)\left(\frac{1}{2}\varepsilon_0 \varepsilon^X E^2\right)} = \frac{d^2}{s^E \varepsilon^X \varepsilon_0}. \tag{2.70}$$

(c) *Definition in Devices*: Though the constitutive equations can be derived from the internal energy in (b), since the key equations are limited depending on the specimen geometry, there are several definitions according to the mode considered (try Example Problem 2.9):

$$\begin{bmatrix} x \\ D \end{bmatrix} = \begin{bmatrix} s^E & d \\ d & \varepsilon_0 \varepsilon^X \end{bmatrix} \begin{bmatrix} X \\ E \end{bmatrix},$$

the electromechanical coupling factor is defined by

$$k^2 = \frac{(\text{Coupling factor})^2}{(\text{Product of the diagonal parameters})} = \frac{(d)^2}{(s^E \varepsilon_0 \varepsilon^X)}. \tag{2.71}$$

(d) From the relations between the E-constant, E-constant elastic compliances, s^E, s^D, stiffness c^E, c^D; and stress-free, strain-free permittivity $\varepsilon_0 \varepsilon^X$, $\varepsilon_0 \varepsilon^x$, inverse permittivity $\kappa_0 \kappa^X$, $\kappa_0 \kappa^x$ (refer to Eqs. (2.20)–(2.25)),

$$1 - k^2 = \frac{s^D}{s^E} = \frac{c^E}{c^D} = \frac{\varepsilon^x}{\varepsilon^X} = \frac{\kappa^X}{\kappa^x}. \tag{2.72}$$

(e) *Dynamic Definition*: In the 4-terminal equivalent circuit (introduced in Section 2.8) in Fig. 2.23, using the electric terminal parameters voltage V and current I, and the mechanical terminal parameters force F and vibration velocity $\dot{u}$ related to each other as

$$\begin{bmatrix} F \\ I \end{bmatrix} = \begin{bmatrix} Z_1 & -\Phi \\ \Phi & Y_1 \end{bmatrix} \begin{bmatrix} \dot{u} \\ V \end{bmatrix}, \tag{2.73}$$

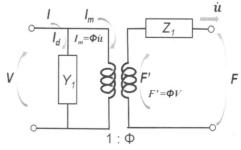

Fig. 2.23: 4-port equivalent circuit model.

the dynamic electromechanical coupling factor k_v^2 is defined by [(complex power in the mechanical branch)/(complex power in the electrical branch)] under short-circuit condition of mechanical terminal, or [(complex power in the electrical branch)/(complex power in the mechanical branch)] under short-circuit condition of electrical terminal, which leads to

$$k_v^2 = \left| \frac{\left(\frac{\Phi^2}{Z_1 Y_1}\right)}{1 + \left(\frac{\Phi^2}{Z_1 Y_1}\right)} \right|. \tag{2.74}$$

Since $Z_1 = jZ_0 \tan(\frac{\omega L}{2v})$, $Y_1 = j\omega C_d$, $\phi = \frac{2d_{31}w}{s_{11}^E}$ and $Z_0 = wb\rho v = \frac{wb}{vs_{11}^E}$ in the k_{31} mode, k_v^2 is ω dependent. By taking $\omega \to 0$, $k_v^2 \to k_{31}^2 = \frac{d_{31}^2}{s^E \varepsilon_0 \varepsilon^X}$. Refer to Example Problem 2.15 for this derivation.

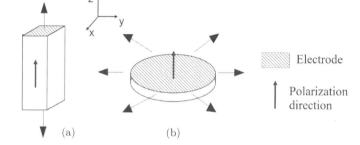

Fig. 2.24:   (a) Longitudinal length extension, (b) planar extension vibration modes of piezoelectric devices.

Example Problem 2.9

Calculate the electromechanical coupling factor k_{ij} of a piezoelectric ceramic vibrator for the following vibration modes (see Fig. 2.24, where vibration // E (a) or $\perp E$ (b)):

(a) Longitudinal length extension mode (// E): k_{33}
(b) Planar extension mode of the circular plate: k_p

Hint:

From the constitutive equations (Definition (c)), the electro-mechanical coupling factor is defined by

$$k^2 = \frac{(\text{Coupling factor})^2}{(\text{Product of the diagonal parameters})} = \frac{(d)^2}{(s^E \varepsilon_0 \varepsilon^X)}. \tag{P2.9.1}$$

Solution.

(a) The relating equations for this k_{33} mode are

$$x_3 = s_{33}^E X_3 + d_{33} E_3,$$

$$D_3 = d_{33} X_3 + \varepsilon_{33}{}^X E_3,$$

$$\to k_{33} = d_{33} / \sqrt{s_{33}^E \cdot \varepsilon_{33}^X} \tag{P2.9.2}$$

(b) The relating equations for the k_p mode are the following three, including 2-D x_1 and x_2 equations:

$$x_1 = s_{11}^E X_1 + s_{12}^E X_2 + d_{31} E_3,$$

$$x_2 = s_{12}^E X_1 + s_{22}^E X_2 + d_{32} E_3,$$

$$D_3 = d_{31} X_1 + d_{32} X_2 + \varepsilon_{33}{}^X E_3.$$

Assuming axial symmetry, $s_{11}^E = s_{22}^E$, $d_{31} = d_{32}$ and $X_1 = X_2 (= X_p)$, the above equations are transformed to the following two equations:

$$x_1 + x_2 = 2(s_{11}^E + s_{12}^E) X_p + 2d_{31} E_3,$$

$$D_3 = 2d_{31} X_p + \varepsilon_{33}{}^X E_3.$$

$\rightarrow U_{ME}$ comes from the d_{31} term as $(1/2)\cdot 2d_{31}E_3 X_p$, U_{MM} comes from the $s_{11}{}^E$ term as $(1/2)\cdot 2(s_{11}{}^E + s_{12}{}^E)X_p{}^2$ and U_{EE} comes from the $\varepsilon_{33}{}^X$ term as $(1/2)\varepsilon_{33}{}^X E_3{}^2$. Thus,

$$k_p = 2d_{31}/\sqrt{2(s_{11}^E + s_{12}^E)\cdot \varepsilon_{33}^X}$$

$$= [d_{31}/\sqrt{s_{11}^E \cdot \varepsilon_{33}^X}]\cdot \sqrt{2/(1-\sigma)} = k_{31}\cdot \sqrt{2/(1-\sigma)}, \tag{P2.9.3}$$

where σ is Poisson's ratio given by

$$\sigma = -s_{12}^E/s_{11}^E. \tag{P2.9.4}$$

Since $\sigma \approx 1/3$, $k_p \approx \sqrt{3}k_{31}$.

2.6.2.2 *The energy transmission coefficient* $\lambda_{\max}$

Not all the stored energy can be actually used, and the actual work done depends on the mechanical load. With zero mechanical load or a complete clamp (no strain), no output work is done. The energy transmission coefficient is defined by

$$\lambda_{\max} = (\text{Output mechanical energy/Input electricalenergy})_{\max} \tag{2.75a}$$

or equivalently,

$$\lambda_{\max} = (\text{Output electrical energy/Input mechanical energy})_{\max} \tag{2.75b}$$

The difference of the above Eq. (2.75) from Eq. (2.66) is "stored" or "output/spent".

Let us consider the case where an electric field E is applied to a piezoelectric under constant external stress X (< 0, because a compressive stress is necessary to work to the outside). This corresponds to the situation that a mass is put suddenly on the actuator, as shown in Fig. 2.25(a). Figure 2.25(b) shows two electric field versus induced strain curves, corresponding to two conditions: under the mass load and no

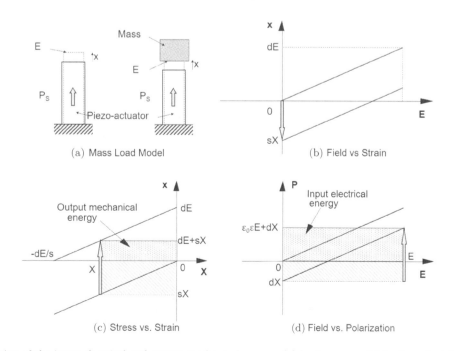

(a) Mass Load Model   (b) Field vs Strain

(c) Stress vs. Strain   (d) Field vs. Polarization

Fig. 2.25: Calculation of the input electrical and output mechanical energy: (a) load mass model for the calculation, (b) electric field vs. induced strain curve, (c) stress vs. strain curve and (d) electric field vs. polarization curve.

mass. Because the area on the field-strain domain does not mean the energy, we should use the stress–strain and field-polarization domains in order to discuss the mechanical and electrical energy, respectively. Figure 2.25(c) illustrates how to calculate the mechanical energy. Note that the mass shrinks the actuator first by sX (s: piezo material's compliance, and $X < 0$). This mechanical energy sX^2 is a sort of "loan" of the actuator credited from the mass, which should be subtracted later. This energy corresponds to the hatched area in Fig. 2.25(c). By applying the step electric field, the actuator expands by the strain level dE under a constant stress condition. This is the mechanical energy provided from the actuator to the mass, which corresponds to $|dEX|$. Like paying back the initial "loan", the output work (from the actuator to the mass) can be calculated as the area subtraction (shown by the dotted area in Fig. 2.25(c))

$$\int (-X)\, dx = -(dE + sX)X. \tag{2.76}$$

Figure 2.25(d) illustrates how to calculate the electrical energy. The mass load X generates the "loan" electrical energy by inducing $P = dX$ (see the hatched area in Fig. 2.25(d)). By applying a sudden electric field E, the actuator (like a capacitor) receives the electrical energy of $\varepsilon_0 \varepsilon E^2$. Thus, the total energy is given by the area subtraction (shown by the dotted area in Fig. 2.25(d))

$$\int (E)dP = (\varepsilon_0 \varepsilon E + dX)E. \tag{2.77}$$

We need to choose a proper load to maximize the *energy transmission coefficient*. From the maximum condition of

$$\lambda = -x \cdot X/P \cdot E = X \int (-X)\, dx/E \int (E)dP$$

$$= -[d(X/E) + s(X/E)^2]/[\varepsilon_0 \varepsilon + d(X/E)]. \tag{2.78}$$

Letting $y = X/E$, then

$$\lambda = -(sy^2 + dy)/(dy + \varepsilon_0 \varepsilon).$$

The maximum λ can be obtained when y satisfies

$$(d\lambda/dy) = [-(2sy + d) \cdot (dy + \varepsilon_0 \varepsilon) + (sy^2 + dy) \cdot d]/(dy + \varepsilon_0 \varepsilon)^2 = 0.$$

Then, from $y_0^2 + 2(\varepsilon_0 \varepsilon/d)y_0 + (\varepsilon_0 \varepsilon/s) = 0$, and

$$y_0 = (\varepsilon_0 \varepsilon/d)[-1 + \sqrt{(1 - k^2)}].$$

Here, $k^2 = d^2/(s \cdot \varepsilon_0 \varepsilon)$. Note that only $y_0 = (\varepsilon_0 \varepsilon/d)[-1 + \sqrt{(1 - k^2)}]$ is valid for realizing the meaningful maximum point, since $y_0 = (\varepsilon_0 \varepsilon/d)[-1 - \sqrt{(1 - k^2)}]$ and $y_0 = (\varepsilon_0 \varepsilon/d)[-1 + \sqrt{(1 - k^2)}]$ provide $(d^2\lambda/dy^2) > 0$ (i.e., minimum point) and < 0 (i.e., maximum point), respectively. By putting $y = y_0$ into $\lambda(y)$, we can get the maximum value of λ:

$$\lambda_{\max} = -s[-2(\varepsilon_0 \varepsilon/d)y_0 - (\varepsilon_0 \varepsilon/s)] + dy_0)/(dy_0 + \varepsilon_0 \varepsilon)$$

We can finally obtain the following two equivalent expressions:

$$\lambda_{\max} = [(1/k) - \sqrt{(1/k^2) - 1}]^2 = [(1/k) + \sqrt{(1/k^2) - 1}]^{-2}. \tag{2.79}$$

Notice that

$$k^2/4 < \lambda_{\max} < k^2/2,$$

for a reasonable k value ($< 90\%$). For a small k, $\lambda_{\max} = k^2/4$, and for a large k, $\lambda_{\max} = k^2/2$.

It is also worth noting that the maximum condition stated above does not agree with the condition which provides the maximum output mechanical energy. The maximum output energy can be obtained when the dotted area in Fig. 2.25(c) becomes maximum under the constraint of the rectangular corner point tracing on the line (from dE on the vertical axis to $-dE/s$ on the horizontal axis). Therefore, the load should be a half of the maximum generative stress (i.e., "Blocking stress") and the mechanical energy: $-[dE - s(dE/2s)](-dE/2s) = (dE)^2/4s$. In this case, since the input electrical energy is given by $[\varepsilon_0 \varepsilon E + d(-dE/2s)]E$,

$$\lambda = 1/2[(2/k^2) - 1], \tag{2.80}$$

which is close to the value $\lambda_{\max}$ when k is small, but has a difference when k is large, as predicted.

2.6.2.3 *The efficiency η*

$$\eta = (\text{Output mechanical energy})/(\text{Consumed electrical energy}) \tag{2.81a}$$

or

$$\eta = (\text{Output electrical energy})/(\text{Consumed mechanical energy}). \tag{2.81b}$$

The difference of the efficiency definition from Eq. (2.75) is "input" energy and "consumed" energy in the denominators. In a work cycle (e.g., an electric field cycle), the input electrical energy is transformed partially into mechanical energy and the remaining is stored as electrical energy (electrostatic energy like a capacitor) in an actuator. In this way, the ineffective electrostatic energy can be returned to the power source, leading to near 100% efficiency, if the loss is small. Typical values of dielectric loss in PZT are about $1 - 3\%$.

Example Problem 2.10.

A paper authored by a mechanical engineering professor described the following: "A *coupling coefficient* is a measure of the *efficiency* with which a piezoelectric material converts the energy in an imposed signal to useful mechanical energy". "By applying 1 J of electric energy to a piezoelectric with an electromechanical coupling factor k, we accumulate k^2 J of mechanical energy in this piezo material. Thus, this actuator can work mechanically up to k^2 J to the outside, and the efficiency is considered to be $k^2\%$". These sentences include two major misconceptions. Describe them and provide their rectifications.

Hint:

First, not all the stored energy can be actually used, and the actual work done depends on the mechanical load. Maximize the work:

$$\int (-X)\, dx = -(dE + sX)X. \tag{P2.10.1}$$

Second, without recovering the stored electric energy, the device efficiency should drop significantly, as only $k^2\%$. Consider how to recover it.

Solution:

Let us consider the simplest case where an electric field E is applied suddenly to a piezoelectric under constant external stress X (< 0, compressive stress) in Fig. 2.25(a). Figure 2.26(a) shows the mechanical work calculation process. The mass shrinks the actuator first by sX (s: piezo material's compliance, and $X < 0$). This mechanical energy sX^2 is a sort of "loan" of the actuator credited from the mass, which should be paid back or subtracted later. By applying the step electric field, the actuator expands by the strain level dE under a constant stress condition. This is the mechanical energy provided from the actuator

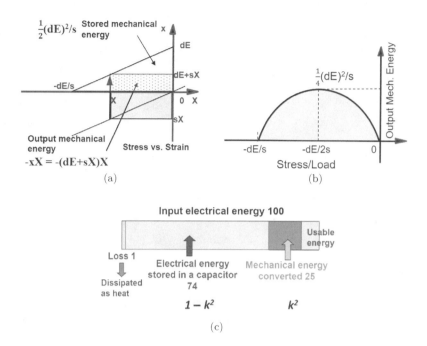

Fig. 2.26: (a) Output mechanical energy calculation process. (b) Output mechanical energy maximization point. (c) Energy conversion rate in a typical piezoelectric.

(electromechanically transduced energy) to the mass, which corresponds to dEX. Like paying back the initial "loan", the output work (from the actuator to the mass) can be calculated as the area subtraction (shown by the dotted area in Fig. 2.26(a))

$$\int (-X)\, dx = -(dE + sX)X \qquad (P2.10.1)$$

Output mechanical energy can be transformed as

$$-(dE + sX)X = -s[X + (1/2)(dE/s)]^2 + (1/4)(dE)^2/s. \qquad (P2.10.2)$$

As Fig. 2.26(b) illustrates, the optimized load should be $(1/2)$ of the maximum generative force (i.e., *Blocking force*), and the maximum output mechanical energy is $(1/4)(dE)^2/s$. Taking into account the stored energy $(1/2)(dE)^2/s$, by choosing the mechanical load (i.e., corresponding to the *mechanical impedance matching*), the work will reach $(1/2)$ of the stored mechanical energy.

Figure 2.26(c) visualizes the energy conversion rate in a typical piezoelectric. Taking an example value $k = 50\%$ for a piezoelectric pseudo-DC device, $k^2 = 25\%$, then the input electrical energy 100 is converted into mechanical energy 25, by remaining $(1 - k^2) = 74$ as stored electrical energy (in a capacitor). Because the loss factor (dielectric loss $\tan \delta'$) is less than 1%, actual loss dissipated as heat is usually less than 1%. Thus, if we can collect the stored electrical energy back to the drive circuit, we can declare that the loss is only 1%, or the efficiency is 99% (very high!).

Because this paper's author, because he is mechanical engineer, may not know how to recover the electrostatic energy stored in the actuator/capacitor. Thus, he releases it by shorting the drive circuit in order to move to the next operation. Of course, in this worst scenario, efficiency less than 25% (equal to k^2) is true in his/her paper. Let us consider how we can recover the electrostatic energy from the piezoelectric capacitor [$(1 - k^2)$ in the damped capacitance]. Examples can be found in dot-matrix/ink-jet printer and diesel injection valve control applications, where multilayer actuators are driven at 1 kHz, much lower than the resonance frequency. Though the k_{33} mode is different, refer to a simpler equivalent circuit for the k_{31} mode shown in Fig. 2.23 (no-loss case is illustrated). In this equivalent circuit, motional current and damped current ($C_d = (1 - k^2)C_0$) should have $k^2 : (1 - k^2)$ ratio under an off-resonance condition. If we insert the inductance L in the driving system so as to create a resonance circuit of L and C_d under the condition of $\omega^2 = 1/LC_d$ (ω: operation cycle such as 1 kHz), the electric energy stored in the damped capacitance C_d starts flip-flopping with L (i.e., energy catchball), without losing this energy (if we neglect the loss or heat generation). When the actuator returns to the original zero-position, the electric energy in the damped capacitance is shifted to the inductor L, so that the next actuation can start synchronously with the electric energy return to the C_d. A *negative capacitance* usage is an alternative solution recently, because the heavy and bulky magnetoelectric inductor is problematic for miniaturization.

2.6.3 *Mechanical Quality Factor Q_M*

The mechanical quality factor, Q_M, is a parameter that characterizes the sharpness of the electromechanical resonance spectrum. When the motional admittance Y_m (see Section 2.7) is plotted around the resonance frequency ω_0, the mechanical quality factor Q_M is defined with respect to the full width [$2\Delta\omega$] at $Y_m/\sqrt{2}$ (i.e., 3 dB down) as

$$Q_M = \omega_0/2\Delta\omega. \tag{2.82}$$

Also note that Q_M^{-1} is equal to the mechanical loss ($\tan\phi'$) in the k_{31} case, which is the primary origin of the heat generation (or efficiency) under the resonance operation. When we define a complex elastic compliance, $s^E = s^{E'} - js^{E''}$, the mechanical loss tangent is provided by $\tan\phi' = s^{E''}/s^{E'}$. The Q_M value is very important in evaluating the magnitude of the resonant displacement and strain. The vibration amplitude at an off-resonance frequency ($dE \cdot L$, L: length of the sample) is amplified by a factor proportional to Q_M at the resonance frequency. For example, a longitudinally vibrating rectangular plate through the transverse piezoelectric effect d_{31} generates the maximum displacement given by $(8/\pi^2)Q_M d_{31}EL$ for the fundamental mode. The details are discussed in Section 2.7.

2.6.4 *Acoustic Impedance Z*

Though this parameter is not unique for piezoelectrics, it is closely associated with the piezoelectric device designing. The acoustic impedance Z is a parameter used for evaluating the acoustic energy transfer between two materials. It is defined, in general, by

$$Z^2 = (\text{pressure/volume velocity}). \tag{2.83}$$

In a solid material,

$$Z = \sqrt{\rho c}, \tag{2.84}$$

where ρ is the density and c is the elastic stiffness of the material.

Acoustic impedance (or mechanical impedance) matching is necessary for transferring mechanical energy from one material to the other. Figure 2.27 shows a conceptual cartoon illustrating two extreme cases. The mechanical work done by one material on the other is evaluated by the product of the applied force F and the displacement ΔL:

$$W = F \times \Delta L. \tag{2.85}$$

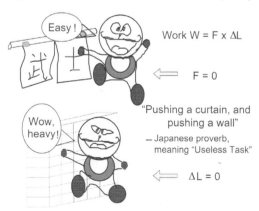

Fig. 2.27: Concept of mechanical impedance matching.

If the material is very soft, the force F can be very small, leading to very small W (practically no work!). This corresponds to "Pushing a curtain", exemplified by the case when the acoustic wave is generated in water directly by an elastically hard PZT transducer. Most of the acoustic energy generated in the PZT is reflected at the interface, and only a small portion of acoustic energy transfers into water. On the other hand, if the material is very hard, the displacement ΔL will be very small, again leading to very small W. This corresponds to "Pushing a wall". Polymer piezoelectric polyvinylidene di-fluoride (PVDF) cannot drive a hard steel part effectively. Therefore, the *acoustic impedance* must be adjusted to maximize the output mechanical power:

$$\sqrt{\rho_1 c_1} = \sqrt{\rho_2 c_2}, \tag{2.86}$$

where ρ is the density and c is the elastic stiffness, and the subscripts 1 and 2 denote the two materials. In practice, an acoustic impedance matching layer (elastically intermediate material between PZT and water, such as a polymer) is inserted between two phases (1 and 2). More precisely, the acoustic impedance Z should be chosen as the geometrical average $\sqrt{Z_1 \cdot Z_2}$ of Z_1 in phase 1 and Z_2 in phase 2, so that the transfer of mechanical energy in the PZT to water will be optimized.

In more advanced discussions, there are three kinds of impedances: specific acoustic impedance (pressure/particle speed), acoustic impedance (pressure/volume speed) and radiation impedance (force/speed). See Ref. [39] for the details.

2.6.5 *Maximum Vibration Velocity* $v_{\max}$

The power density of a piezoelectric is measured by different figures of merit (FOM) for different applications:[40]

(1) Off-resonance actuator applications — positioners

$$\text{FOM} = d\,(\text{piezoelectric constant}).$$

(2) Resonance actuator applications — ultrasonic motors

$$\text{FOM} = v\,(vibration\ velocity) \approx Q_m \cdot d(\text{for low level excitation}).$$

(3) Resonance transducer applications — piezoelectric transformers, sonars (transmitters and receivers)

$$\text{FOM} = k \cdot v(\text{k : electromechanical coupling factor}).$$

In order to obtain a large mechanical output power, the ceramics are driven under a high vibration level, namely, under a relatively large AC electric field around the electromechanical resonance frequency. Though the vibration velocity (i.e., first derivative of the vibration displacement in respect to time) is almost proportional to the applied AC electric field under a relatively small field range, with increasing the electric field, the induced vibration velocity will saturate above a certain critical field. This originates from the sudden intensive elastic loss increase, or the reduction of the mechanical quality factor above this critical field. Thus, heat generation becomes significant, as well as a degradation in piezoelectric properties. Therefore, the high-power device such as an ultrasonic motor requires a very "hard" piezoelectric with a high *mechanical quality factor* Q_m (i.e., low elastic loss) in order to suppress heat generation. The Q_m is defined as an inverse value of the intensive elastic loss factor, $\tan \phi'$. It is also notable that the actual mechanical vibration velocity at the resonance frequency is directly proportional to this Q_m value (i.e., *displacement amplification factor*).

In order to analyze the various piezoelectric parameter changes as a function of vibration level, we occasionally use vibration velocity, instead of the applied electric field. Though the *vibration amplitude* may be used, the *vibration velocity* is used more popularly in the discussion. The reason is to eliminate the size effect; that is, when the vibration amplitude is small and proportional to the applied electric field and the length L, it should be expressed by

$$\Delta L = (8/\pi^2)Q_m \ d_{31}L \ E_3 \sin(\omega_R t) \tag{2.87}$$

for a d_{31}-type rectangular piezoelectric plate. Since the vibration velocity at the edge of the plate sample is the first derivative of amplitude in respect of t, we obtain

$$v = (8/\pi^2)Q_m d_{31}LE_3\omega_R \cos(\omega_R t). \tag{2.88}$$

Taking into account the fundamental resonance frequency $f_R = (1/\sqrt{\rho s_{11}})/2L$, the vibration velocity at the plate edge can be transformed as

$$v = (8/\pi)Q_m d_{31}E_3(1/\sqrt{\rho s}_{11}) \cos(\omega_R t). \tag{2.89}$$

Note that the vibration amplitude is sample size L dependent, but that the vibration velocity is not. Because the vibration velocity is proportional to the electric field, and the proportional constant is given primarily by $Q_m d_{31}/\sqrt{\rho s}_{11} = Q_m d_{31}v_{11}$, which is sample size independent, we use it as a measure of the vibration level.

As explained above, with increasing the electric field, the induced vibration velocity will saturate above a certain critical field, and heat generation is associated. Figure 2.28 shows vibration velocity dependence of the mechanical quality factors Q_A and Q_B, and corresponding temperature rise for A (resonance) and B (antiresonance) type resonances of a longitudinally vibrating PZT ceramic transducer k_{31}.[41] Since the additional electric power is converted mostly to heat, rather than the vibration velocity increase, we define the *maximum vibration velocity* as the v_{max} under which the piezoelectric plate shows 20°C temperature rise at the nodal point (i.e., the specimen center part) above the room temperature. The RMS (root mean square) value of v_{max} of popular hard PZT rectangular plates

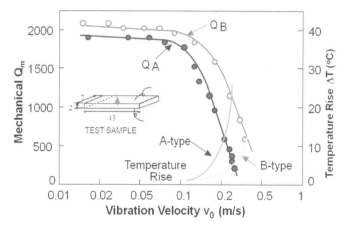

Fig. 2.28: Vibration velocity dependence of the mechanical quality factors Q_A and Q_B, and corresponding temperature rise for A (resonance) and B (antiresonance) type resonances of a longitudinally vibrating PZT ceramic transducer k_{31}.[41]

ranges from 0.3 m/s (Fig. 2.28 case) to 0.6 m/s, which is a sort of material's constant, important parameter for the high-power applications. When we consider the high-power performance in a wide variety of piezo materials such as Pb-free and PZT, the *maximum mechanical energy* density is suitable by taking into account the *mass density*, which is defined as $(1/2)\rho v_{rms}^2$ under the maximum vibration velocity condition. Current top data range from 1000–1500 J/m^3. By multiplying the resonance frequency f on the mechanical energy density, we can obtain the maximum vibration power density, the top data of which range 30–40 W/cm^3 in PZTs.

2.7 Piezoelectric Resonance and Antiresonance

When the field E is alternating, mechanical vibration is caused in a piezoelectric device, and if the drive frequency is adjusted to a mechanical resonance frequency of the device, large amplified resonating strain is generated. This phenomenon can be understood as a strain amplification due to synchronous accumulation of the input energy with time (i.e., amplification in terms of time), and is called *piezoelectric resonance*.

The amplification factor is proportional to the mechanical quality factor Q_M (inversely proportional to the elastic loss). Piezoelectric resonance is very useful for realizing energy trap devices, filters, actuators, medical and underwater transducers, piezo transformers, etc. We consider in this section the electromechanical resonance under an AC external electric field theoretically. Two popular sample geometry k_{31} and k_{33} modes are discussed without and with integrating loss factors. The key difference among these geometries can be found in E-constant and D-constant constraint conditions. The detailed discussion is found in Ref. 40 by introducing three losses (i.e., dielectric, elastic and piezoelectric losses).

2.7.1 *The k_{31} Longitudinal Vibration Mode (Loss-Free)*

2.7.1.1 *Piezoelectric dynamic equation*

Let us consider the longitudinal mechanical vibration of a piezo -ceramic plate through the transverse piezoelectric effect (d_{31}), as shown in Fig. 2.29. Sinusoidal electric field E_z (angular frequency ω) is applied along the polarization P_z direction. If the polarization is in the z-direction and $x - y$ planes are the planes of the electrodes, the extensional vibration in the x (length) direction is represented by the following dynamic equations (when the length L is more than 4–6 times the width w or the thickness b,

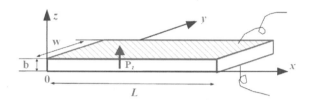

Fig. 2.29: Longitudinal vibration through the transverse piezoelectric effect (d_31) in a rectangular plate ($L \gg w \gg b$).

we can neglect the coupling modes with width or thickness vibrations):

$$\rho(\partial^2 u/\partial t^2) = F = (\partial X_{11}/\partial x) + (\partial X_{12}/\partial y) + (\partial X_{13}/\partial z), \qquad (2.90)$$

where ρ is the density of the piezo ceramic, u is the displacement of a small volume element in the ceramic plate in the x-direction.

We integrate the piezoelectric constitutive equations, where strain x and electric displacement D are controlled by the intensive parameters, electric field E and stress X. The relations between stress, electric field (only $E_z = E_3$ exists) and the induced strain in 3D expression are given by

$$
\begin{bmatrix} x_1 \\ x_2 \\ x_3 \\ x_4 \\ x_5 \\ x_6 \end{bmatrix} =
\begin{bmatrix}
s_{11} & s_{12} & s_{13} & 0 & 0 & 0 \\
s_{12} & s_{11} & s_{13} & 0 & 0 & 0 \\
s_{13} & s_{13} & s_{33} & 0 & 0 & 0 \\
0 & 0 & 0 & s_{44} & 0 & 0 \\
0 & 0 & 0 & 0 & s_{44} & 0 \\
0 & 0 & 0 & 0 & 0 & s_{66}
\end{bmatrix}
\begin{pmatrix} X_1 \\ X_2 \\ X_3 \\ X_4 \\ X_5 \\ X_6 \end{pmatrix} +
\begin{pmatrix}
0 & 0 & d_{31} \\
0 & 0 & d_{31} \\
0 & 0 & d_{33} \\
0 & d_{15} & 0 \\
d_{15} & 0 & 0 \\
0 & 0 & 0
\end{pmatrix}
\begin{bmatrix} E_1 \\ E_2 \\ E_3 \end{bmatrix}, \qquad (2.91a)
$$

$$
\begin{bmatrix} D_1 \\ D_2 \\ D_3 \end{bmatrix} =
\begin{bmatrix}
0 & 0 & 0 & 0 & d_{15} & 0 \\
0 & 0 & 0 & d_{15} & 0 & 0 \\
d_{31} & d_{31} & d_{33} & 0 & 0 & 0
\end{bmatrix}
\begin{bmatrix} X_1 \\ X_2 \\ X_3 \\ X_4 \\ X_5 \\ X_6 \end{bmatrix} +
\varepsilon_0
\begin{bmatrix}
\varepsilon_{11} & 0 & 0 \\
0 & \varepsilon_{11} & 0 \\
0 & 0 & \varepsilon_{33}
\end{bmatrix}
\begin{bmatrix} E_1 \\ E_2 \\ E_3 \end{bmatrix}. \qquad (2.92a)
$$

Note $s_{66} = 2(s_{11} - s_{12})$ in the ∞ symmetry like random ceramics. Refer to Eq. (2.15).

Equation (2.91a) is transformed as follows, by putting $E_1 = E_2 = 0$:

$$x_1 = s_{11}^E X_1 + s_{12}^E X_2 + s_{13}^E X_3 + d_{31} E_3,$$

$$x_2 = s_{12}^E X_1 + s_{11}^E X_2 + s_{13}^E X_3 + d_{31} E_3,$$

$$x_3 = s_{13}^E X_1 + s_{13}^E X_2 + s_{33}^E X_3 + d_{33} E_3,$$

$$x_4 = s_{44}^E X_4,$$

$$x_5 = s_{44}^E X_5,$$

$$x_6 = 2(s_{11}^E - s_{12}^E) X_6. \tag{2.91b}$$

When the plate is very long and thin, X_2 and X_3 may be set equal to zero through the plate (i.e., mechanical boundary conditions). Since shear stress will not be generated by the electric field $E_z(= E_3)$, Eq. (2.91b) is reduced to only one equation:

$$x_1 = s_{11}^E X_1 + d_{31} E_3, \quad \text{or} \quad X_1 = x_1/s_{11}^E - (d_{31}/s_{11}^E) E_z \tag{2.93}$$

Introducing Eq. (2.93) into Eq. (2.90), and allowing for strain definition $x_1 = \partial u/\partial x$ (non-suffix x corresponds to the Cartesian coordinate, and x_1 is the strain along the 1 (x) direction) and $\partial E_z/\partial x = 0$ (due to the equal potential on each electrode), leads to a *harmonic vibration* equation, $u = u_1(x)e^{j\omega t} + u_2(x)e^{-j\omega t}$:

$$\rho(\partial^2 u/\partial t^2) = (1/s_{11}^E)(\partial x_1/\partial x), \quad \text{or} \quad -\omega^2 \rho s_{11}^E u = \partial^2 u/\partial x^2. \tag{2.94}$$

Here, ω is the angular frequency of the sinusoidal drive field E_z and the displacement u. Supposing the displacement u also vibrates with the frequency of ω, a general solution of Eq. (2.94) is expressed by

$$u(x) = A \sin\left(\frac{\omega}{v_{11}^E} x\right) + B \cos\left(\frac{\omega}{v_{11}^E} x\right) \tag{2.95}$$

where v_{11}^E is the *sound velocity* along the length x direction in the piezo ceramic plate, which is expressed by

$$v_{11}^E = 1/\sqrt{\rho s_{11}^E}. \tag{2.96}$$

From Eq. (2.93) and $x_1(x)$ given by

$$x_1(x) = \frac{\partial u}{\partial x} = A \frac{\omega}{v_{11}^E} \cos\left(\frac{\omega}{v_{11}^E} x\right) - B \frac{\omega}{v_{11}^E} \sin\left(\frac{\omega}{v_{11}^E} x\right), \tag{2.97}$$

with the boundary condition $X_1 = 0$ at $x = 0$ and L (sample length) (due to the mechanically free condition at both plate ends), we can determine A and B:

$$A \frac{\omega}{v_{11}^E} = d_{31} E_z$$

$$A \frac{\omega}{v_{11}^E} \cos\left(\frac{\omega}{v_{11}^E} L\right) - B \frac{\omega}{v_{11}^E} \sin\left(\frac{\omega}{v_{11}^E} L\right) = d_{31} E_z$$

Thus,

$$\begin{cases} A = \left(\frac{v_{11}^E}{\omega}\right) d_{31} E_z \\ B = -\left(\frac{v_{11}^E}{\omega}\right) d_{31} E_z \dfrac{\sin\left(\frac{\omega}{2v_{11}^E} L\right)}{\cos\left(\frac{\omega}{2v_{11}^E} L\right)} \end{cases}$$

$$\text{(Displacement)} \quad u(x) = \left(\frac{v_{11}^E}{\omega}\right) d_{31} E_z \frac{\sin\left[\frac{\omega(2x-L)}{2v_{11}^E}\right]}{\cos\left(\frac{\omega L}{2v_{11}^E}\right)} \tag{2.98}$$

$$\text{(Strain)} \quad \partial u/\partial x = x_1 = d_{31} E_z \left(\frac{\cos\left[\frac{\omega(2x-L)}{2v_{11}^E}\right]}{\cos\left(\frac{\omega L}{2v_{11}^E}\right)}\right) \tag{2.99}$$

First, the displacement and strain are proportional to the external electric field E_z, which is an interesting contrast to the k_{33} mode, where these are proportional to the electric displacement D_z (refer to Section 2.7.2). Second, their distributions in terms of x in Eqs. (2.98) and (2.99) are anti-symmetrically and symmetrically sinusoidal in respect to $x = L/2$ position (the numerator becomes minimum, 0 or maximum, 1), and the maximum strain (i.e., *nodal line*) exists on this line. Note that $\omega \to 0$ (i.e., pseudo-DC) makes Eq. (2.99) to $x_1 = d_{31} E_z$, that is, uniform strain distribution on the whole piezo-plate.

2.7.1.2 *Admittance around resonance and antiresonance*

When the specimen is utilized as an electrical component such as a filter or a vibrator, the electrical admittance ((induced current)/(applied voltage) ratio) or impedance ((applied voltage)/(induced current)) plays an important role. Now, we use another set of constitutive equations with respect to electric displacement D in Eq. (2.92a). Taking into account non-zero components, X_1, E_3, ($X_2 = X_3 = 0$ due to thin size, though x_1, x_2 are not zero) only the following equation is eligible:

$$D_3 = d_{31} X_1 + \varepsilon_0 \varepsilon_{33}{}^X E_3. \tag{2.92b}$$

Refresh your memory on that the external current flow into the specimen by the surface free charge increment is equal to the negative of the internal electric displacement-based current, $i = \partial D_3/\partial t (= j\omega D_3)$. Also note that the externally applied voltage V is related with the internal electric field as $E_z = -\text{grad}(V) = -(V/b)$ (when the electric field is uniform along the z direction). Though the electric field is uniform in the sample along the x direction due to the surface electrode, the electric displacement D is not uniform because of the stress X distribution along the x direction, maximum at the nodal line ($x = L/2$). The total current i is given by integrating $\partial D_3/\partial t$ with respect to the top area:

$$i = j\omega w \int_0^L D_3 dx = j\omega w \int_0^L (d_{31} X_1 + \varepsilon_0 \varepsilon_{33}{}^X E_z) dx$$

$$= j\omega w \int_0^L [d_{31}\{x_1/s_{11}{}^E - (d_{31}/s_{11}{}^E) E_z\} + \varepsilon_0 \varepsilon_{33}{}^X E_z] dx. \tag{2.100}$$

w is the plate width. Using strain distribution in Eq. (2.99), the admittance for the mechanically free sample is calculated to be [above i is the internal displacement current, which is considered $(-i)$ from the external power supply]

$$Y = (-i/V) = (i/E_z \cdot b)$$

$$= (j\omega w L/E_z b) \int_0^L [(d_{31}^2/s_{11}^E) \left(\frac{\cos\left[\frac{\omega(L-2x)}{2v_{11}^E}\right]}{\cos\left(\frac{\omega L}{2v_{11}^E}\right)}\right) E_z + [\varepsilon_0 \varepsilon_{33}{}^X - (d_{31}^2/s_{11}^E)] E_z] dx$$

$$= (j\omega w L/b)\varepsilon_0 \varepsilon_{33}{}^{LC}[1 + (d_{31}{}^2/\varepsilon_0 \varepsilon_{33}{}^{LC} s_{11}^E)(\tan(\omega L/2v_{11}^E)/(\omega L/2v_{11}^E)]$$

$$= (j\omega wL/b)\varepsilon_0\varepsilon_{33}{}^X[(1-k_{31})+k_{31}{}^2 \cdot (\tan(\omega L/2v_{11}^E)/(\omega L/2v_{11}^E)]$$

$$= j\omega C_d \left[1 + \frac{k_{31}^2}{1-k_{31}^2}\frac{\tan(\Omega_{11})}{\Omega_{11}}\right]$$

$$= j\omega C_0 \left[(1-k_{31}^2)+k_{31}^2\frac{\tan(\Omega_{11})}{\Omega_{11}}\right], \tag{2.101}$$

where w is the width, L the length, b the thickness of the rectangular piezo sample, and V is the applied voltage $(=-E_z \cdot b)$. $k_{31}{}^2 = (d_{31}{}^2/\varepsilon_0\varepsilon_{33}^X s_{11}^E)$. We adopt the following notations for making the formulas simpler:

- ε_{33}^{LC} is the permittivity in a "longitudinally clamped" (LC) sample, which is given by

$$\varepsilon_0\varepsilon_{33}{}^{LC} = \varepsilon_0\varepsilon_{33}{}^X - (d_{31}^2/s_{11}^E), \quad \text{or}$$

$$= \varepsilon_0\varepsilon_{33}{}^X(1-k_{31}{}^2). \tag{2.102}$$

$$k_{31}^2 = \frac{d_{31}^2}{\varepsilon_0\varepsilon_{33}^X s_{11}^E} \tag{2.103}$$

Note here that this ε_{33}^{LC} is different from extensive permittivity ε_{33}^x, precisely speaking. ε_{33}^{LC} is the permittivity in the sample mechanically clamped only along the x (or 1, length) direction, free along z (or 3, polarization direction) or y directions, while ε_{33}^x means the permittivity clamped completely in the three directions.

- Free capacitance C_0 and damped capacitance C_d are defined by mechanical free and 1D clamped ones, respectively:

$$C_0 = \varepsilon_0\varepsilon_{33}^X \frac{Lw}{b}, \tag{2.104}$$

$$C_d = \varepsilon_0\varepsilon_{33}^{LC}\frac{Lw}{b}. \tag{2.105}$$

- Normalized frequency Ω_{11}:

$$\Omega_{11} = (\omega L/2v_{11}^E). \tag{2.106}$$

The first term $(j\omega wL/b)\varepsilon_0\varepsilon_{33}^{LC}$ of admittance Eq. (2.101) is the *damped capacitance* (longitudinally clamped condition), $j\omega C_d$, and the second term is characterized by $[\tan(\Omega_{11})/(\Omega_{11})]$, which is the *motional capacitance* (relating with the geometrical length change). For small ω, since $[\tan(\Omega_{11})/(\Omega_{11})] \approx 1$ and $[(1-k_{31})+k_{31}^2 \cdot (\tan(\Omega_{11})/(\Omega_{11})] \approx 1$, Y approaches $(j\omega wL/b)\, \varepsilon_0\varepsilon_{33}{}^X$, that is, free capacitance, $j\omega C_0$. With increasing ω, $\tan(\Omega_{11})$ increases significantly (up to $+\infty$) until $\Omega_{11} = \omega L/2v_{11}^E$ reaches $\pi/2$ with the mechanical vibration amplitude increase. The piezoelectric resonance is achieved where the admittance becomes infinite or the impedance is zero. The resonance frequency f_R is, thus, calculated from Eq. (2.106) (by putting $\omega L/2v_{11}^E = \pi/2$ for infinite admittance), and the fundamental frequency is given by

$$f_R = \omega_R/2\pi = v_{11}^E/2L = 1/(2L\sqrt{\rho s_{11}{}^E}). \tag{2.107}$$

On the other hand, the antiresonance state is generated for zero admittance or infinite impedance:

$$(\omega_A L/2v_{11}^E)\cot(\omega_A L/2v_{11}^E) = -d_{31}^2/\varepsilon_{33}{}^{LC}s_{11}{}^E = -k_{31}{}^2/(1-k_{31}{}^2), \tag{2.108}$$

where ω_A is the antiresonance frequency, and the final transformation is provided by the k_{31} definition.

Figure 2.30 shows an example admittance magnitude and phase spectra for a rectangular piezo ceramic plate $(L = 20)$ for a fundamental longitudinal mode (k_{31}) through the transverse piezoelectric effect (d_{31}).

on the basis of Eq. (2.101). Note that the shown data include losses, because the loss-free spectrum with infinite admittance is practically difficult to draw, and the 3 dB down method to obtain mechanical quality factors Q_m is also inserted in advance (see Section 2.7.2).

2.7.1.3 *Resonance and antiresonance vibration modes*

The resonance and antiresonance states are both mechanical resonance states with amplified strain/displacement states, but they are very different from the driving viewpoints. When we excite the vibration of the piezoelectric plate in Fig. 2.29 mechanically, the resonance and antiresonance modes can actually be observed as the natural mechanical resonance under electrical short-circuit and open-circuit conditions, respectively. The mode difference is described by the following intuitive model

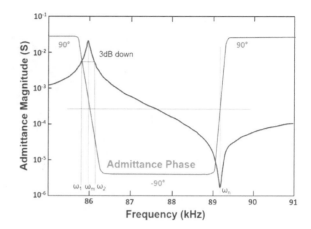

Fig. 2.30: Admittance magnitude and phase spectra for a rectangular (k_{31}) piezoceramic plate via the transverse piezoelectric effect (d_{31}).

under electric field excitation below. In a high electromechanical coupling material with k almost equal to 1, the resonance or antiresonance states appear for $\tan(\omega L/2v) = \infty$ or 0 [i.e., $\omega L/2v = (m - 1/2)\pi$ or $m\pi$ (m: integer)], respectively. The strain amplitude x_1 distribution for each state (calculated using Eq. (2.99)) is illustrated in Fig. 2.31 (the figure is slightly off-resonance to escape from the infinite amplitude). In the resonance state, the strain distribution is basically sinusoidal with the maximum at the center of the plate ($x = L/2$) (see the numerator). When ω is close to ω_R, $(\omega_R L/2v) = \pi/2$, leading to the denominator cos ($\omega_R L/2v$) → 0. Significant strain amplification is obtained. It is worth noting that the stress X_1 is zero at the plate ends ($x = 0$ and L), but the strain x_1 is not exactly zero, but is equal to $d_{31}E_z$ (practically invisible due to large strain amplification along the x direction). According to this large strain amplitude, large capacitance changes (called *motional capacitance*) are induced, and under a constant applied voltage the current can easily flow into the device (i.e., admittance Y is infinite). To the contrary, at the antiresonance, the strain induced in the device compensates completely (because extension and compression are compensated in one wave on the specimen length, and the plate ends become the node), resulting in no motional capacitance change, and the current cannot flow easily into the sample (i.e., admittance Y zero). Thus, for a high k_{31} material, the first antiresonance frequency f_A should be almost twice as large as the first resonance frequency f_R.

It should be pointed out again that both resonance and antiresonance states are in the mechanical resonance, which can create large strain in the sample under small input electrical energy (not exactly minimum, precisely speaking). When we use a constant voltage supply, the specimen vibration is excited only at the resonance mode, as indicated from Eq. (2.99), because the electrical power is very small at the

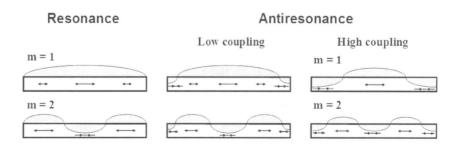

Fig. 2.31: Strain distribution in the resonance and antiresonance states for a k_{31}-type piezoelectric plate.

antiresonance mode. This provides a common misconception to junior engineers that "the antiresonance is not a mechanical resonance". In contrast, when we use a constant current supply, the vibration is excited only at the antiresonance, instead, because the impedance $1/Y$ shows the maximum at the antiresonance, leading to significant enhancement of input electrical energy. The stress X_1 at the plate ends ($x = 0$ and L) is supposed to be zero in both cases. However, though the strain x_1 at the plate ends is near zero/very small (precisely, $d_{31}E_z$, because of low-voltage and high-current drive) for the resonance, the strain x_1 is not zero (actually the maximum) for the antiresonance (because of high-voltage and low-current drive). This means that there is only one vibration node at the plate center for the resonance (top-left in Fig. 2.31), and there are additional two nodes at both plate ends for the first antiresonance (top-right in Fig. 2.31). The reason is from the antiresonance drive, i.e., high-voltage/low-current (minimum power) drive due to the high impedance. The converse piezo effect strain under E directly via d_{31} (uniform strain in the sample) superposes on the mechanical resonance strain distribution (distributed strain with nodes in the sample), two strains of which have exactly the same level theoretically at the antiresonance for $k_{31} \approx 1$.

In a typical PZT case, where $k_{31} = 0.3$, the antiresonance state varies from the previously mentioned mode and becomes closer to the resonance mode (top-center in Fig. 2.31). The low-coupling material exhibits an antiresonance mode where the capacitance change due to the size change (i.e., *motional capacitance*) is compensated completely by the current required to charge up the static capacitance (called *damped capacitance*). Note that above the resonance frequency, the motional capacitance phase changes by 180° owing to $\tan(\omega L/2v) < 0$. Thus, the antiresonance frequency f_A will approach the resonance frequency f_R.

2.7.2 *The k_{33} Longitudinal Vibration Mode (Loss-Free)*

2.7.2.1 *Piezoelectric dynamic equation*

Let us consider now the longitudinal vibration k_{33} mode in comparison with the k_{31} mode. Note that the vibration direction is in parallel to the spontaneous polarization direction P_S. When the resonator is long in the z direction and the electrodes are deposited on each end of the rod, as shown in Fig. 2.32, the following conditions are satisfied:

$$X_1 = X_2 = X_4 = X_5 = X_6 = 0 \text{ and } X_3 \neq 0.$$

Thus, the necessary constitutive equations (Eqs. (2.91a) and (2.92a)) are just two for this configuration:

$$X_3 = (x_3 - d_{33}E_z)/s_{33}^E, \quad (2.109)$$

$$D_3 = \varepsilon_0\varepsilon_{33}{}^X E_z + d_{33}X_3. \quad (2.110)$$

Assuming a local displacement u_3 in the z direction, from Eq. (2.109), the dynamic equation is given $\left(x_3 = \frac{\partial u_3}{\partial z}\right)$:

$$\rho\frac{\partial^2 u_3}{\partial t^2} = \frac{1}{s_{33}^E}\left[\frac{\partial^2 u_3}{\partial z^2} - d_{33}\frac{\partial E_z}{\partial z}\right] \quad (2.111)$$

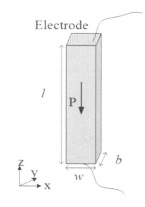

Fig. 2.32: Longitudinal vibration via the piezoelectric effect (d_{33}) in a rod ($L \gg w \approx b$).

The important notion is the *depolarization field* in the case when the longitudinal vibration direction $//$ P_S. In the previous subsection, the k_{31} mode was considered, where $\partial E_z/\partial x = 0$ was used because the electrode covers the whole vibration direction. However, the electrical condition for the longitudinal k_{33} vibration is *not* $\partial E_z/\partial z = 0$, but rather $\partial D_z/\partial z = 0$, because of no electrode along the vibration direction. Remember the Gauss Law, $\text{div}(D_3) = \sigma$ (charge) or $\text{div}(E_3) = \sigma - \frac{1}{\varepsilon_0\varepsilon_{33}^E}\text{div}(P_3)$, and no free charge σ on the piezo rod between the top and bottom electrode gap. Thus, $\partial D_z/\partial z = 0$ is obvious. The stress distribution X_3 generates the induced polarization distribution P_3 in the piezo rod, which should generate the internal electric field distribution $E_3 = -\frac{P_3}{\varepsilon_0 \varepsilon_{33}^X}$ to satisfy $D_z = $ constant (Recall $D = dX + \varepsilon_0 \varepsilon^X E$). Because of the negative electric field with respect to the polarization, this is called "depolarization field".

From Eqs. (2.109) and (2.110), $D_3 = \varepsilon_0 \varepsilon_{33}^X E_z + d_{33}[(x_3 - d_{33}E_z)/s_{33}{}^E]$, then taking $(\partial/\partial z)$ on both sides

$$0 = \varepsilon_0 \varepsilon_{33}^X \frac{\partial E_Z}{\partial z} + \frac{d_{33}}{s_{33}^E}\left[\left(\frac{\partial^2 u_3}{\partial z^2}\right) - d_{33}\left(\frac{\partial E_Z}{\partial z}\right)\right], \quad \text{or}$$

$$\varepsilon_0 \varepsilon_{33}^X (1 - k_{33}^2)\left(\frac{\partial E_Z}{\partial z}\right) = -\frac{d_{33}}{s_{33}^E}\left(\frac{\partial^2 u_3}{\partial z^2}\right)\left[k_{33}^2 = \frac{d_{33}^2}{s_{33}^E \varepsilon_0 \varepsilon_{33}^X}\right] \tag{2.112}$$

Thus, Eq. (2.112) is transformed into the following dynamic equation:

$$\rho \frac{\partial^2 u_3}{\partial t^2} = \frac{1}{s_{33}^D}\frac{\partial^2 u_3}{\partial z^2} \tag{2.113}$$

$$s_{33}^D = (1 - k_{33}^3)\, s_{33}^E \tag{2.114}$$

or for the harmonic oscillation with $u_3 e^{-i\omega t}$,

$$\frac{\partial^2 u_3}{\partial z^2} + \frac{\omega^2}{v_{33}^{D\,2}} = 0 \quad (v_{33}^D = 1/\sqrt{\rho s_{33}^D}) \tag{2.115}$$

Compared with the sound velocity $v_{11}{}^E = 1/\sqrt{\rho\, s_{11}{}^E})$ in Eq. (2.96) with the surface electrode (E-constant) sample along the vibration direction, non-electrode k_{33} (D-constant) sample exhibits $v_{33}{}^D = 1/\sqrt{\rho\, s_{33}{}^D}$, which is faster (elastically stiffened) than that in E constant condition.

Since the derivative equation formula is the same as the k_{31} case, the solving process is also similar. Assuming the solution of Eq. (2.115) as

$$\begin{cases} u_3 = A\sin\left(\frac{\omega z}{v_{33}^D}\right) + B\cos\left(\frac{\omega z}{v_{33}^D}\right) \\ x_3 = \frac{\partial u_3}{\partial z} = \left(\frac{\omega}{v_{33}^D}\right)A\cos\left(\frac{\omega z}{v_{33}^D}\right) - \left(\frac{\omega}{v_{33}^D}\right)B\sin\left(\frac{\omega z}{v_{33}^D}\right). \end{cases} \tag{2.116}$$

And the boundary conditions $X_3 = 0$ at $z = 0, L$ into Eqs. (2.109) and (2.110) lead to $\frac{\partial u_3}{\partial z} = x_3 = d_{33}E_z = (d_{33}/\varepsilon_0 \varepsilon_{33}{}^X)D_3$. Therefore,

$$\begin{cases} \left(\frac{\omega}{v_{33}^D}\right)A = \frac{d_{33}}{\varepsilon_0 \varepsilon_{33}^X}D_3 \\ \left(\frac{\omega}{v_{33}^D}\right)\left[A\cos\left(\frac{\omega L}{v_{33}^D}\right) - B\sin\left(\frac{\omega L}{v_{33}^D}\right)\right] = \frac{d_{33}}{\varepsilon_0 \varepsilon_{33}^X}D_3 \end{cases}$$

give the solutions for A and B:

$$\begin{cases} A = \frac{d_{33}}{\varepsilon_0 \varepsilon_{33}^X}\frac{v_{33}^D}{\omega}D_3 \\ B = -\frac{d_{33}}{\varepsilon_0 \varepsilon_{33}^X}\frac{v_{33}^D}{\omega}D_3 \tan\left(\frac{\omega L}{2v_{33}^D}\right) \end{cases}$$

Finally, we obtain

$$[\text{Displacement}] \quad u_3 = \frac{d_{33}}{\varepsilon_0 \varepsilon_{33}^X}\frac{v_{33}^D}{\omega}D_3\left[\sin\left(\frac{\omega}{2v_{33}^D}(2z - L)\right)\bigg/\cos\left(\frac{\omega L}{2v_{33}^D}\right)\right], \tag{2.117}$$

$$[\text{Strain}] \quad x_3 = \frac{d_{33}}{\varepsilon_0 \varepsilon_{33}^X}D_3\left[\cos\left(\frac{\omega}{2v_{33}^D}(2z - L)\right)\bigg/\cos\left(\frac{\omega L}{2v_{33}^D}\right)\right] \tag{2.118}$$

In comparison with the k_{31} mode, where the displacement and strain are proportional to the electric field E_3, in the k_{33} mode, those are proportional to the electric displacement D_3.

2.7.2.2 Admittance around resonance and antiresonance

Under the electric field E_z, current can be calculated (D_3 is constant along z-direction in the k_{33} case!) by

$$i = j\omega \int D_3 dx dy = j\omega D_3 wb \tag{2.119}$$

From the relationship

$$D_3 = \varepsilon_0\varepsilon_{33}{}^X(1 - k_{33}^2)E_z + (d_{33}/s_{33}^E)x_3,$$

integration along z axis provides

$$\int_0^L D_3 dz = \varepsilon_0\varepsilon_{33}^X(1 - k_{33}^2)\int_0^L E_z dz + \frac{d_{33}}{s_{33}^E}\int_0^L x_3 dz.$$

Then,

$$D_3 L = \varepsilon_0\varepsilon_{33}^{LC}(-V) + \left(\frac{d_{33}}{s_{33}^E}\right)\frac{d_{33}}{\varepsilon_0\varepsilon_{33}^X}\frac{2v_{33}^D}{\omega}D_3\left[\tan\left(\frac{\omega L}{2v_{33}^D}\right)\right],$$

and

$$D_3 = \varepsilon_0\varepsilon_{33}^{LC}\left(\frac{-V}{L}\right)\bigg/\left[1 - k_{33}^2\left\{\frac{\tan\left(\frac{\omega L}{2v_{33}^D}\right)}{\left(\frac{\omega L}{2v_{33}^D}\right)}\right\}\right] \tag{2.120}$$

where $v_{33}^D = 1/\sqrt{\rho s_{33}^D}$, and $s_{33}^D = s_{33}^E(1 - k_{33}^2)$ (i.e., stiffened elasticity).

We adopt the following notations similar to the k$_{31}$ mode:

- ε_{33}^{LC} is the permittivity in a "longitudinally clamped" (LC) sample, which is given by

$$\varepsilon_0\varepsilon_{33}^{LC} = \varepsilon_0\varepsilon_{33}^X - (d_{33}^2/s_{33}^E), \quad \text{or} \tag{2.121}$$
$$= \varepsilon_0\varepsilon_{33}^X(1 - k_{33}^2).$$
$$k_{33}^2 = d_{33}^2/\varepsilon_0\varepsilon_{33}^X s_{33}^E \tag{2.122}$$

Note here again that the $\varepsilon_{33}{}^{LC}$ is different from extensive permittivity $\varepsilon_{33}{}^x$. $\varepsilon_{33}{}^{LC}$ is the permittivity in the sample mechanically clamped only along the z (or 3, length) direction, free along x or y directions, while the accurate definition of ε_{33}^x means the permittivity clamped completely in the three directions.

- Free capacitance C_0 and damped capacitance C_d are defined by mechanical free and 1D clamped ones, respectively:

$$C_0 = \varepsilon_0\varepsilon_{33}^X\frac{wb}{L} \tag{2.123}$$

$$C_d = \varepsilon_0\varepsilon_{33}^{LC}\frac{wb}{L} \tag{2.124}$$

- Normalized frequency Ω_{33}:

$$\Omega_{33} = (\omega L/2v_{33}^D) \tag{2.125}$$

Combining Eqs. (2.119) and (2.120), the admittance $Y(=-i/V)$ is calculated as follows:

$$Y = \frac{i}{-V} = \frac{j\omega\varepsilon_0\varepsilon_{33}^{LC}\left(\frac{wb}{L}\right)}{\left[1 - k_{33}^2\left\{\frac{\tan\left(\frac{\omega L}{2v_{33}^D}\right)}{\left(\frac{\omega L}{2v_{33}^D}\right)}\right\}\right]} = j\omega C_d + \frac{j\omega C_d}{\left[-1 + 1/k_{33}^2\left\{\frac{\tan(\Omega_{33})}{(\Omega_{33})}\right\}\right]}$$

$$= j\omega C_d + \frac{1}{\left[-\frac{1}{j\omega C_d} + 1/j\omega C_d k_{33}^2\left\{\frac{\tan(\Omega_{33})}{(\Omega_{33})}\right\}\right]} \tag{2.126}$$

Here, we used $\varepsilon_{33}^{LC} = \varepsilon_{33}^X(1-k_{33}^2)$, $s_{33}^D = s_{33}^E(1-k_{33}^2)$, $k_{33}^2 = \frac{d_{33}^2}{\varepsilon_0\varepsilon_{33}^X s_{33}^E}$, $v_{33}^D = 1/\sqrt{\rho s_{33}^D}$ and $C_d = \varepsilon_0\varepsilon_{33}^{LC}\left(\frac{wb}{L}\right)$. The second expression is to show the *damped admittance* and the *motional admittance*, separately, and the final expression is explicitly to reveal that the *motional admittance* branch should include the *negative capacitance* (with exactly the same damped capacitance value) in series with the pure vibration related contribution proportional to $\tan(\Omega_{33})/(\Omega_{33})$.

The resonance frequency is obtained from $Y = \infty$, that is,

$$\left(\frac{\omega L}{2v_{33}^D}\right) = k_{33}^2\tan\left(\frac{\omega L}{2v_{33}^D}\right) \tag{2.127}$$

To the contrary, the antiresonance frequency is obtained from $Y = 0$, that is,

$$\tan\left(\frac{\omega L}{2v_{33}^D}\right) = \infty, \quad \text{or} \quad \frac{\omega L}{2v_{33}^D} = \frac{\pi}{2}, \quad \text{leading to } f_A = \frac{v_{33}^D}{2L} \tag{2.128}$$

Different from the k_{31} case, the k_{33} mode exhibits the antiresonance as a natural mechanical resonance frequency with a half wave length exactly on the rod length under the sound velocity of v_{33}^D (i.e., stiffened vibration), and the resonance is a subsidiary vibration mode associated with the electromechanical coupling.

2.7.2.3 *Resonance and antiresonance vibration modes (comparison among k_{33} and k_{31})*

In comparison with the resonance/antiresonance strain distribution status in the k_{31} mode in Fig. 2.31, Fig. 2.33 illustrates the strain distribution status in the k_{33} mode. Because k_{31} and k_{33} modes possess E-constant and D-constant constraints, respectively, in k_{31}, the resonance frequency is directly related with v_{11}^E or s_{11}^E, while in k_{33}, the antiresonance frequency is a half-wave of the sound wave directly related with v_{33}^D or s_{33}^D (c_{33}^D), which are stiffer (higher frequency) than E-constant status. The antiresonance in k_{31}

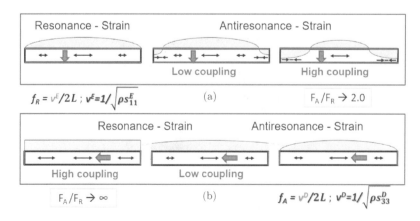

Fig. 2.33: Strain distribution in the resonance and antiresonance states. Longitudinal vibration through the transverse d_{31} (a) and longitudinal d_{33} (b) piezoeffect in a rectangular plate.

and the resonance in k_{33} are subsidiary, originated from the electromechanical coupling factors. It is also worth noting that with increasing the k value toward 1, the ratio f_A/f_R approaches 2 in k_{31}, while it can reach ∞ in k_{33} (since $f_R \to 0$), and that the strain distribution becomes almost flat or uniform in k_{33}, though the stress distributes sinusoidally with zero at the plate ends.

2.7.3 *Resonance/Antiresonance Dynamic Equations with Losses*

2.7.3.1 *Loss and mechanical quality factor in k_{31} mode*[40]

2.7.3.1.1 Resonance Q_A

Now, we introduce the three losses via the complex parameters into the dynamic admittance equation (2.102) around the resonance frequency:[41] $\varepsilon_3^{X^*} = \varepsilon_3^X(1 - j\tan\delta_{33}')$, $s_{11}^{E^*} = s_{11}^E(1 - j\tan\phi_{11}')$ and $d_{31}^* = d_{31}(1 - j\tan\theta_{31}')$.

$$Y = Y_d + Y_m = j\omega C_d(1 - j\tan\delta_{33}''') + j\omega C_d K_{31}^2[(1 - j(2\tan\theta_{31}'$$
$$- \tan\phi_{11}')][(\tan(\omega L/2v_{11}^E *)/(\omega L/2v_{11}^{E*})], \tag{2.129}$$

where

$$C_0 = (wL/t)\varepsilon_0\varepsilon_{33}{}^X \text{(free electrostatic capacitance, real number)} \tag{2.130}$$

$$C_d = (1 - k_{31}^2)C_0 \text{(damped/longitudinary clamped capacitance, realnumber)}, \tag{2.131}$$

$$K_{31}^2 = \frac{k_{31}^2}{1 - k_{31}^2}. \tag{2.132}$$

Note that the loss for the first term ("damped/clamped" admittance) is represented by the dielectric loss $\tan\delta'''$:

$$\tan\delta_{33}''' = [1/(1 - k_{31}^2)][\tan\delta_{33}' + k_{31}^2(\tan\phi_{11}' - 2\tan\theta_{31}')] \tag{2.133}$$

Though the formula is identical to Eq. (2.53), $\tan\delta_{33}'''$ is not exactly to the "extensive" non-prime loss $\tan\delta$, because the extensive loss should be defined under three-dimensionally clamped condition, not by just 1D longitudinal clamp. Taking into account

$$v_{11}^{E^*} = \frac{1}{\sqrt{\rho s_{11}^E(1 - j\tan\phi_{11}')}} = v_{11}^E\left(1 + j\frac{\tan\phi_{11}'}{2}\right), \tag{2.134}$$

we further calculate $1/[\tan(\omega L/2v^*)]$ with an expansion-series approximation around the A-type resonance frequency $(\omega_A L/2v) = \pi/2$, taking into account that the resonance state is defined in this case for the minimum impedance point. Using normalized frequency parameters,

$$\Omega_A = \omega_A L/2v_{11}{}^E = \pi/2, \quad \Delta\Omega_A = \Omega - \pi/2(<< 1), \tag{2.135}$$

and

$$\frac{1}{\tan\Omega^*} = \cot\left(\frac{\pi}{2} + \Delta\Omega_A - j\frac{\pi}{4}\tan\phi_{11}'\right) = \Delta\Omega_A - j\frac{\pi}{4}\tan\phi_{11}'$$

the *motional admittance* Y_m is approximated around the first resonance frequency ω_A by

$$Y_m = j(8/\pi^2)\,\omega_A\,C_d\,K_{31}^2\,[(1 - j(2\tan\theta_{31}' - \tan\phi_{11}')]/[(4/\pi)\Delta\Omega_A - j\tan\phi_{11}']. \tag{2.136}$$

The maximum Y_m is obtained at $\Delta\Omega_A = 0$:

$$Y_m^{\max} = (8/\pi^2)\,\omega_A\,C_d\,K_{31}^2\,(\tan\phi_{11}')^{-1} = (8/\pi^2)\,\omega_A\,C_0\,k_{31}^2 Q_A, \qquad (2.137)$$

The mechanical quality factor for A-type resonance $Q_A = (\tan\phi_{11}')^{-1}$ can be proved as follows: Q_A is defined by $Q_A = \omega_A/2\Delta\omega$, where $2\Delta\omega$ is a full width of the 3 dB down (i.e., $1/\sqrt{2}$, because $20\log_{10}(1/\sqrt{2}) = -3.01$) of the maximum value $Y_m^{\max}$ at $\omega = \omega_A$. Since $|Y| = |Y|^{\max}/\sqrt{2}$ can be obtained when the "conductance = susceptance"; $\Delta\Omega_A = (\pi/4)\tan\phi_{11}'$ [see the denominator of Eq. (2.136)],

$$Q_A = \Omega_A/2\Delta\,\Omega_A = (\pi/2)/2(\pi/4)\tan\phi_{11'} = (\tan\phi_{11'})^{-1}. \qquad (2.138)$$

Similarly, the maximum displacement $u^{\max}$ is obtained at $\Delta\Omega = 0$:

$$u^{\max} = (8/\pi^2)\,Q_A\,d_{31}\,E_Z L. \qquad (2.139)$$

The maximum displacement at the resonance frequency is $(8/\pi^2)Q_A$ times larger than that at a non-resonance frequency, $d_{31}E_Z L$. Under the constant voltage/field drive, the displacement is amplified at the resonance frequency, while under the constant current drive, the displacement u and the impedance Z are amplified at the antiresonance frequency by the factor of $(8/\pi^2)Q_B$, as explained in **Example Problem 2.11**.

Example Problem 2.11.

Under pseudo-DC operation, the input electric energy is split into the converted mechanical energy by k^2 and the stored electric energy by $(1 - k^2)$, leading to the damped and motional capacitance ratio $(1 - k^2)$ vs. k^2. However, under the resonance drive, though the damped admittance is provided by $\omega_A C_0(1 - k_{31}^2)$, the maximum of the motional admittance for the fundamental resonance frequency is described by

$$Y_m^{\max} = \frac{8}{\pi^2}\omega_A C_0 k_{31}^2 Q_A. \qquad (P2.11.1)$$

The calibration factor $(8/\pi^2)$ (≈ 0.81) is required for the fundamental resonance frequency, rather than just 1. Explain why this calibration factor is required for the fundamental resonance condition.

Hint:

Calculate the motional admittance for higher order resonance harmonics. Only one fundamental resonance mode does not spend all mechanically convertible energy. You may also use the relation

$$\Sigma\left[\frac{1}{(2m-1)^2}\right] = \left(\frac{\pi^2}{8}\right) \qquad (P2.11.2)$$

Solution:

We start from Eq. (2.129):

$$Y_m = j\omega C_d\,K_{31}^2[(1 - j(2\tan\theta_{31'} - \tan\phi_{11}'))][(\tan(\omega L/2v_{11}^{E^*})/(\omega L/2v_{11}{}^{E^*})], \qquad (P2.11.3)$$

Note that the A-type resonance is obtained at $\left(\frac{\omega_{A,n}L}{2v_{11}^E}\right) = n(\frac{\pi}{2})$, where $n = 1, 3, 5, \ldots$ (the nth higher-order harmonics) and $v_{11}^{E\,2} = 1/\rho s_{11}^E$. Then, taking into account the complex elastic compliance

$$v_{11}^{E^*} = \frac{1}{\sqrt{\rho s_{11}^E(1 - j\tan\phi_{11}')}} = v_{11}^E\left(1 + j\frac{\tan\phi_{11}'}{2}\right), \qquad (P2.11.4)$$

we further calculate $1/[\tan(\omega L/2v^*)]$ with an expansion-series approximation around the A-type resonance frequency $(\omega_A L/2v_{11}^E) = n(\pi/2)$, taking into account that the resonance state is defined in this case for the minimum impedance (maximum admittance) point. Using new frequency parameters,

$$\Omega_A = \omega_A L/2v_{11}{}^E = n(\pi/2), \quad \Delta\Omega_A = \Omega - n(\pi/2)(<< 1), \tag{P2.11.5}$$

and

$$\frac{1}{\tan \Omega^*} = \cot\left(n\frac{\pi}{2} + \Delta\Omega_A - j\frac{1}{2}\left(n\frac{\pi}{2}\right)\tan \phi'_{11}\right) = \Delta\Omega_A - j\frac{n\pi}{4}\tan \phi'_{11} \tag{P2.11.6}$$

the *motional admittance* Y_m is approximated around the n-th resonance frequency $\omega_{A,n}$ by

$$Y_{m,n} = j(8/\pi^2 n^2)\,\omega_{A,n}C_0 k_{31}^2/[(4/n\pi)\,\Delta\Omega_{A,n} - j\tan \phi'_{11}]. \tag{P2.11.7}$$

The maximum $Y_{m,n}$ is obtained at $\Delta\Omega_{A,n} = 0$:

$$Y_{m,n}^{max} = (8/\pi^2 n^2)\,\omega_{A,n}C_0 k_{31}^2\,(\tan \phi'_{11})^{-1} = (8/\pi^2 n^2)\,\omega_{A,n}C_0 k_{31}^2 Q_m \tag{P2.11.8}$$

Supposing that the intensive elastic loss $\tan \phi'_{11}$ or the mechanical quality factor Q_m is insensitive to the frequency difference among the higher-order harmonic resonance frequencies, we can understand that each harmonic mode originates from the effective motional capacitance equal to $(8/\pi^2 n^2)C_0 k_{31}^2$, and the admittance is enhanced by the factor of Q_m. Under the resonance, the input cyclic electric energy will excite the mechanical vibration and motional capacitance synchronously by a factor of Q_m by spending the cyclic excitation number proportional to Q_m. The motional capacitance is proportional to $(1/n^2)$ for the n-th order harmonic resonance mode. Knowing a general relationship $\Sigma\left[\frac{1}{(2m-1)^2}\right] = \left(\frac{\pi^2}{8}\right)$ (m: positive integer), when we add motional capacitances for all harmonic resonance modes,

$$(8/\pi^2)\,C_0 k_{31}^2 \sum\nolimits_{n=1,\,3,\,5,\dots}\left(\frac{1}{n^2}\right) = C_0 k_{31}^2 \tag{P2.11.9}$$

Since the total motional capacitance for all harmonic resonance modes corresponds exactly to the free capacitance minus damped capacitance, the calibration factor $(8/\pi^2 n^2)$ can be understood as the distribution ratio of the mechanical energy to all n-th harmonic modes. The above concept on the higher-order harmonic modes will be used in the Equivalent Circuit Model explained in Section 2.8.

2.7.3.1.2 Antiresonance Q_B

On the other hand, a higher-quality factor at the antiresonance is usually observed in comparison with that at the resonance point,[42,43] the reason of which was interpreted by Mezheritsky from the combination of three loss factors.[43] In this subsection, we provide an alternative and, more importantly, a user-friendly formula to determine piezoelectric losses by analyzing the admittance/impedance spectra at resonance and antiresonance.[36] The antiresonance corresponds to the minimum admittance of Eq. (2.129).

$$Y = j\omega C_d(1 - j\tan \delta_{33}''') + j\omega C_d\left(\frac{k_{31}^2}{1 - k_{31}^2}\right)[(1 - j(2\tan \theta'_{31} - \tan \phi'_{11})]$$

$$\times\,[(\tan(\omega L/2v_{11}{}^{E*})/(\omega L/2v_{11}{}^{E*})]$$

In the resonance discussion, we neglected the damped admittance, because the motional admittance is significantly large due to $\tan(\omega L/2v_{11}{}^{E*}) \nearrow \infty$. On the contrary, in the antiresonance discussion, we consider basically the subtraction between the damped and motional admittances (i.e., motional admittance phase changes 90° to −90° above the resonance frequency); that is, the admittance should be exactly zero when the loss is not included, or is only the minimum when we consider the losses (i.e., integration of the complex parameters) in Eq. (2.102).

Using again a normalized frequency parameter, $\Omega_{11} = \frac{\omega L}{2v_{11}^E}$, we introduce the normalized admittance Y' (in terms of $j\omega C_0$) for further calculation:

$$Y' = 1 - k_{31}{}^2 + k_{31}^2 \frac{\tan(\omega l/2v_{11}^E)}{\omega l/2v_{11}^E} = 1 - k_{31}^2 + k_{31}^2 \frac{\tan(\Omega_{11})}{\Omega_{11}}. \tag{2.140}$$

Since the expansion series of $\tan \Omega_{11}$ is convergent in this case, taking into account

$$1/v_{11}^{E^*} = \sqrt{\rho s_{11}^E(1 - j\tan\phi_{11}')} = v_{11}^E\left(1 - j\frac{\tan\phi_{11}'}{2}\right) \tag{2.141}$$

we can apply the following expansion approximation in terms of $\tan\phi_{11}'$:

$$\tan(\Omega_{11}^*) = \tan\left(\Omega_{11} - j\frac{\Omega_{11}\tan\phi_{11}'}{2}\right) = \tan\Omega_{11} - j\frac{\Omega_{11}\tan\phi_{11}'}{2\cos^2\Omega_{11}}.$$

Introducing losses for the parameters in Eq. (2.129) leads to

$$Y' = 1 - k_{31}{}^2\left[1 - j\left(2\tan\theta_{31}' - \tan\delta_{33}' - \tan\phi_{11}'\right)\right]$$
$$+ k_{31}^2\left[1 - j\left(2\tan\theta_{31}' - \tan\delta_{33}' - \tan\phi_{11}'\right)\right]\frac{\tan\Omega_{11}^*}{\Omega_{11}^*}. \tag{2.142}$$

Note that the *electromechanical coupling loss* $(2\tan\theta_{31}' - \tan\delta_{33}' - \tan\phi_{11}')$ contributes significantly in this antiresonance discussion. We separate Y' into conductance G (real part) and susceptance B (imaginary part) as $Y' = G + jB$:

$$G = 1 - k_{31}^2 + k_{31}^2\frac{\tan\Omega_{11}}{\Omega_{11}}. \tag{2.143}$$

$$B = \left(k_{31}^2 - k_{31}^2\frac{\tan\Omega_{11}}{\Omega_{11}}\right)\left(2\tan\theta_{31}' - \tan\delta_{33}' - \tan\phi_{11}'\right) - \frac{k_{31}^2}{2}\left(\frac{1}{\cos^2\Omega_{11}} - \frac{\tan\Omega_{11}}{\Omega_{11}}\right)\tan\phi_{11}'. \tag{2.144}$$

The antiresonance frequency Ω_b should satisfy $G = 0$ (i.e., definition from Eq. (2.101)), and

$$1 - k_{31}^2 + k_{31}^2\frac{\tan\Omega_B}{\Omega_B} = 0. \tag{2.145}$$

Using new parameters,

$$\Omega_{11} = \Omega_B + \Delta\Omega_B, \tag{2.146}$$

similar to $\Delta\Omega_A$ for the resonance, $\Delta\Omega_B$ is also a small number, and the first-order approximation can be utilized.

$$\frac{\tan\Omega_{11}}{\Omega_{11}} = \frac{\tan\Omega_B}{\Omega_B} + \frac{1}{\Omega_B}\left(\frac{1}{\cos^2\Omega_B} - \frac{\tan\Omega_B}{\Omega_B}\right)\Delta\Omega_B.$$

Neglecting the high-order term which has two or more small factors (loss factor or $\Delta\Omega_B$)

$$G = \frac{k_{31}^2}{\Omega_B}\left(\frac{1}{\cos^2\Omega_B} - \frac{\tan\Omega_B}{\Omega_B}\right)\Delta\Omega_B. \tag{2.147}$$

$$B = \left(2\tan\theta_{31}' - \tan\delta_{33}' - \tan\phi_{11}'\right) - \frac{k_{31}^2}{2}\left(\frac{1}{\cos^2\Omega_B} - \frac{\tan\Omega_B}{\Omega_B}\right)\tan\phi_{11}'. \tag{2.148}$$

Consequently, the minimum absolute value of admittance can be achieved when $\Delta\Omega_B$ is 0. The antiresonance frequency Ω_B is determined by Eq. (2.145). In order to find the 3dB-up point, let $G = B$, where

$\sqrt{G^2 + B^2} = \sqrt{2}B$ is satisfied:

$$\frac{k_{31}^2}{\Omega_B} \left(\frac{1}{\cos^2 \Omega_B} - \frac{\tan \Omega_B}{\Omega_B} \right) \Delta\Omega_B = \left(2\tan\theta_{31}' - \tan\delta_{33}' - \tan\phi_{11}' \right)$$

$$- \frac{k_{31}^2}{2} \left(\frac{1}{\cos^2 \Omega_B} - \frac{\tan \Omega_B}{\Omega_B} \right) \tan\phi_{11}'. \tag{2.149}$$

Further, since the antiresonance quality factor is given by

$$Q_{B,31} = \frac{\Omega_B}{2|\Delta\Omega_B|}, \tag{2.150}$$

Eq. (2.149) can be represented as

$$\frac{k_{31}^2}{2Q_{B,31}} \left(\frac{1}{\cos^2 \Omega_B} - \frac{\tan \Omega_B}{\Omega_B} \right) = -\left(2\tan\theta_{31}' - \tan\delta_{33}' - \tan\varphi_{11}' \right) + \frac{k_{31}^2}{2} \left(\frac{1}{\cos^2 \Omega_B} - \frac{\tan \Omega_B}{\Omega_B} \right) \tan\varphi_{11}'.$$

We can now obtain the result as

$$\frac{1}{Q_{B,31}} = \tan\phi_{11}' - \frac{2}{k_{31}^2} \left(2\tan\theta_{31}' - \tan\delta_{33}' - \tan\phi_{11}' \right) / \left(\frac{1}{\cos^2 \Omega_B} - \frac{\tan \Omega_B}{\Omega_B} \right).$$

Noting the following relation

$$\frac{1}{\cos^2 \Omega_B} - \frac{\tan \Omega_B}{\Omega_B} = \frac{(1 - k_{31}^2)^2 \Omega_B^2 + k_{31}^2}{k_{31}^4},$$

we obtain the final formula:

$$\begin{cases} \dfrac{1}{Q_{A,31}} = \tan\phi_{11}' \\[2mm] \dfrac{1}{Q_{B,31}} = \dfrac{1}{Q_{A,31}} - \dfrac{2}{1 + \left(\frac{1}{k_{31}} - k_{31} \right)^2 \Omega_B^2} \left(2\tan\theta_{31}' - \tan\delta_{33}' - \tan\phi_{11}' \right) \end{cases} \tag{2.151}$$

You may understand that in k_{31} mode, where the wave propagation direction with the electrode is perpendicular to the spontaneous polarization direction, the primary mechanical resonance (a half wave length vibration of the plate length) corresponds to the "resonance" mode with the sound velocity $s_{11}{}^E$ and the "antiresonance" mode corresponds to the subsidiary mode via the electromechanical coupling. Accordingly, the mechanical quality factor Q_B at the antiresonance is modified from the resonance Q_A by the electromechanical coupling loss $(2\tan\theta_{31}' - \tan\delta_{33}' - \tan\phi_{11}')$. Also from the Eq. (2.151), we can understand that when $(2\tan\theta_{31}' - \tan\delta_{33}' - \tan\phi_{11}') > 0, Q_{B,31} > Q_{A,31}$, while $(2\tan\theta_{31}' - \tan\delta_{33}' - \tan\phi_{11}') < 0, Q_{B,31} < Q_{A,31}$. The fact $Q_{B,31} > Q_{A,31}$ in PZTs indicates the largest contribution of piezoelectric loss $\tan\theta_{31}'$ in comparison with other losses.

2.7.3.2 *Loss and mechanical quality factor in k_{33} mode*[43]

The length extensional mode is shown in Fig. 2.32, where $L >> w, b$. Taking into account the dynamic equation given by Eq. (2.126), we can derive the impedance expression of the k_{33} mode bar in a similar fashion to k_{31} mode.[44]

$$Z(\omega) = \frac{1}{j\omega C_d} \left(1 - k_{33}^2 \frac{\tan \Omega_{33}}{\Omega_{33}} \right), \tag{2.152}$$

where

$$C_d = \frac{wb}{L}\varepsilon_0\varepsilon_{33}^X\left(1 - k_{33}^2\right), \tag{2.153}$$

$$\Omega_{33} = \frac{\omega L}{2}\sqrt{\rho s_{33}^D} = \frac{\omega L}{2}\sqrt{\rho s_{33}^E\left(1 - k_{33}^2\right)}, \tag{2.154}$$

$$k_{33}^2 = \frac{d_{33}^2}{\varepsilon_0\varepsilon_{33}^X s_{33}^E}. \tag{2.155}$$

By introducing the complex parameters,

$$(k_{33}^2)^* = k_{33}^2(1 - j\chi_{33}), \tag{2.156}$$

$$\chi_{33} = 2\tan\theta_{33}' - \tan\delta_{33}' - \tan\phi_{33}' \text{ ("electromechanical coupling loss")}, \tag{2.157}$$

$$C_d^* = C_d(1 - j\tan\delta_{33}''') \text{ (damped capacitance loss)}, \tag{2.158}$$

$$\tan\delta_{33}''' = \frac{1}{1 - k_{33}^2}[\tan\delta_{33}' - k_{33}^2(2\tan\theta_{33}' - \tan\phi_{33}')], \tag{2.159}$$

$$\Omega^* = \Omega\sqrt{1 - j\tan\phi_{33}'''}, \tag{2.160}$$

$$\tan\phi_{33}''' = \frac{1}{1 - k_{33}^2}[\tan\phi_{33}' - k_{33}^2(2\tan\theta_{33}' - \tan\delta_{33}')]. \tag{2.161}$$

Note again that the parameters in k_{33} mode have similar forms as k_{31} mode, and the difference is that the loss factors by $-\chi_{33}$, $\tan\delta_{33}'''$, $\tan\phi_{33}'''$, which shows identical forms to the "extensive" loss parameters in terms of "intensive" losses, but are not the same, strictly speaking. The difference between the extensive (non-prime) losses and these triple-prime losses comes from the 3D or 1D mechanically clamped conditions. Refer to the k_t mode in the next subsection, where the loss factors are purely "extensive losses" since the elastic stiffness c_{33}^D is the primary parameter, and the mechanical 3D clamp is practically satisfied. Therefore, a similar derivation process to the k_{31} mode can be applied, and the results are given by

$$Q_{B,33} = \frac{1}{\tan\phi_{33}'''} = \frac{1 - k_{33}^2}{\tan\phi_{33}' - k_{33}^2(2\tan\theta_{33}' - \tan\delta_{33}')}, \tag{2.162}$$

$$\frac{1}{Q_{A,33}} = \frac{1}{Q_{B,33}} + \frac{2}{k_{33}^2 - 1 + \Omega_A^2/k_{33}^2}(2\tan\theta_{33}' - \tan\delta_{33}' - \tan\phi_{33}'). \tag{2.163}$$

Different from the k_{31} mode, you may approximately understand that in k_{33} and k_t modes, where the wave propagation direction is parallel to the spontaneous polarization direction, the primary mechanical resonance (a half wave length vibration of the length or thickness) corresponds to the "antiresonance" mode with the sound velocity s_{33}^D or c_{33}^D, respectively (D-constant with the depolarization field), and the "resonance" state is the subsidiary mode via the piezoelectric coupling. Also from the Eq. (2.163), we can understand that when $(2\tan\theta_{33}' - \tan\delta_{33}' - \tan\phi_{33}') > 0$, $Q_{B,33} > Q_{A,33}$, while $(2\tan\theta_{33}' - \tan\delta_{33}' - \tan\phi_{33}') < 0$, $Q_{B,31} < Q_{A,31}$. From Eq. (2.163) and setting $Z = 0$, the resonance frequency is provided simply by

$$\Omega_A = k_{33}^2\tan\Omega_A. \tag{2.164}$$

Note again that a half-wave length of the vibration is longer than the rod length at the resonance frequency (more uniform strain distribution than the sinusoid); a half-wave length is realized at the antiresonance frequency.

2.7.3.3 *Loss and mechanical quality factor in other modes*

To obtain the loss factor matrix, five vibration modes need to be characterized in PZT ceramics with ∞mm crystallographic symmetry (10 independent dielectric, elastic and piezoelectric loss components for intensive and extensive parameters), as summarized in Fig. 2.34 and Table 2.4. The methodology is based on the equations of quality factors Q_A (resonance) and Q_B (antiresonance) in various modes with regard to loss factors and other properties.[44] We can measure Q_A and Q_B for each mode by using the 3 dB-up/down method (or quadrantal frequency method) in the impedance/admittance spectra (see Fig. 2.30). In addition to some derivations based on fundamental relations of the material properties, all the 20 loss factors can be obtained for piezoelectric ceramics. We derived the relationships between mechanical quality factors Q_A (resonance) and Q_B (antiresonance) in all required five modes shown in Table 2.4. The results are summarized below:[44]

(a) k_{31} mode: (intensive elastic loss)

$$Q_{A,31} = \frac{1}{\tan \phi'_{11}},$$

$$\frac{1}{Q_{B,31}} = \frac{1}{Q_{A,31}} - \frac{2}{1 + \left(\frac{1}{k_{31}} - k_{31}\right)^2 \Omega_{B,31}^2}(2\tan\theta'_{31} - \tan\delta'_{33} - \tan\phi'_{11})$$

$$\Omega_{A,31} = \frac{\omega_a l}{2v_{11}^E} = \frac{\pi}{2}\left[v_{11}^E = 1/\sqrt{\rho s_{11}^E}\right], \quad \Omega_{B,31} = \frac{\omega_b l}{2v_{11}^E}, \quad 1 - k_{31}^2 + k_{31}^2 \frac{\tan\Omega_B}{\Omega_B} = 0$$

(b) k_t mode: (extensive elastic loss)

$$Q_{B,t} = \frac{1}{\tan \phi_{33}},$$

$$\frac{1}{Q_{A,t}} = \frac{1}{Q_{B,t}} - \frac{2}{k_t^2 - 1 + \Omega_{A,t}^2/k_t^2}(2\tan\theta_{33} - \tan\delta_{33} - \tan\phi_{33})$$

$$\Omega_{B,t} = \frac{\omega_b l}{2v_{33}^D} = \frac{\pi}{2}\left[v_{33}^D = 1/\sqrt{\rho/c_{33}^D}\right], \quad \Omega_{A,t} = \frac{\omega_a l}{2v_{33}^D}, \Omega_{A,t} = k_t^2 \tan\Omega_{A,t}$$

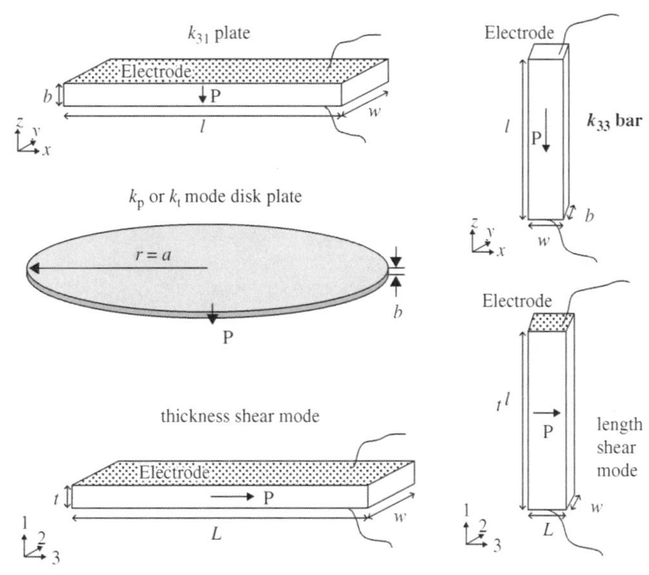

Fig. 2.34: Sketches of the sample geometries for five required vibration modes.

Table 2.4: The characteristics of various piezoelectric resonators with different shapes and sizes.

	Factor	Boundary Conditions	Resonator Shape	Definition
a	k_{31}	$X_1 \neq 0,\ X_2 = X_3 = 0$ $x_1 \neq 0,\ x_2 \neq 0,\ x_3 \neq 0$		$\dfrac{d_{31}}{\sqrt{s_{11}^E \varepsilon_0 \varepsilon_{33}^X}}$
b	k_{33}	$X_1 = X_2 = 0,\ X_3 \neq 0$ $x_1 = x_2 \neq 0,\ x_3 \neq 0$	 **Fundamental Mode**	$\dfrac{d_{33}}{\sqrt{s_{33}^E \varepsilon_0 \varepsilon_{33}^X}}$
c	k_p	$X_1 = X_2 \neq 0,\ X_3 = 0$ $x_1 = x_2 \neq 0,\ x_3 \neq 0$	 **Planar/Radial**	$k_{31}\sqrt{\dfrac{2}{1-\sigma}}$
d	k_t	$X_1 = X_2 \neq 0,\ X_3 \neq 0$ $x_1 = x_2 = 0,\ x_3 \neq 0$	 **Thickness Mode**	$k_{33}\sqrt{\dfrac{\varepsilon_0 \varepsilon_{33}^x}{c_{33}^D}}$
e	$k_{24} = k_{15}$	$X_1 = X_2 = X_3 = 0,\ X_4 \neq 0$ $x_1 = x_2 = x_3 = 0,\ x_5 \neq 0$	 **Shear Mode**	$\dfrac{d_{15}}{\sqrt{s_{55}^E \varepsilon_0 \varepsilon_{11}^X}}$

(c) k_{33} mode:

$$Q_{B,33} = \frac{1}{\tan \phi'''_{33}} = \frac{1 - k_{33}^2}{\tan \phi'_{33} - k_{33}^2 (2 \tan \theta'_{33} - \tan \delta'_{33})}$$

$$\frac{1}{Q_{A,33}} = \frac{1}{Q_{B,33}} + \frac{2}{k_{33}^2 - 1 + \Omega_A^2/k_{33}^2}\left(2\tan\theta'_{33} - \tan\delta'_{33} - \tan\phi'_{33}\right)$$

$$\Omega_{B,33} = \frac{\omega_b l}{2 v_{33}^D} = \frac{\pi}{2}\left[v_{33}^D = 1/\sqrt{\rho s_{33}^D}\right], \quad \Omega_{A,33} = \frac{\omega_a l}{2 v_{33}^D}, \Omega_{A,33} = k_{33}^2 \tan\Omega_{A,33}$$

(d) k_{15} mode (constant E – length shear mode): (intensive elastic loss)

$$Q_{A,15}^E = \frac{1}{\tan \phi'_{55}}$$

$$\frac{1}{Q_{B,15}^E} = \frac{1}{Q_{A,15}^E} - \frac{2}{1 + (\frac{1}{k_{15}} - k_{15})^2 \Omega_B^2}\left(2\tan\theta'_{15} - \tan\delta'_{11} - \tan\phi'_{55}\right)$$

$$\Omega_B = \frac{\omega_b L}{2 v_{55}^E} = \frac{\omega_b L}{2 v_{55}^E}\sqrt{\rho s_{55}^E}, \quad 1 - k_{15}^2 + k_{15}^2 \frac{\tan\Omega_B}{\Omega_B} = 0$$

(e) k_{15} mode (constant D – thickness shear mode): (extensive elastic loss)

$$Q_{B,15}^D = \frac{1}{\tan \phi_{55}}$$

$$\frac{1}{Q_{A,15}^D} = \frac{1}{Q_{B,15}^D} - \frac{2}{k_{15}^2 - 1 + \Omega_A^2/k_{15}^2}\left(2\tan\theta_{15} - \tan\delta_{11} - \tan\phi_{55}\right)$$

$$\Omega_A = \frac{\omega_a t}{2 v_{55}^D} = \frac{\omega_a t}{2}\sqrt{\frac{\rho}{c_{55}^D}}, \quad \Omega_A = k_{15}^2 \tan\Omega_A$$

Note again that because k_{31} and k_{33}/k_t modes possess E-constant and D-constant constraints, respectively, in k_{31}, the resonance frequency is directly related with v_{11}^E or s_{11}^E as $f_A = \frac{v_{11}^E}{2L} = 1/2L\sqrt{\rho s_{11}^E}$, while in

k_{33}/k_t, the antiresonance frequency is directly related with ${v_{33}}^D$ or ${s_{33}}^D$, ${c_{33}}^D$ as $f_B = \frac{{v_{33}}^D}{2L} = 1/2L\sqrt{\rho {s_{33}}^D}$ or $1/2b\sqrt{\rho/{c_{33}}^D}$. It is important to distinguish k_{33} ($X_1 = X_2 = 0$, $x_1 = x_2 \neq 0$) from k_t ($X_1 = X_2 \neq 0$, $x_1 = x_2 = 0$) from the boundary conditions. Note the relations $s_{33}^D = s_{33}^E(1-k_{33}^2)$ and $c_{33}^E = c_{33}^D(1-k_t^2)$, and $k_{33} > k_t$, in general, and also c_{33} is not just an inverse of s_{33}, but $\frac{1}{s_{33}} = c_{33} - \frac{2c_{13}^2}{c_{11}+c_{12}}$ or $\frac{1}{c_{33}} = s_{33} - \frac{2s_{13}^2}{s_{11}+s_{12}}$ in $\infty\, mm$ symmetry. The pure "extensive" loss $tan\phi_{33}$ is obtained from the loss relating with c_{33}^D from the definition, that is, in the k_t mode. When the length of a rod k_{33} is not very long, the mode approaches the k_t, and $c_{33}^D \approx 1/s_{33}^D$. The antiresonance in k_{31} and the resonance in k_{33}/k_t are subsidiary, originated from the electromechanical coupling factors. We also remind the reader of the relation for the *electromechanical coupling factor losses* from Eq. (2.57):

$$\left(2\tan\theta' - \tan\delta' - \tan\phi'\right) = -\left(2\tan\theta - \tan\delta - \tan\phi\right) \tag{2.165}$$

Since the side is not clamped ($x_1 = x_2 \neq 0$) in the k_{33} mode (different from the k_t mode), the triple-prime losses in the previous subsection are not exactly equal to non-prime extensive losses. 3D-clamped k_t mode exhibits the purely "extensive" non-prime losses, though the Q_m formulas for the k_{33} mode seem to be rather close to the extensive losses transformed from the intensive losses.

2.7.4 Q_A and Q_B in the IEEE Standard

It is also important to discuss the assumption in the IEEE Standard,[45] where the difference of the mechanical quality factors among the resonance and antiresonance modes is neglected, that is, $Q_A = Q_B$. This historically originates from the neglect of the *coupling loss* (i.e., piezoelectric loss) and the assumption of $\tan\phi' \gg \tan\delta'$ around the resonance region, leading to only one loss factor, that is, the intensive elastic loss. However, if we adopt our three-loss model, this situation ($Q_A = Q_B$) occurs only when $(2\tan\theta' - \tan\delta' - \tan\phi') = 0$, or $\tan\theta' = (\tan\delta' + \tan\phi')/2$. The IEEE Standard corresponds to only when the piezoelectric loss is equal to the average value of the dielectric and elastic losses, which exhibits a serious contradiction to the well-known PZT experimental results, that is, $Q_A < Q_B$. To the contrary, if we totally neglect the piezoelectric loss $\tan\theta'$, $Q_A > Q_B$ is inevitably derived, which is further contradictory to the experiments. As we can realize in Fig. 2.30 from the peak sharpness, the PZTs always exhibit Q_A (resonance) $< Q_B$ (antiresonance), irrelevant to the vibration mode (Fig. 2.30 is an example of the k_{31} mode). This concludes that $(\tan\delta'_{33} + \tan\phi'_{11} - 2\tan\theta'_{31}) < 0$, or $(\tan\delta'_{33} + \tan\phi'_{11})/2 < \tan\theta'_{31}$ for k_{31}, and $(\tan\delta_{33} + \tan\phi_{33} - 2\tan\theta_{33}) > 0$, or $(\tan\delta_{33} + \tan\phi_{33})/2 > \tan\theta_{33}$ for k_t. It is worth noting that the intensive piezoelectric loss is larger than the average of the dielectric and elastic intensive losses in Pb-contained piezo ceramics.

2.8 Equivalent Circuits with Piezoelectric Vibrators

Mechanical and electrical systems are occasionally equivalent from the mathematical formula's viewpoint. Therefore, an electrician tries to understand a mechanical system behavior from a more familiar LCR electrical circuit analysis; or vice versa, a mechanic uses a mass-spring-damper model to understand an electric circuit. However, two important notes must be taken into account: (1) mechanical loss is handled as "viscous damping" in the equivalent circuit, though the losses in piezoelectric ceramics may be "solid damping" in reality; (2) equivalent circuit approach is almost successful, as long as we consider a steady sinusoidal (harmonic) vibration. When we consider a transient response, such as a pulse drive of a mechanical system with finite specimen size, the equivalent circuit analysis generates a significant discrepancy. We will discuss the impulse force application on the piezoelectric energy harvesting device in the next chapter.

The equivalent (electric) circuit (EC) is a widely used tool which can greatly simplify the process of design and analysis of the piezoelectric devices, in which the circuit, in a standard form,[46] can only graphically characterize the mechanical loss by applying a resistor (and dielectric loss sometimes). Different from a pure mechanical system, a piezoelectric vibration exhibits an "antiresonance" mode in addition to a "resonance" mode, due to the existence of the damped capacitance (i.e., only the partial of the input electric energy is transduced into the mechanical energy). As discussed in Section 2.7, without introducing the "*piezoelectric loss*", it is difficult to explain the difference of the mechanical quality factors at the resonance and antiresonance modes. Damjanovic therefore introduced an additional branch into the standard circuit, which is used to present the influence of the piezoelectric loss.[47] However, concise and decoupled formulas of three (dielectric, elastic and piezoelectric) losses have not been derived, which can be used for the measurements of losses in piezoelectric materials as a user-friendly method. We consider new equivalent circuits of piezoelectric devices with these three losses in this chapter.

2.8.1 *Equivalency between Mechanical and Electrical Systems*

There are two classifications of an LCR electrical circuit: series connection and parallel connection. Though both circuits are equivalent, in general, focused usage is different.

2.8.1.1 *LCR series connection equivalent circuit*

The dynamic equation for a pure mechanical system composed of a mass, a spring and a damper illustrated in Table 2.5(a) is expressed by

$$M(d^2u/dt^2) + \zeta(du/dt) + cu = F(t), \quad \text{or} \tag{2.166a}$$

$$M(dv/dt) + \zeta v + c \int_0^t v dt = F(t) \tag{2.166b}$$

where u is the displacement of a mass M, v is the velocity ($= du/dt$), c spring constant, ζ damping constant of the dash-pot and F is the external force. Note that when a continuum elastic material is considered, the actual damping may be "solid damping", but we consider or approximate here the "*viscous damping*", where the damping force is described in proportion to the velocity, from a mathematical simplicity viewpoint.

On the other hand, the dynamic equation for an electrical circuit composed of a series connection of an inductance L, a capacitance C, and a resistance R illustrated in Table 2.5(b) is expressed by

$$L(d^2q/dt^2) + R(dq/dt) + (1/C)q = V(t), \quad \text{or} \tag{2.167a}$$

$$L(dI/dt) + RI + (1/C) \int_0^t I dt = V(t) \tag{2.167b}$$

where q is charge, I is the current ($= dq/dt$) and V is the external voltage. Taking into account the equation similarity, the engineer introduces an equivalent circuit; that is, consider a mechanical system with using an equivalent electrical circuit, which is intuitively simpler for an electrical engineer. In contrast, consider an electrical circuit with using an equivalent mechanical system, which is intuitively simpler for a mechanical engineer. Equivalency between these two systems is summarized in the center column in Table 2.5.

When we consider steady harmonic (i.e., sinusoidal) vibrations of the system at the frequency $\omega(V(t) = V_0 e^{j\omega t}, I(t) = I_0 e^{i\omega t - \delta})$, Eq. (2.167) can be transformed into

$$[j\omega L + R + (1/j\omega C)]I = V, \text{or} \tag{2.168a}$$

$$Y = I/V = [j\omega L + R + (1/j\omega C)]^{-1}. \tag{2.168b}$$

Table 2.5: Equivalency between mechanical and electrical systems, composed of M (mass), c (spring constant), ζ (viscous damper); L (inductance), C (capacitance), R (resistance). (a) mechanical system, (b) LCR series connection and (c) LCR parallel connection.

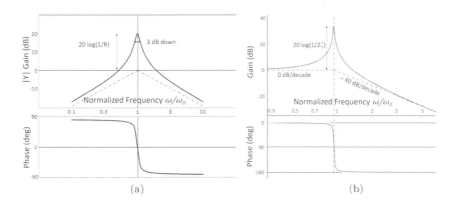

	Mechanical	Electrical (F – V)	Electrical (F – I)
	Force F(t)	Voltage V(t)	Current I(t)
	Velocity v/ů	Current I	Voltage V
	Displacement u	Charge q	--
	Mass M	Inductance L	Capacitance C
	Spring Compliance 1/c	Capacitance C	Inductance L
	Damping ζ	Resistance R	Conductance G

Fig. 2.35: Bode diagram for a series LCR circuit: (a) admittance, and (b) charge(second-order system).

Under a certain constant voltage (such as 1 V), the current (A) behavior provides the frequency dependence of the *circuit admittance*. Thus, this series connection equivalent circuit (EC) is useful to discuss the piezo-electric resonance mode with the admittance maximum peak. We consider the *Bode plot* of Eq. (2.168b). Referring to Fig. 2.35(a), the admittance $|Y|$ *gain* (current under unit voltage) is plotted as a function of frequency ω in both logarithmic scales. The steady state oscillation plot exhibits the following:

(1) 20 dB/decade ($\propto \omega C$) asymptotic curve with 90° phase in the low-frequency region.
(2) The peak at ω_0, resonance angular frequency for zero damping, given by $\omega_0 = 1/\sqrt{LC}$, with the peak height $|Y|_{\max} = (1/R)$, and $Q = \sqrt{L/C}/R$, which corresponds to the quality factor.
(3) −20 dB/decade ($\propto 1/\omega L$) asymptotic curve with −90° phase in the high-frequency region.

Let us calculate the quality factor Q in the LCR circuit defined by $\omega_R/2\Delta\omega_R$, where $\Delta\omega_R$ is the half width of the admittance frequency spectrum to provide the $1/\sqrt{2}$ (i.e., 3 dB down) of the maximum admittance (1/R) at the resonance frequency ω_R (see Fig. 2.35(a)). Since these cut-off frequencies ω_c are provided by

$$\frac{1}{\sqrt{2}} = \frac{1}{\sqrt{\left[\left(\frac{\omega_c L}{R}\right) - \left(\frac{1}{\omega_c RC}\right)\right]^2 + 1}}.$$

then, two roots for the cut-off frequency ω_C are given by

$$\omega_{c1,c2} = \mp \frac{R}{2L} + \sqrt{\left(\frac{R}{2L}\right)^2 + \left(\frac{1}{LC}\right)},$$

Since $2\Delta\omega_R = \omega_{c2} - \omega_{c1}$ and $\omega_R = 1/\sqrt{LC}$, the quality factor is expressed by

$$Q = \omega_R/2\Delta\omega_R = (1/\sqrt{LC})/(R/L) = \sqrt{L/C}/R. \tag{2.169}$$

When we consider the charge q, rather than current I under the voltage (Eq. (2.167a)),

$$L(d^2q/dt^2) + R(dq/dt) + (1/C)q = V(t).$$

Taking the harmonic oscillation, the above equation is transformed to

$$[-\omega^2 L + j\omega R + (1/C)]q = V, \quad \text{or} \tag{2.170a}$$

$$q/V = [-\omega^2 L + j\omega R + (1/C)]^{-1}. \tag{2.170b}$$

The *Bode plot* of Eq. (2.170), well known as the *second-order system*, is drawn in Fig. 2.35(b), where the *gain* of charge q is plotted as a function of frequency ω in both logarithmic scale. Note that the displacement vs. electric field also exhibits the second-order system response. The steady state oscillation plot exhibits the following:

(1) 0 dB/decade asymptotic curve in the low-frequency region.
(2) The peak at ω_0, resonance angular frequency for zero damping, given by $\omega_0 = 1/\sqrt{LC}$, with the peak height $(1/2\zeta) = \sqrt{L/C}/R = Q$, which corresponds to the quality factor.
(3) -40 dB/decade asymptotic curve in the high-frequency region.

Try the following Example Problem 2.12 to understand mechanical analysis of a piezo actuator system.

Example Problem 2.12.

We consider a simple vibration system of a mass, M, actuated by a multilayer piezoelectric actuator spring in Fig. 2.36, which has an elastic stiffness, c, piezoelectric strain coefficient, d, a length, L, and a cross-sectional area, A. Displacement of the mass, $u(t)$, is induced in a piezoelectric actuator by an applied electric field, E. The oscillation of the mass is sustained by the piezoelectric force (product of strain $d \cdot E$ and stiffness c), which is generated by the applied electric field and described by $(Acd)E$. The dynamic equation is written as follows:

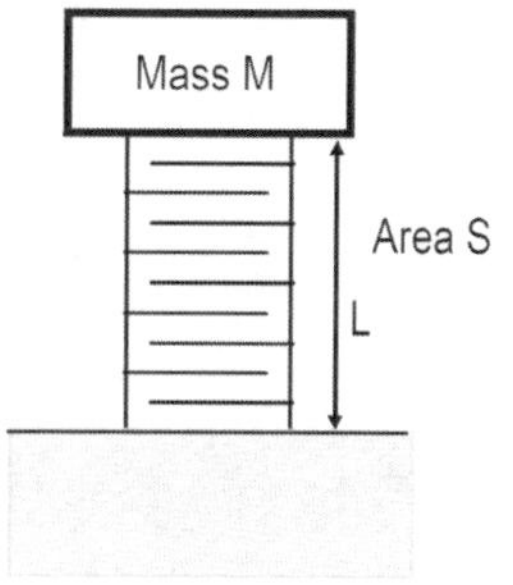

$$M\frac{d^2u}{dt^2} + \zeta\frac{du}{dt} + \left(\frac{ac}{L}\right)u = (Acd)E \tag{P2.12.1}$$

Fig. 2.36: A simple actuator and mass model.

where ζ represents the *damping effect*, which is originated from the material's elastic loss. Using the Laplace transformation, Eq. (P2.12.1) becomes

$$Ms^2U + \zeta sU + \left(\frac{Ac}{L}\right)U = Acd\tilde{E} \tag{P2.12.2}$$

Here, U and $\tilde{E}$ are the Laplace transforms of u and E, respectively. We may now define the following:

$$U(s) = G_2(s)\tilde{E}(s) \tag{P2.12.3}$$

where the function $G(s)$ is given by

$$G_2(s) = Acd/(Ms^2 + \zeta s + Ac/L) \tag{P2.12.4}$$

This function, which essentially relates the input $\tilde{E}(s)$ to the output $U(s)$, is called the *transfer function*. When the denominator of the transfer function includes s^2 term, the transfer function is called the *second-order transfer function*, popularly analyzed in many applications. By replacing s in Eq. (P2.12.4) with $s = j\omega$, we obtain the exciting electric field frequency dependence of the induced displacement. Notice the fundamental resonance frequency of the system as

$$\omega_0^2 = (Ac/ML) - (\zeta^2/4M^2), \quad \text{and} \tag{P2.12.5}$$

$$T = 1/\omega_0 \tag{P2.12.6}$$

The transfer function (equivalent to the displacement vibration) can be simplified as the following standard *second-order system*:

$$G(j\omega) = \frac{1}{[(-\omega^2 T^2) + 2\zeta j\omega T + 1]} \tag{P2.12.7}$$

Describe the *Bode plot* for the standard *second-order system*. The *Bode plot* is a representation of the transfer function (amplitude and phase) as a function of frequency (unit: $\omega/\omega_0 = \omega T$) on a logarithmic scale.

Solution:

First, let us consider approximate curves for the low- and high-frequency regions.

- For $\omega \to 0$, $G(j\omega) \to 1$. Thus, gain $|G(j\omega)| = 1$, so that in decibels

$$dB = 20\log_{10}(1) = 0 \tag{P2.12.8}$$

0 dB/decade, i.e., flat frequency dependence. Regarding the phase, the real number corresponds to 0°. Gain and phase Bode plots are in Fig. 2.37. Note that this Bode diagram is the same as in Fig. 2.35(b), as long as it is the same second-order system.
- For $\omega \to \infty$, $G(j\omega) \to 1/(-\omega^2 T^2)$
  Gain:

$$|G(j\omega)| = \left| \frac{1}{(-\omega^2 T^2)} \right| \tag{P2.12.9a}$$

so that in decibels

$$dB = -20log_{10}(\omega T)^2 = -40\log_{10}(\omega T) \tag{P2.12.9b}$$

40 dB down/decade with frequency. Regarding the phase, negative real number corresponds to $-180°$ as indicated by the gain and phase curves in Fig. 2.37.

The low- and high-frequency portions of the gain curve can be approximated with two straight lines as shown in Fig. 2.37. We consider next the resonance region.
- Resonance range: we will now consider the deviation from these two lines around the bend-point frequency, which is obtained from the relation $\omega T = 1$. Substituting $\omega T = 1$ in Eq. (P2.12.7) yields

$$G(j\omega) = \frac{1}{(2\zeta j\omega T)} = \frac{1}{(2\zeta j)} \tag{P2.12.10}$$

so that the gain and phase become $[1/(2\zeta)]$ and $-90°$, respectively. The constant ζ corresponds to the damping constant. If $\zeta = 0$, an infinite amplitude will occur at the bend-point frequency (i.e., the resonance frequency). When ζ is large ($>1/2$), however, the resonance peak will disappear, and monotonous decrease in amplitude is observed. It is notable that $(1/\zeta) \propto$ mechanical quality factor Q_m. Precisely, $2\zeta = \pi^2/(8Q_m)$. The displacement is amplified as $\Delta L = (8/\pi^2)Q_m \cdot d_{31}EL$ at its fundamental resonance frequency.

Finally, we will consider the displacement curve under a cyclic electric field in a piezoelectric actuator system. In a low-frequency range, the displacement $\Delta L = dEL$ with zero phase lag, where the piezoelectric constant d is d_{33} or d_{31}, depending on the setting. In a piezo multilayer actuator's case, it is d_{33}. ΔL exhibits a linear line with a slope proportional to the piezoelectric constant, depicted in Fig. 2.38(a) (this is for the d_{33} case, where $d_{33} > 0$, thus positive slope). In a high-frequency range, the ΔL amplitude decreases dramatically as 40 dB down/decade. Because the phase is $-180°$, ΔL exhibits a linear line with a negative slope (opposite to the piezoelectric constant sign) (Fig. 2.38(c)). Since the displacement magnitude changes significantly with the operating frequency in this

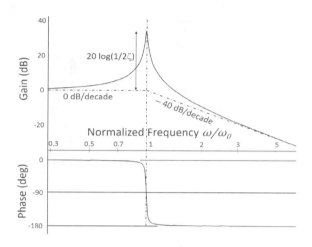

Fig. 2.37:   Bode plot for a standard second-order system.

region, the displacement scale is arbitrary in this figure. It is most intriguing at the resonance frequency, where ΔL is amplified significantly by a factor of the mechanical quality factor $(8/\pi^2)Q_m$ (e.g., 1000 times in a hard PZT), and the displacement curve does not exhibit a linear line, but an elliptic Lissajous curve because of $-90°$ phase lag (counter-clockwise rotation due to positive d_{33}) (Fig. 2.38(b)). Note that applied electric field is much lower than the E_{max} at the low (off-resonance) frequency. Due to the displacement amplification around the resonance frequency, cracking occurs at the nodal point in the piezo ceramic, if we keep a high electric field.

2.8.1.2 *LCR parallel connection equivalent circuit*

Let us now consider the dynamic equation for an electrical circuit (EC) composed of a parallel connection of an inductance L_B, a capacitance C_B and a conductance G_B illustrated in Table 2.5(c):

$$C_B(dV/dt) + G_B V + (1/L_B) \int_0^t V dt = I(t) \tag{2.171}$$

where I is the current from the current supply, and V is the same voltage applied on three components. In comparison with Eq. (2.168), equivalency between the two mechanical and electrical systems is summarized in the last column in Table 2.5. When we consider steady sinusoidal vibrations of the system at the frequency

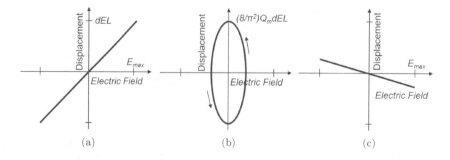

Fig. 2.38:   Displacement vs. electric field Lissajous curves: (a) low frequency, (b) resonance frequency and (c) high frequency.

$\omega(I(t) = I_0 e^{j\omega t}, V(t) = V_0 e^{i\omega t - \delta})$, Eq. (2.172) can now be transformed into

$$[j\omega C_B + G_B + (1/j\omega L_B)]V = I, \quad \text{or}$$

$$Z = V/I = [j\omega C_B + G_B + (1/j\omega L_B)]^{-1}. \tag{2.172}$$

Under a certain constant current (such as 1 A), the voltage (V) behavior provides the frequency dependence of the *circuit impedance*. Thus, this parallel connection EC is preferred to discuss the piezoelectric "antiresonance mode" (i.e., B-type resonance) with the impedance maximum peak, in practice. To the contrary, the LCR series connection in Table 2.5(b) is popularly used to discuss the resonance mode with the admittance peak.

Example Problem 2.13.

Two equivalent circuits in Table 2.5(b) and 2.5(c) are modeled for the same mechanical system in Table 2.5(a). Therefore, we can expect the mutual relationships between these inductance, capacitance and resistance/conductance values. Obtain the mutual relationships.

Hint:

Since the voltage – current behavior should be equivalent in these series and parallel connection circuits, the admittance in Eq. (2.169) should be an inverse of the impedance in Eq. (2.172).

Solution:

Equating Z with $1/Y$

$$Z = [j\omega C_B + G_B + (1/j\omega L_B)]^{-1} = 1/Y = [j\omega L + R + (1/j\omega C)], \tag{P2.13.1}$$

we obtain the following equation:

$$C_B\left(\frac{1}{C} - \omega^2 L\right) + \frac{1}{L_B}\left(L - \frac{1}{\omega^2 C}\right) + RG_B + j\left[G_B\left(\omega L - \frac{1}{\omega C}\right) + R\left(\omega C_B - \frac{1}{\omega L_B}\right)\right] = 1 \tag{P2.13.2}$$

In order to keep the same resonance frequency in both circuits,

$$\omega_0^2 = \frac{1}{LC} = \frac{1}{L_B C_B} \tag{P2.13.3}$$

should be maintained, which also satisfies the imaginary term = 0 in Eq. (P2.13.2). Thus, Eq. (P2.13.2) indicates another equation,

$$RG_B = 1. \tag{P2.13.4}$$

Though we have the circuit component flexibility, as long as $LC = L_B C_B$, the simplest solution is the utilization of the same L, C and R for L_B, C_B and $1/G_B$.

2.8.2 *Equivalent Circuit (Loss-Free) of the k_{31} Mode*

We introduce the equivalent circuit (EC) of piezoelectric devices, a widely used tool which can greatly simplify the process of designing the devices. We consider the k_{31}-type piezo-plate specimen in Fig. 2.29 again. The key in a piezoelectric is illustrated schematically in Fig. 2.39.

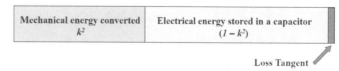

Fig. 2.39: Energy conversion in a piezoelectric.

where the input electric energy is partially converted to the output mechanical energy by the factor or k^2, while the remaining energy $(1 - k^2)$ is stored in a capacitor (so-called *damped capacitance*). The loss observed as heat generation is usually small (around a couple of %), which is proportional to the loss tangent/dissipation factor. Different from the previous section of a simple LCR series connection, where only the resonance mode shows up, when we include the damped capacitance, the antiresonance mode appears, where the damped and motional capacitances are basically canceled out. In other words, the existence of the damped capacitance is essential to generate the antiresonance mode. The admittance Y for the k_{31} mode piezo-plate was analyzed around the resonance/antiresonance modes in detail in Section 2.7.1.2. You are reminded of the admittance equations:

$$Y = (j\omega wL/b)\varepsilon_0\varepsilon_{33}{}^{LC}[1 + (d_{31}^2/\varepsilon_0\varepsilon_{33}{}^{LC}s_{11}^E)(\tan(\omega L/2v_{11}{}^E)/(\omega L/2v_{11}{}^E)]$$

$$= (j\omega wL/b)\varepsilon_0\varepsilon_{33}{}^X[(1 - k_{31}) + k_{31}^2 \cdot (\tan(\omega L/2v_{11}{}^E)/(\omega L/2v_{11}{}^E)]$$

$$= j\omega C_d\left[1 + \frac{k_{31}^2}{1 - k_{31}^2}\frac{\tan(\Omega_{11})}{\Omega_{11}}\right] = j\omega C_0\left[(1 - k_{31}^2) + k_{31}^2\frac{\tan(\Omega_{11})}{\Omega_{11}}\right] \qquad (2.173)$$

where w is the width, L the length, b the thickness of the rectangular piezo sample. We adopt the following notations to make the formulas simpler:

$$k_{31}^2 = d_{31}^2/\varepsilon_0\varepsilon_{33}^X s_{11}^E$$

$$\varepsilon_0\varepsilon_{33}{}^{LC} = \varepsilon_0\varepsilon_{33}{}^X - (d_{31}^2/s_{11}^E) = \varepsilon_0\varepsilon_{33}^X(1 - k_{31}^2).$$

$$C_0 = \varepsilon_0\varepsilon_{33}^X\frac{Lw}{b}(\text{free capacitance})$$

$$C_d = \varepsilon_0\varepsilon_{33}^{LC}\frac{Lw}{b}(\text{damped capacitance})$$

$$\Omega_{11} = (\omega L/2v_{11}^E).$$

Equation (2.173) indicates that the first term is originated from the *clamped capacitance* [proportional to $(1 - k^2)$], and the second term is the *motional capacitance* associated with the mechanical vibration [proportional to k^2]. By splitting Y into the damped admittance Y_d and the motional part Y_m,

$$Y_d = j\omega C_d, \qquad (2.174a)$$

$$Y_m = j\omega C_d\left[\frac{k_{31}^2}{1 - k_{31}^2}\frac{\tan(\Omega_{11})}{\Omega_{11}}\right]. \qquad (2.174b)$$

The damped branch can be represented by a capacitor with *damped capacitance* C_d in Fig. 2.40(a). We connect the motional branch (mechanical vibration) in "parallel" to this damped capacitance, because Eq. (2.173) indicates the summation of these two admittances.

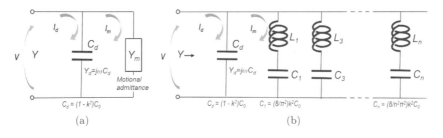

Fig. 2.40: Equivalent circuit for the k_{31} mode (loss-free): (a) conceptual EC and (b) LC series EC.

One comment is made here in conjunction with the 4-terminal equivalent circuit in the following section. We will consider the motional current I_m and the vibration velocity $\dot{u}(x = L)$ at the k_{31} plate end. Refresh your memory to what we learned in Section 2.7.1.2 on that the external current flow into the specimen by the surface free charge increment is equal to the negative of the internal electric displacement-based current, $i = \partial D_3/\partial t \ (= j\omega D_3)$. The total current i is given by integrating $\partial D_3/\partial t$ with respect to the top area, using $D_3 = d_{31}X_1 + \varepsilon_0 \varepsilon_{33}^X E_3$:

$$i = j\omega w \int_0^L D_3 dx = j\omega w \int_0^L (d_{31}X_1 + \varepsilon_0 \varepsilon_{33}^X E_z)dx$$

$$= j\omega w \int_0^L \left[(d_{31}/s_{11}^E)x_1 + \varepsilon_0 \varepsilon_{33}{}^X(1 - k_{31}^2)E_z \right] dx. \tag{2.175}$$

Since $j\omega \int_0^L x_1 dx = 2\dot{u}(x = L)(x = L/2$ is the nodal point!) from $x_1 = (\frac{\partial u}{\partial x})$, the motional current (first term in Eq. (2.175)) can be expressed by

$$I_m = \frac{2wd_{31}}{s_{11}^E}\dot{u}(x = L) = \Phi\dot{u}(x = L)$$

$$\Phi = \frac{2wd_{31}}{s_{11}^E} \tag{2.176}$$

Equation (2.176) indicates the relationship between the motional current and the vibration velocity at the plate end, and the correlation factor Φ is called *force factor*, which will be discussed again in Section 5.5.1.

2.8.2.1 *Resonance mode*

Because the maximum Y corresponds to the resonance mode, we can analyze merely the motional Y_m, which is much larger than Y_d. Since $Y_m = j\omega C_d \left[\frac{k_{31}^2}{1-k_{31}^2} \frac{\tan(\Omega_{11})}{\Omega_{11}} \right]$ is infinite (∞) around the A-type resonance frequency, that is, $\omega_A L/2v_{11}^E = \Omega_{11,A,n} = n\pi/2(n = 1, 3, 5, \ldots)$, taking *Mittag-Leffler's theorem* of $\frac{\tan(\Omega_{11})}{\Omega_{11}}$ around $\omega_{A,n}$, we get

$$\frac{\tan(\Omega_{11})}{\Omega_{11}} = \sum_{n:odd}^{\infty} \left(\frac{8}{n^2\pi^2} \right) /[1 - (\Omega_{11}/\Omega_{11,A,n})^2]. \tag{2.177}$$

When we use an LC series connection equivalent circuit (EC) model in Table 2.5(b) on the motional branch just around the resonance frequency peak region, we convert the mass contribution to L and elastic compliance to C, and create L_n and C_n series connections as shown in Fig. 2.40(b). Note here that a similar approach can be made for the parallel connection EC, because of the equivalency among the series and parallel ECs (refer to Example Problem 2.13). Each pair of $(L_1, C_1), (L_3, C_3), \ldots (L_n, C_n)$ contributes to the fundamental, second and the nth resonance vibration mode, respectively. Remember that n is only for the odd-number, or even-number n does not show up in the piezoelectric resonance, which corresponds basically to the antiresonance mode. Though each branch is activated only at its own nth resonance frequency, the capacitance's contribution remains even at an inactive frequency range, in particular, at a low-frequency range (note the impedance of capacitance, $1/j\omega C$, and inductance, $j\omega L$. Under a low-frequency region, the inductance contribution will disappear). Neglecting the damped admittance, the motional impedance of this LC circuit around the resonance $\omega_{A,n}$ is approximated by

$$1/Y_{m,n} = j\omega L_n + 1/j\omega C_n \approx j(L_n + 1/\omega_{A,n}^2 C_n)(\omega - \omega_{A,n}) \tag{2.178}$$

where $\omega_{A,n}^2 = 1/L_n C_n$. Using Eqs. (2.177) and (2.178), we can obtain the following equation:

$$Y_m = j\omega C_d \left[\frac{k_{31}^2}{1 - k_{31}^2} \sum_{n:odd}^{\infty} \left(\frac{8}{n^2 \pi^2} \right) / \left\{ 1 - \left(\frac{\Omega_{11}}{\Omega_{11,A,n}} \right)^2 \right\} \right]$$

$$= \sum_{n,odd}^{\infty} [j/(L_n + 1/\omega_{A,n}^2 C_n)(\omega_{A,n} - \omega)]$$

Taking into account further approximation, $\dfrac{1}{\left\{ 1 - \left(\frac{\Omega_{11}}{\Omega_{11,A,n}} \right)^2 \right\}} \approx \dfrac{\omega_{11,A,n}}{2(\omega_{11,A,n} - \omega_{11})}$ for each *n-th* branch, we

can obtain the following two equations which express the L_n, C_n in terms of the transducer's physical parameters:

$$L_n = (bL s_{11}^E / 4 v_{11}^{E2} w d_{31}^2)/2 = (\rho/2)(Lbw)(s_{11}^E / 2wd_{31})^2 = (\rho/2)(Lbw)/\Phi^2 \tag{2.179a}$$

$$C_n = 1/\omega_{A,n}^2 L_n = (L/n\pi v_{11}^E)^2 (8/\rho)(w/Lb)(d_{31}^2/s_{11}^{E2})$$

$$= (2/n^2 \pi^2)(L/bw)(2wd_{31}/s_{11}^E)^2 s_{11}^E = (2/n^2 \pi^2)(L/bw) s_{11}^E \Phi^2 \tag{2.179b}$$

$$\omega_{A,n} = 1/\sqrt{L_n C_n} = n\pi/L\sqrt{\rho s_{11}^E} \tag{2.180}$$

Note initially that L_n is a constant, irrelevant to n. All harmonics have the same L, which originates from the ceramic density ρ or the specimen mass M. C_n is proportional to $1/n^2$ and the elastic compliance s_{11}^E. Note that the parameter $(2wd_{31}/s_{11}^E)$ is distinguished in both Eqs. (2.179a) and (2.179b), which is a *force factor* $\Phi = 2wd_{31}/s_{11}^E$. The total motional capacitance $\sum_n C_n$ is calculated as follows, using an important

general relationship, $\Sigma \left[\frac{1}{(2m-1)^2} \right] = (\frac{\pi^2}{8})$:

$$\sum_n C_n = \sum_n \frac{1}{n^2} \left(\frac{8}{\pi^2} \right) \left(\frac{Lw}{b} \right) \left(\frac{d_{31}^2}{s_{11}^E} \right) = \left(\frac{Lw}{b} \right) \left(\frac{d_{31}^2}{s_{11}^E} \right) = k_{31}^2 C_0 \tag{2.181}$$

Therefore, we can understand that the total capacitance $C_0 = (wL/b)\varepsilon_0 \varepsilon_{33}^X$ is split into the damped capacitance $C_d = (1 - k_{31}^2)C_0$ and the total motional capacitance $k_{31}^2 C_0$, which is reasonable from the energy conservation viewpoint.

2.8.2.2 *Antiresonance mode*

We consider next the antiresonance mode at the *n-th* mode, where the total admittance $Y = 0$ in Eqs. (2.173), (2.179a), and (2.179b):

$$Y = j\omega C_d \left[1 + \frac{k_{31}^2}{1 - k_{31}^2} \frac{\tan(\Omega_{11})}{\Omega_{11}} \right] = j\omega C_d + \sum_n \frac{1}{\frac{1}{j\omega C_n} + j\omega L_n} = 0$$

This admittance corresponds to the closed-circuit admittance under externally open-circuit condition (i.e., smallest admittance condition). Accordingly, we obtain the *n*th antiresonance (B-type) frequency $\omega_{B,n}$ in terms of the EC circuit parameters:

$$\omega_{B,n} = \sqrt{\left(1 + \frac{C_n}{C_d} \right) / L_n C_n} = \sqrt{\left(\frac{1}{C_n} + \frac{1}{C_d} \right)/L}. \tag{2.182}$$

2.8.3 *Equivalent Circuit of the k_{31} Mode with Losses*

2.8.3.1 *IEEE standard equivalent circuit*

Figure 2.41 shows the IEEE Standard equivalent circuit (EC) for the k_{31} mode with only one elastic loss $(\tan \phi')$.[45] This elastic loss introduction in the mechanical branch is based on the assumption that the elastic loss in a piezoelectric material follows a "viscous damping" model, merely from the mathematical simplicity viewpoint. The dielectric or piezoelectric losses are neglected. In this EC model, in addition to Eqs. (2.179a), (2.179b) and (2.180) in the loss-free EC, the circuit analysis provides the following R and Q (electrical quality factor, which corresponds to the mechanical quality factor Q_m in the piezo-plate) relation [refer to Eq. (2.169)]:

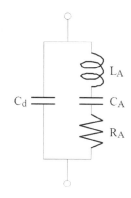

$$Q = \sqrt{L_A/C_A}/R_A. \tag{2.183}$$

Fig. 2.41: Equivalent circuit for the k_{31} mode (IEEE).

In order to demonstrate the usefulness of the equivalent circuit model for the piezoelectric device analysis, a simulation tool is introduced. PSpice is a popular circuit analysis software code for simulating the performance of electric circuits, which is widely distributed for students in the university Electrical Engineering department. EMA Design Automation, Inc. in the United States http://www.ema-eda.com is distributing a free-download OrCAD Capture, this schematic design solution. The reader can access the download site: http://www.orcad.com/products/orcad-lite-overview?gclid=COaXitWJp9ECFcxKDQodCGMB0w.

Figure 2.42 shows the PSpice simulation process of the IEEE Standard k_{31} type. Figure 2.42(a) shows an equivalent circuit for the k_{31} mode. L, C and R values were calculated for PZT4 with $40 \times 6 \times 1\,\mathrm{mm}^3$ (Eqs. (2.179a), (2.179b) and (2.183)), and Fig. 2.42(b) plots the simulation results on the currents under $1\,V_{\mathrm{ac}}$, that is, admittance magnitude and phase spectra. IPRINT1 (current measurement), IPRINT2 and IPRINT3 are the measure of the total admittance ($\square$ line), motional admittance ($\circ$ line) and damped admittance (∇ line), respectively. First, the damped admittance shows a slight increase with the frequency ($j\omega C_d$) with $+90°$ phase in a full frequency range. Second, the motional admittance shows a peak at the resonance frequency, where the phase changes from $+90°$ (i.e., capacitive) to $-90°$ (i.e., inductive). In other words, the phase is exactly zero at the resonance frequency. The admittance magnitude decreases above the resonance frequency with a rate of $-20\,\mathrm{dB}$ down in a *Bode plot*. Third, by adding the above two, the total admittance is obtained. The admittance magnitude shows two peaks, maximum and minimum, which correspond to the resonance and antiresonance points, respectively. You can find that the peak sharpness (i.e., the mechanical quality factor) is the same for both peaks, because only one loss is included in the equivalent circuit. The antiresonance frequency is obtained at the intersection of the damped and motional admittance curves. Because of the phase difference between the damped ($+90°$) and motional ($-90°$) admittance, the phase is exactly zero at the antiresonance, and changes to $+90°$ above the antiresonance frequency because of the larger magnitude of the damped capacitance. Remember that the phase is $-90°$ (i.e., inductive) at a frequency between the resonance and antiresonance frequencies.

2.8.3.2 *Equivalent circuit with three losses*

2.8.3.2.1 Hamilton's Principle

In Section 2.7.3, we derived the difference between the mechanical quality factor Q_A at the resonance and Q_B at the antiresonance in the k_{31} specimen, based on the three dielectric, elastic and piezoelectric loss factors, $\tan \delta'$, $\tan \phi'$ and $\tan \theta'$. We now consider how to generate an EC with these three losses in order to realize the difference between Q_A and Q_B even in the EC. We start from *Hamilton's Principle*, a powerful

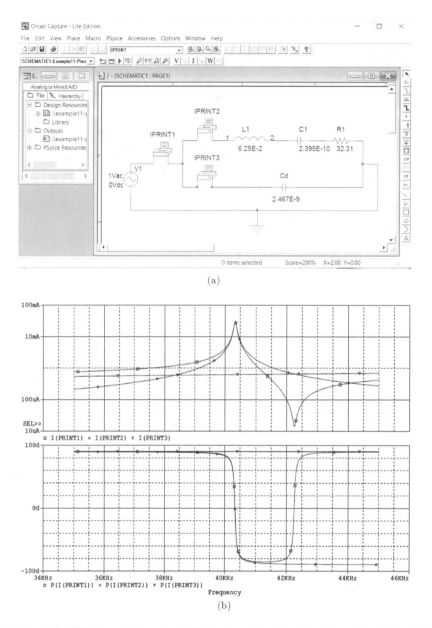

(a)

(b)

Fig. 2.42: PSpice simulation of the IEEE-type k_{31} mode. (a) Equivalent circuit for the k_{31} mode. L, C and R values were calculated for PZT4 with $40 \times 6 \times 1\,\mathrm{mm}^3$. (b) Simulation results on admittance magnitude and phase spectra.

tool for "mechanics" problem solving, which can transform a physical system model to a *variational problem* solving. We integrate loss factors directly into Hamilton's Principle for a piezoelectric k_{31} plate (Fig. 2.29).[48] Refer to Ref. 48 for the detailed derivation process. The following admittance expression can be derived, which is equivalent to Eq. (2.129) in the previous Section 2.7.3:

$$Y^* = j\omega \cdot \frac{lw}{b} \cdot \left(\varepsilon_0 \varepsilon_{33}^{X'} - Re\left[\frac{(d_{31}^*)^2}{s_{11}^{E\,*}}\right]\right) + \omega \cdot \frac{lw}{b} \cdot \left(\varepsilon_0\,\varepsilon_{33}^{X''} + Im\left[\frac{(d_{31}^*)^2}{s_{11}^{E\,*}}\right]\right)$$

$$+ j\omega \cdot \frac{8lw}{b\pi^2} \cdot Re\left[\frac{(d_{31}^*)^2}{s_{11}^{E\,*}}\right] \cdot \frac{\frac{\pi^2}{l^2 \rho s_{11}^{E\,*}}}{\frac{\pi^2}{l^2 \rho s_{11}^{E\,*}} - \omega^2} + j\omega \cdot \frac{8lw}{b\pi^2} \cdot \left(jIm\left[\frac{(d_{31}^*)^2}{s_{11}^{E\,*}}\right]\right) \cdot \frac{\frac{\pi^2}{l^2 \rho s_{11}^{E\,*}}}{\frac{\pi^2}{l^2 \rho s_{11}^{E\,*}} - \omega^2} \quad (2.184)$$

Among the above four terms in Eq. (2.184), the first and second terms correspond to the damped capacitance and its dielectric loss (i.e., the "extensive"-like dielectric loss (tan δ''') in Eq. (2.133) in the previous section), respectively, while the third and fourth terms correspond to the motional capacitance and the losses combining with "intensive" elastic and piezoelectric losses.

2.8.3.2.2 k_{31} Equivalent Circuit with Three Losses

Damjanovic[47] introduced a motional branch to describe the third term in Eq. (2.184), which contains a motional resistor, a motional capacitor and a motional inductor. Meanwhile, an additional branch is also injected into the classical circuit[46] to pictorially express the last term in Eq. (2.184) to present the influence of the piezoelectric loss, where the new resistance, capacitance and inductance are all proportional to corresponding motional elements with the proportionality constant being $jIm\left[\frac{(d_{31}^*)^2}{s_{11}^{E\,*}}\right]/Re\left[\frac{(d_{31}^*)^2}{s_{11}^{E\,*}}\right]$.

Shi *et al.* proposed a concise *equivalent circuit* (EC) shown in Fig. 2.43(a) with three losses.[4] Compared with the IEEE Standard EC with only one elastic loss or Damjanovic's EC with a full set of L, C, R, only one additional electrical element G'_m is introduced into the classical circuit.[46] The new coupling conductance can reflect the coupling effect between the elastic and the piezoelectric losses. The admittance of this new EC can be mathematically expressed as

$$Y^* = G_d + j\omega C_d + \frac{G'_m + j\omega C_m}{(1 + G'_m/G_m - \omega^2 L_m C_m) + j\left(\omega C_m/G_m + \omega L_m G'_m\right)}. \tag{2.185}$$

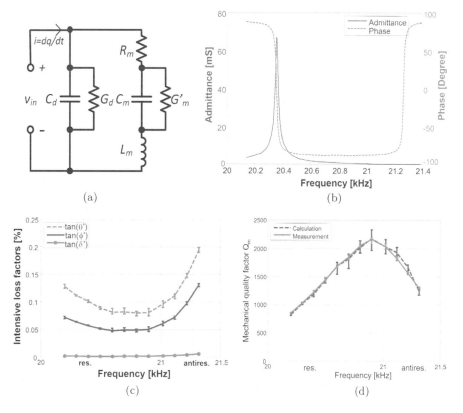

Fig. 2.43: (a) New equivalent circuit proposed with three intensive loss factors, (b) admittance spectrum to be used in the simulation, (c) frequency spectra of intensive loss factors (i.e., dielectric, elastic and piezoelectric losses) obtained from the admittance spectrum (b) fitting, and (d) calculated mechanical quality factor Q_m as a function frequency around the resonance and antiresonance frequencies.

The parameters of the new EC can therefore be obtained by comparing Eq. (5.24) with Eq. (5.25) as new expressions of three "intensive" loss factors:

$$\tan \phi' = \omega C_m / G_m \quad [G_m = 1/R_m], \tag{2.186a}$$

$$\tan \theta' = \tan(\phi' - \beta'), \tag{2.186b}$$

$$\tan \delta' = k_{31}^2 \tan(2\theta' - \phi') + \frac{G_d}{\omega C_d}, \tag{2.186c}$$

where the phase delay $\tan \beta' = \frac{\omega C_m}{G'_m} - \sqrt{\left(\frac{\omega C_m}{G'_m}\right)^2 + 1}$ denotes the disparity between the piezoelectric and elastic components. The value of β' generally holds negative or approaches to zero (when $G'_m \to 0$), which implies that the piezoelectric loss is persistently larger or equal to the elastic component. The significance of the piezoelectric loss has been therefore verified in theory from the equivalent circuit viewpoint.

Using the experimental data in Fig. 2.43(b), almost frequency-independent circuit parameters as $C_d = 3.2\,nF$, $C_m = 0.29\,nF$, $L_m = 210\,mH$ and $G_d = 0$ (extensive dielectric loss $\tan \delta'''$ is small) can be obtained, and the frequency-dependent parameters (G_m and G'_m).[48] By manipulating Eqs. (2.186)–(2.186c), we determined intensive dielectric, elastic and piezoelectric losses as a function of frequency, as shown in Fig. 2.43(c).

2.8.3.2.3 Quality Factor in the Equivalent Circuit

The mechanical quality factor, Q_m, is always applied to evaluate the effect of losses. When arriving at steady state, it can be expressed by

$$Q_m = 2\pi \cdot \frac{\text{energy stored/cycle}}{\text{energy lost/cycle}}. \tag{2.187}$$

The denominator is supposed to compensate the dissipation, w_{loss}; that is, $\int_V w_I dV = \frac{\pi}{2}|v_3^*||q_3^*|\cos\varphi$, where the phase difference between current and input voltage, φ, ranges within $[-\frac{\pi}{2}, \frac{\pi}{2}]$. Meanwhile, the reactive portion of the input energy returns to the amplifier and is neither used nor dissipated. Furthermore, the maximum stored and kinetic energies also get equilibrium in an electric cycle. With definitions of energy items and appropriate substitutions, Q_m can be calculated as[49]

$$Q_m = \frac{\omega_a^2 - \omega_r^2}{\cos\varphi} \cdot \frac{\omega^2}{|\omega^2 - (\omega_r^*)^2||\omega^2 - (\omega_a^*)^2|}. \tag{2.188}$$

As ω^2 approaches ω_r^2 or ω_a^2, the phase difference will approach zero. Therefore, for low k_{31}^2 materials, with substituting Eq. (2.188), mechanical quality factors at the resonance and antiresonance frequencies can be calculated as

$$Q_A = \frac{1}{\tan \phi'} = \frac{1}{R_m}\sqrt{\frac{L_m}{C_m}}, \tag{2.189a}$$

$$Q_B = \frac{1}{\tan \phi' + \frac{8K_{31}^2}{\pi^2}[\tan \phi' + \tan \delta' - 2\tan\theta']} \quad [K_{31}^2 = k_{31}^2/(1 - k_{31}^2)]. \tag{2.189b}$$

Equations (2.189a) and (2.189b) obtained from a new equivalent circuit are basically the same as we derived analytically in the previous subsection Eq. (2.151). Hence, the calculation of Q_m at these special frequencies has been verified by the well-accepted conclusion. Not only at these frequencies, Eq. (2.188) also infers an advanced calculation method of Q_m for a wide bandwidth. Figure 2.43(d) shows the frequency spectrum of the mechanical quality factor Q_m calculated from Eq. (2.188). You can clearly find that (1) Q_B

at antiresonance is larger than Q_A at resonance, and (2) the maximum Q_m (i.e., the highest efficiency) can be obtained at a frequency between the resonance and antiresonance frequencies. This frequency can theoretically be obtained by taking the first derivative of Eq. (2.188) in terms of ω to be equal to zero, which suggests the best operating frequency of the transducer to realize the maximum efficiency.

2.8.4 *Equivalent Circuit of the k_{33} Mode with Losses*

Remember that the k_{33} mode is governed by the sound velocity v^D, not by v^E, and that the antiresonance is the primary mechanical resonance given by $f = v^D/2L$, and the resonance is the subsidiary mode originating from the electromechanical coupling factor k_{33}. The difference primarily originates from the "*depolarization field*" created in the "longitudinal" piezoelectric effect oscillator (k_{33}, k_t), in comparison with the "transversal" piezoelectric effect oscillator (k_{31}). Let us consider formulating the EC for the k_{33} mode, as shown in Fig. 2.44(a).

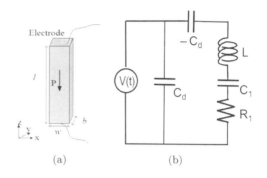

(a)　　　　(b)

Fig. 2.44: (a) k_{33} mode piezoceramic rod; (b) equivalent circuit for the k_{33} mode.

2.8.4.1 *Admittance calculation*

Referring to the derivation process introduced in Section 2.7.2, we just summarize the key formulas first.

- The constitutive equations

$$X_3 = (x_3 - d_{33}E_z)/s_{33}^E \tag{2.190a}$$

$$D_3 = \varepsilon_0\varepsilon_{33}{}^X E_z + d_{33}X_3 \tag{2.190b}$$

- Dynamic equation

$$\rho\,\frac{\partial^2 u}{\partial t^2} = \frac{1}{s_{33}^D}\frac{\partial^2 u}{\partial z^2}\quad (s_{33}^D = (1 - k_{33}^3)s_{33}^E) \tag{2.191}$$

- Admittance is expressed as

$$Y = \frac{j\omega\varepsilon_0\varepsilon_{33}^{LC}\left(\frac{wb}{L}\right)}{\left[1 - k_{33}^2\left\{\frac{\tan\left(\frac{\omega L}{2v_{33}^D}\right)}{\left(\frac{\omega L}{2v_{33}^D}\right)}\right\}\right]} = j\omega C_d + \frac{j\omega C_d}{[-1 + 1/k_{33}^2\{\frac{\tan(\Omega_{33})}{(\Omega_{33})}\}]} \tag{2.192}$$

Here, we used $\Omega_{33} = \left(\frac{\omega L}{2v_{33}^D}\right)$, $\varepsilon_{33}^{LC} = \varepsilon_{33}^X(1 - k_{33}^2)$, $s_{33}^D = s_{33}^E(1 - k_{33}^2)$, $k_{33}^2 = \frac{d_{33}^2}{\varepsilon_0\varepsilon_{33}^X s_{33}^E}$, $v_{33}^D = 1/\sqrt{\rho s_{33}^D}$ and $C_d = \varepsilon_0\varepsilon_{33}^{LC}\left(\frac{wb}{L}\right)$. The second expression is to show the *damped admittance* and the *motional admittance*, separately.

2.8.4.2 *Resonance/antiresonance of the k_{33} mode*

When we consider the resonance condition, $Y = \infty$, the resonance frequency is obtained from Eq. (2.192) as

$$\left(\frac{\omega L}{2v_{33}^D}\right)\cot\left(\frac{\omega L}{2v_{33}^D}\right) = k_{33}^2 \quad \left[v_{33}^D = 1/\sqrt{\rho s_{33}^D}\right] \tag{2.193}$$

Since the resonance is the subsidiary mode, the resonance frequency of the k_{33} mode depends strongly on the electromechanical coupling factor k_{33} value.

The antiresonance mode is obtained by putting $Y = 0$, which provides the condition, $\tan\left(\frac{\omega L}{2v_{33}^D}\right) = \infty$. Thus, the antiresonance frequency is determined by $n\left(\frac{v_{33}^D}{2L}\right)$ ($n = 1, 3, 5, \ldots$), and the vibration mode shows an exact half-wave-length on the specimen with length L under sound velocity v_{33}^D, while the resonance is the subsidiary vibration mode as discussed above. This provides intriguing contrast to the k_{31} mode, where the resonance mode is the primary vibration with a half-wave-length of the specimen of L, and the antiresonance is the subsidiary vibration mode.

2.8.4.3 *Equivalent circuit of the k_{33} mode*

We can rewrite the Eq. (2.192) as follows:

$$Y = j\omega C_d + \cfrac{1}{-\cfrac{1}{j\omega C_d} + \cfrac{1}{j\tan\left(\frac{\omega L}{2v_{33}^D}\right)\frac{2bwd_{33}^2}{\rho v_{33}^D L^2 s_{33}^E{}^2}}}, \tag{2.194}$$

where $C_d = \left(\frac{\varepsilon_0 \varepsilon_{33}^{x_3} bw}{L}\right)$. $\varepsilon_0 \varepsilon_{33}^{x_3}$ is the same as the longitudinally clamped permittivity, $\varepsilon_{33}^{x_3} = \varepsilon_{33}^X(1 - k_{33}^2)$.

From Eq. (2.194), we can understand that the equivalent circuit of the k_{33} mode is composed of the first term "*damped capacitance*" and the second term "*motional admittance*". Also, the motional branch is obtained by a series connection of so-called "negative capacitance – C_d" (exactly the same absolute value of the damped capacitance in the electric branch) and the pure motional admittance $j\tan\left(\frac{\omega L}{2v_{33}^D}\right)\frac{2bwd_{33}^2}{\rho v_{33}^D L^2 s_{33}^E{}^2}$. Figure 2.44(b) illustrates the fundamental mode equivalent circuit (EC) by translating the motional admittance with only a pair of L and C. The IEEE Standard model includes only one resistance R_1, which corresponds to the elastic loss $\tan\phi'''$ in the material's parameter. The admittance should be the minimum at the antiresonance frequency, where the pure mechanical resonance status is realized, because the damped capacitance should be compensated by this negative capacitance $-C_d$ in the closed loop circuit. On the contrary, at the resonance, the admittance should be the maximum, and the effective motional capacitance in the motional branch is provided by $1/(\frac{1}{C_1} + \frac{1}{-C_d})$, which provides $s_{33}^D (= s_{33}^E(1 - k_{33}^2))$, rather than s_{33}^E (i.e., origin of C_1). The reader can understand intuitively that the negative capacitance comes from the "depolarization field", or the D-constant status of the k_{33} vibration mode, different from the k_{31} E-constant mode. Figure 2.44(b) integrates a resistance R_1 in series with L_1 and C_1 in the pure mechanical branch. In comparison with Eqs. (2.179a), (2.179b) and (2.183) in the k_{31} mode, the EC components L, C and R of the k_{33} mode can be denoted as

$$L_n = (bLs_{33}^D/4v_{33}^D{}^2 wd_{33}^2)/2 = (\rho/8)(Lb/w)(s_{33}^{D2}/d_{33}^2), \tag{2.195a}$$

$$C_n = 1/\omega_{A,n}^2 L_n = (L/n\pi v_{33}^D)^2(8/\rho)(w/Lb)(d_{33}^2/s_{33}^{D2}) \tag{2.195b}$$

$$= (8/n^2\pi^2)(Lw/b)(d_{33}^2/s_{33}^{D2})s_{33}^D,$$

$$R_n = \sqrt{L_n/C_n}/Q. \tag{2.195c}$$

Here, $s_{33}^D = s_{33}^E(1 - k_{33}^2)$, $k_{33}^2 = \frac{d_{33}^2}{\varepsilon_0\varepsilon_{33}^X s_{33}^E}$ and $Q = \tan\phi'''$ as the material's constants.

2.8.5 *4- and 6-Terminal Equivalent Circuits (EC) — k_{31} Case*

Though the 2-terminal EC is useful for the basic no-load piezoelectric samples, we need to extend it to 4- and 6-terminal EC models in order to consider the mechanical load effect for practical transducer/actuator applications with composite structures such as Langevin transducers.

2.8.5.1 *4-terminal equivalent circuit*

(a) 4-terminal Equivalent Circuit (Zero Loss)
We consider again the k_{31}-type piezoelectric plate, whose admittance is described by Eq. (2.101). When we consider the damped electric branch and the motional mechanical branch together, we can generate a 2-terminal equivalent circuit (EC) as shown in Fig. 2.45(a). However, since the electric and mechanical branches are physically different, it is more reasonable to discuss these branches separately, which intuitively create a 4-terminal (or 2-port) equivalent circuit,

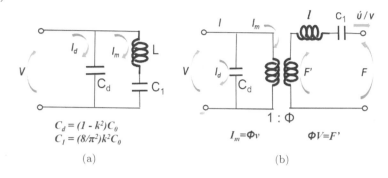

$$C_d = (1 - k^2)C_0$$
$$C_1 = (8/\pi^2)k^2C_0$$

(a)

$$I_m = \Phi v \qquad \Phi V = F'$$

(b)

Fig. 2.45: (a) 2- and (b) 4-terminal EC's for k31 mode (zero loss).

as exemplified in Fig. 2.45(b). The electric branch (left-hand side) is separated from the mechanical branch (right-hand side) by a *transformer*, which transforms voltage and current (electrical energy) to force and vibration velocity (mechanical energy), respectively, with the transformation ratio of Φ and $1/\Phi$, in order to change the unit from the electric to mechanical parameters. This Φ is called *force factor*. In this case, the port on the mechanical branch can be mechanically loaded (symmetrically in the 4-terminal model), depending on the piezoelectric composite structure.

Let us formulate electric component parameters of the 4-terminal EC of the k_{31} plate in Fig. 2.29. The motional current I_m is given by

$$I_m = E_z b Y_m = E_z b j \omega C_d \left(\frac{k_{31}^2}{1 - k_{31}^2}\right) \frac{\tan(\omega L/2v_{11})}{(\omega L/2v_{11})} \tag{2.196}$$

while the vibration velocity $\dot{u}$ at the plate edge is described from Eq. (2.98) as

$$(\partial u/\partial t)_{x=L} = j d_{31} E_z v_{11} \tan(\omega L/2v_{11}).$$

Taking into account the definition of the transformer ratio Φ in terms of current and vibration velocity by

$$I_m = \Phi \dot{u} = \Phi(\partial u/\partial t)_{x=L}, \tag{2.197}$$

Φ can be obtained as

$$\Phi = \frac{2w d_{31}}{s_{11}^E} \tag{2.198}$$

Note the general relations: $F' = \Phi V$ and $\dot{u} = I_m/\Phi$. Using the former in terms of voltage V and force F' we obtain

$$\Phi E_z b = \left(j\omega l + \frac{1}{j\omega c_1}\right)\left(\frac{\partial u}{\partial t}\right)_{x=L} \tag{2.199}$$

Note here that mechanical force F' is obtained by the product of vibration velocity $\left(\frac{\partial u}{\partial t}\right)_{x=L}$ and the mechanical impedance $(j\omega l + \frac{1}{j\omega c_1})$. Since the voltage is given by the product of motional current I_m and the impedance in the 2-terminal model,

$$V = E_z b = \left(j\omega L + \frac{1}{j\omega C_1}\right)I_m \tag{2.200}$$

we obtain the relationship between the L, C_1 in the 1-port model and l, c_1 in the 2-port model:

$$\left(j\omega l + \frac{1}{j\omega c_1}\right) = \left(j\omega L + \frac{1}{j\omega C_1}\right)\Phi^2$$

We finally obtain the following relations in terms of the force factor Φ:

$$\Phi^2 L = l, \, C_1/\Phi^2 = c_1 \tag{2.201}$$

The force factor $\Phi = 2wd_{31}/s_{11}^E$ has a practical value around 0.1 for PZTs.

Example Problem 2.14.

Figure 2.46 shows a composite piezoelectric oscillator, which is composed of the k_{31}-type piezoelectric plate and two metal plates bonded on both ends of the piezo-plate. Supposing the piezo-plate length L and the metal length are $L/2$ symmetrically on both ends with the same cross-section area, consider the equivalent circuit of this composite oscillator to analyze the vibration mode.

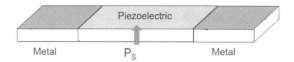

Fig. 2.46: Composite piezoelectric oscillator.

Hint:

4-terminal (2-port) equivalent circuit (EC) for the k_{31} mode is given by Fig. 2.47(a). Consider the elastic material's equivalent circuit.

Solution:

Figure 2.47(a) shows a 2-port equivalent circuit for the k_{31} mode, on which mechanical load can be applied. The resistance connected in series corresponds to the mechanical loss. We consider some load application cases.

Recall the parameters in the 2-terminal model:

$$L_n = (\rho/8)(Lb/w)(s_{11}^{E2}/d_{31}^2)$$

$$C_n = (8/n^2\pi^2)(Lw/b)\left(d_{31}^2/s_{11}^{E2}\right)s_{11}^E$$

$$R_n = \sqrt{L_n/C_n}/Q$$

The force factor, inductance, capacitance and resistance on the mechanical branch can be obtained as

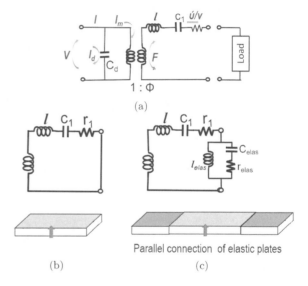

Fig. 2.47: (a) 4-terminal equivalent circuit for the k_{31} mode. (b) No load (short-circuit) condition. (c) Elastic plates attached in parallel.

$$\Phi = \frac{2wd_{31}}{s_{11}^E} \tag{P2.14.1}$$

$$l_n = \Phi^2 L_n = (\rho/2)\,(Lbw) \tag{P2.14.2}$$

$$c_n = C_n/\Phi^2 = (2/n^2\pi^2)(L/wb)s_{11}^E$$

$$r_n = \Phi^2 R_n = (l_n/c_n)^{1/2}/Q$$

Note the difference from the 2-terminal model: the L, C components in the 2-terminal EC include the piezoelectric constant explicitly, but the l, c components in the 4-terminal EC above do not, because the electromechanical coupling is defined in the "force factor" of the transformer, and the mechanical branch parameters should be only pure elastic parameters.

When the two terminals on the mechanical branch are short-circuited (Fig. 2.47(b)), i.e., mechanically free condition (the force F on the piezo-plate ends is zero), this will be transformed to the 2-terminal EC. On the contrary, when the two terminals on the mechanical branch are open-circuited, this condition corresponds to completely clamped (strain-free) on the both plate ends.

Figure 2.47(c) shows the model where the metal plate with the same width, thickness, length L is bonded symmetrically by cutting a half on the both ends of the piezoelectric plate. The load is modeled by the LC "parallel" connection in this case with the parameters, l_{elast}, c_{elast}, and r_{elast}, where ρ, s_{metal} and Q are the metal's density, elastic compliance and the inverse elastic loss, respectively:

$$l_{\text{elast}} = (\rho)\,(Lbw)$$

$$c_{\text{elast}} = \left(1/n^2\pi^2\right)(L/wb)\,s_{\text{metal}}$$

$$r_{\text{elast}} = (l_{\text{elast}}/c_{\text{elast}})^{1/2}/Q \tag{P2.14.3}$$

For the reader's reference, if the metal plate is bonded symmetrically by cutting a half of the thickness on the both top and bottom surfaces of the piezo-plate, the load is modeled by the above LC components in "series" connection. You may understand this situation by taking into account the mechanical impedance series or parallel connection.

Example Problem 2.15.

In the 4-terminal equivalent circuit in Fig. 2.48, the electric terminal parameters voltage V and current I, and the mechanical terminal parameters force F and vibration velocity $\dot{u}$ are related to each other as

$$\begin{bmatrix} F \\ I \end{bmatrix} = \begin{bmatrix} Z_1 & -\Phi \\ \Phi & Y_1 \end{bmatrix}\begin{bmatrix} \dot{u} \\ V \end{bmatrix}. \tag{P2.15.1}$$

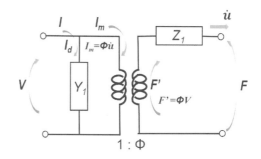

Fig. 2.48: 4-port equivalent circuit model.

The dynamic electromechanical coupling factor k_v^2 is defined by [(Complex power in the mechanical branch)/(complex power in the electrical branch)] under short-circuit condition of mechanical terminal, or [(Complex power in the electrical branch)/(complex power in the mechanical branch)] under short-circuit condition of electrical terminal. Derive the following result, and discuss that this dynamic k_v^2 approaches the material's electromechanical coupling factor $k^2 = \frac{(d)^2}{(s^E \varepsilon_0 \varepsilon^X)}$ with $\omega \to 0$:

$$k_v^2 = \left| \frac{\left(\frac{\Phi^2}{Z_1 Y_1}\right)}{1 + \left(\frac{\Phi^2}{Z_1 Y_1}\right)} \right|. \tag{P2.15.2}$$

Solution:

Figure 2.48 provides the relationships among the electric terminal voltage V and current I, and the mechanical terminal force F and vibration velocity $\dot{u}$ as Eq. (P2.15.1). Since this provides the following equations,

$$F = -\Phi V + Z_1 \dot{u} \tag{P2.15.3}$$

$$Y_1 V = I_d = I - I_m = I - \Phi\dot{u} \tag{P2.15.4}$$

we obtain

$$\begin{bmatrix} F \\ \dot{u} \end{bmatrix} = \begin{bmatrix} -(\Phi + \frac{Z_1 Y_1}{\Phi}) & \frac{Z_1}{\Phi} \\ -\frac{Y_1}{\Phi} & \frac{1}{\Phi} \end{bmatrix}\begin{bmatrix} V \\ I \end{bmatrix}. \tag{P2.15.5}$$

The dynamic electromechanical coupling factor k_v^2 is defined by [(Complex power in the mechanical branch)/(complex power in the electrical branch)] under short-circuit condition of mechanical terminal (i.e., $F = 0$), which leads to

$$\frac{Z_1}{\Phi} I = \left(\Phi + \frac{Z_1 Y_1}{\Phi} \right) V \tag{P2.15.6}$$

Because the mechanical power can be calculated by $Z_1 \dot{u}^2$ and the electrical input power is given be $I \cdot V$, the dynamic k_v^2 is obtained by

$$k_v^2 = \frac{Z_1 \dot{u}^2}{I \cdot V} = \frac{Z_1 (-\frac{Y_1}{\Phi} V + \frac{1}{\Phi} I)^2}{I \cdot V} = \left| \frac{(\frac{\Phi^2}{Z_1 Y_1})}{1 + (\frac{\Phi^2}{Z_1 Y_1})} \right|. \tag{P2.15.7}$$

We used Eq. (P2.15.6) for the final equation derivation. Since $Z_1 = j Z_0 tan(\frac{\omega L}{2v})$, $Y_1 = j\omega C_d$, $\Phi = \frac{2d_{31}w}{s_{11}^E}$ and $Z_0 = wb\rho v = \frac{wb}{vs_{11}^E}$ in the k_{31} mode, k_v^2 is ω dependent. By taking $\omega \to 0$, $k_v^2 \to k_{31}^2 = \frac{d_{31}^2}{s^E \varepsilon_0 \varepsilon^X}$.

2.8.5.1.1 4-Terminal Equivalent Circuit with Three Losses

Uchino proposes a new *4-terminal equivalent circuit* for a k_{31} mode plate, including elastic, dielectric and piezoelectric losses (Fig. 2.49(a)), which can handle symmetrical external mechanical losses.[50] The 4-terminal EC includes an ideal transformer with a voltage step-up ratio Φ to connect the electric (damped capacitance) and the mechanical (motional capacitance) branches, where $\Phi = 2wd_{31}/s_{11}^E$, called *"force factor"* in this *"electromechanical transformer"*. Components l, c_1, r_1 in the mechanical branch are related with L_1, C_1 and R_1 in the 2-terminal EC given in Eqs. (2.179a), (2.179b) and (2.183):

$$l = \Phi^2 L; c_1 = C_1/\Phi^2; r_1 = \Phi^2 R_1, \tag{2.202}$$

where Φ is a real number. Regarding the three losses, as shown in Fig. 2.49(a) in a k_{31} piezo-plate, in addition to the IEEE standard "elastic" loss r_1 and "dielectric" loss R_d, we introduce the *coupling loss* r_{cpl} in the force factor ($\Phi = 2wd_{31}/s_{11}^E$) as inversely proportional to $(\tan \phi'_{11} - \tan \theta'_{31})$, which can be either positive or negative, depending on the $\tan \theta'_{31}$ magnitude. Figure 2.49(b) shows the PSpice software simulation results for three values of r_{cpl}. We can find that (1) the resonance Q_A does not change with changing r_{cpl}, (2) when $r_{cpl} = 100k\Omega$ (i.e., $\tan \theta'_{31} \approx 0$), $Q_A > Q_B$, (3) when $r_{cpl} = 1G\Omega$ (i.e., $\tan \phi'_{11} - \tan \theta'_{31} \approx 0$), $Q_A = Q_B$, and (4) when $r_{cpl} = -100k\Omega$ (i.e., $\tan \phi'_{11} - \tan \theta'_{31} < 0$), $Q_A < Q_B$. Taking into account a typical PZT's case, where $\tan \theta' > (1/2)(\tan \delta' + \tan \phi')$, the well-known experimental result $Q_A < Q_B$ can be expected from the negative r_{cpl}. Thus, the large piezoelectric loss $\tan \theta'_{31}$ in PZTs is the key to exhibit the negative force factor loss, which leads to the experimental fact $Q_A < Q_B$.

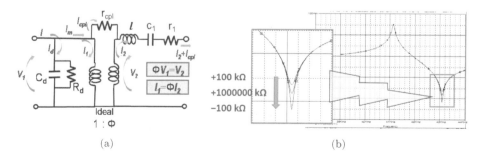

(a)   (b)

Fig. 2.49:   (a) 4-terminal (2-port) EC for a k_{31} plate, including three losses (r_1, R_d, and r_{cpl}). (b) PSpice simulation results on admittance for a PZT4 $40 \times 6 \times 1 \, mm^3$ plate.

2.8.5.2 *6-terminal equivalent circuit*

2.8.5.2.1 Mason's Equivalent Circuit

Mason introduced a famous 6-terminal (3-port) equivalent circuit (EC) model, relating with *distributed element model*.[51] As illustrated in Fig. 2.50, two ports in the mechanical branch of the 6-terminal (3-port) EC for the k_{31} piezoelectric plate correspond to the two edges of the plate, on which different mechanical loads can individually be applied, exemplified by a Langevin transducer with difference head and tail masses.

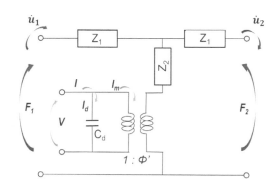

Fig. 2.50: 6-terminal equivalent circuit for k_{31} mode.

Let us determine the electronic components, Z_1, Z_2 and the *force factor* Φ' in the 6-terminal EC model. We denote the displacement along the length of a plate specimen (Fig. 2.29), u, force and vibration velocity on the edge, F_1 and $\dot{u}_1$ at $x = 0$, F_2 and $\dot{u}_2$ at $x = L$, in addition to the input voltage V and motional current I_m.

Supposing a general solution of the displacement as

$$u = A\cos(\omega x/v) + B\sin(\omega x/v) \tag{2.203}$$

and the boundary condition at $x = 0$,

$$A = u_1, B = (\partial u/\partial x)_1 (v/\omega),$$

we obtain

$$\partial u/\partial x = -(\omega/v)\,u_1\sin(\omega x/v) + (\partial u/\partial x)_1\cos(\omega x/v) \tag{2.204}$$

$$\partial u/\partial t = \dot{u} = j\omega[u_1\cos(\omega x/v) + (\partial u/\partial x)_1(v/\omega)\sin(\omega x/v)] \tag{2.205}$$

Now, we consider the force on the cross-section

$$F = wbX_1. \tag{2.206}$$

Since strain is given as $\partial u/\partial x = d_{31}E_z + s_{11}^E X_1$,

$$F = -\frac{wb}{s_{11}^E}[(\partial u/\partial x) - d_{31}E_z] \tag{2.207}$$

or

$$F - \Phi'V = -\frac{wb}{s_{11}^E}(\partial u/\partial x), \tag{2.208}$$

where the force constant Φ' is given by a half of Φ on the 4-terminal EC, because the mechanical branch in the 4-terminal model is the combination of Φ' of the two ports in the 6-terminal model.

$$\Phi' = \Phi/2 = \frac{wd_{31}}{s_{11}^E} \tag{2.209}$$

When we adopt the boundary conditions at x = 0,

$$(\partial u/\partial x)_1 = -\frac{s_{11}^E}{wb}(F_1 - \varphi'V) \tag{2.210}$$

$$\dot{u}_1 = jwu_1 \tag{2.211}$$

Now Eq. (2.205) becomes

$$\partial u/\partial t = \dot{u}_1 \cos(\omega x/v) - j\frac{vs_{11}^E}{wb}(F_1 - \Phi'V)\sin(\omega x/v) \tag{2.212}$$

Also at $x = L$,

$$F_2 - \Phi'V = -\frac{wb}{s_{11}^E}(\partial u/\partial x)_2$$

$$= \left(-\frac{wb}{s_{11}^E}\right)[-(\omega/v)u_1\sin(\omega L/v) + \dot{u}_1\cos(\omega L/v)] \tag{2.213}$$

Now, we can rewrite the relationship among F_1, F_2, $\dot{u}_1$, $\dot{u}_2$:

$$\dot{u}_2 == \dot{u}_1\cos(\omega L/v) - j\frac{1}{Z_0}(F_1 - \Phi'V)\sin(\omega L/v) \tag{2.214}$$

$$F_2 - \Phi'V = (F_1 - \Phi'V)\cos(\omega L/v) - j\dot{u}_1 Z_0\sin(\omega L/v) \tag{2.215}$$

$$I = j\omega C_d V + \Phi'(\dot{u}_2 - \dot{u}_1) \tag{2.216}$$

Note that the motional current is given by $\Phi'(\dot{u}_2 - \dot{u}_1)$. Now, we can construct the 6-terminal EC as shown in Fig. 2.50. In order to satisfy Eqs. (2.214) – (2.216), we obtain all the components including Z_1, Z_2:

$$C_d = \frac{Lw\varepsilon_0\varepsilon_{33}^X(1 - k_{31}^2)}{b} \tag{2.217}$$

$$Z_0 = wb\rho v = wb\left(\frac{\rho}{s_{11}^E}\right)^{1/2} = \frac{wb}{v_{11}^E s_{11}^E} \tag{2.218}$$

$$Z_1 = jZ_0\tan\left(\frac{\omega L}{2v_{11}^E}\right) \tag{2.219}$$

$$Z_2 = \frac{Z_0}{j\sin\left(\frac{\omega L}{v_{11}^E}\right)} \tag{2.220}$$

$$\Phi' = \frac{wd_{31}}{s_{11}^E} \tag{2.221}$$

Note that though Mason's equivalent circuit includes the frequency-dependent Z_1 and Z_2, these impedances can be translated into a pair of L and C for each individual fundamental or higher-order harmonic mode, if required.

2.8.5.2.2 Application of 6-Terminal EC

Dong *et al.* constructed a 6-terminal equivalent circuit with three (dielectric, elastic and piezoelectric) losses, which can handle symmetric external loads for a k_{31} mode plate[52] and Langevin transducer by integrating the head and tail mass loads,[53] then estimate the optimum (i.e., minimum required input electrical energy) driving frequency at which we can drive the transducer, as demonstrated with the highest efficiency.

In order to verify the feasibility of the EC circuit, a partial electrode configuration was designed (Fig. 2.51 top), which reflects intensive and extensive loss behavior on the electrode (center) and non-electrode (side) parts, respectively. The center part was electrically excited, which actuates the side non-electrode elastic load, then the vibration status can be monitored from the admittance from this center portion. The non-electrode side portions are merely the mechanical load. Figure 2.51 bottom shows a combination of

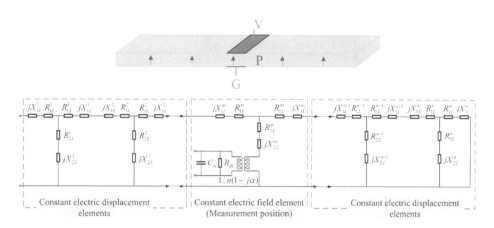

Fig. 2.51: A partial electrode configuration (top) and its EC (bottom) of a combination of 6-terminal ECs which model the center constant E element (i.e., intensive losses) and the side constant D elements (i.e., extensive losses) by integrating loss factors.

6-terminal ECs which model the center constant E element (i.e., intensive losses, $\tan\phi'_{11}$, $\tan\delta'_{33}$, $\tan\theta'_{31}$) and the side constant D elements (i.e., extensive losses, $\tan\phi_{11}$, $\tan\delta_{33}$, $\tan\theta_{31}$) by integrating loss factors into Eqs. (2.217)–(2.21). Note that the non-electrode part was segmented into 20 parts on each side to calculate the voltage distribution generated on the surface during the center actuation. The resonance and antiresonance frequencies and their corresponding mechanical quality factors derived from the circuits are compared with the actual sample with the load and boundary conditions.[52] The voltage distribution of non-electrode sample is simulated with the proposed equivalent circuit (Fig. 2.52) and matches the experimental result on the actual sample. The voltage simulation results have the same sinusoidal distribution trend as the experiments, and the admittance curves show a good agreement between the simulation and the measurements.

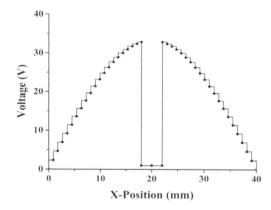

Fig. 2.52: Voltage distribution of non-electrode sample simulated with the new 6-terminal equivalent circuit.

2.8.6 *4- and 6-Terminal Equivalent Circuits (EC) — k_{33} Case*

The k_{33} mode requires a negative capacitance inclusion in the equivalent circuit (EC) in order to reflect D-constant condition, or "depolarization field". Since the sound velocity of the k_{33} mode is given by $1/\sqrt{\rho s_{33}^D}$, which is larger than $1/\sqrt{\rho s_{11}^E}$ of the k_{31} mode, the k_{33} mode is occasionally called a "stiffened mode". Equivalent circuits for the k_{33} mode are summarized for (a) 2-terminal model, (b) 4-terminal model and (c) 6-terminal model in Fig. 2.53. Top and bottom show the difference with the negative capacitor installation; when the negative capacitance is installed in the electrical branch, $-C_d$ is directly inserted, while when it is installed in the mechanical branch, $-C_d/\Phi^2$ or $-C_d/\Phi'^2$ is inserted in series with other electrical components.

Note the difference between Φ and Φ' in the 4- and 6-terminal models:

$$\Phi' = \Phi/2 = \frac{wd_{33}}{s_{33}^D}. \tag{2.222}$$

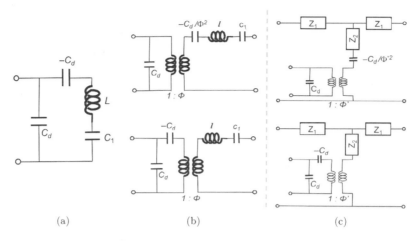

Fig. 2.53: Equivalent circuits for the k_{33} mode: (a) 2-terminal model, (b) 4-terminal model and (c) 6-terminal model. Top and Bottom show the difference with the negative capacitor installation

Z_1 and Z_2 in the 6-terminal model are described as follows:

$$C_d = \frac{Lw\varepsilon_0\varepsilon_{33}^X(1-k_{33}^2)}{b},$$ (2.223)

$$Z_0 = wb\rho v = wb\left(\frac{\rho}{s_{33}^D}\right)^{1/2} = \frac{wb}{v_{33}^D s_{33}^D},$$ (2.224)

$$Z_1 = jZ_0\tan\left(\frac{\omega L}{2v_{33}^D}\right),$$ (2.225)

$$Z_2 = \frac{Z_0}{j\sin\left(\frac{\omega L}{v_{33}^D}\right)}.$$ (2.226)

We can integrate three losses, $\varepsilon_{33}^{X*} = \varepsilon_{33}^X(1-j\tan\delta_{33}')$, $s_{33}^{E*} = s_{33}^E(1-j\tan\phi_{33}')$, $d_{33}^* = d_{33}(1-j\tan\theta_{33}')$, and the electromechanical coupling factor k_{33} loss in the above 6-terminal circuit components, then simulate admittance/impedance response from the circuit. Or, we may integrate three losses as R_{1L}, R_{2L} separately from X_{1L}, X_{2L}, as shown in Fig. 2.51.

Chapter Essentials

1. Origins of the field-induced strains:

    a. *Inverse Piezoelectric Effect*: $x = dE$
    b. *Electrostriction*: $x = ME^2$
    c. *Domain reorientation*: strain hysteresis – This is mostly discussed in the textbook.
    d. *Phase Transition* (antiferroelectric ↔ ferroelectric): strain "jump"

2. The electrostriction equation:

$$x = \qquad QP_S^2 \qquad + \quad 2Q\varepsilon_o\epsilon^X P_S E \quad + \quad Q\varepsilon_o^2\epsilon^{X2}E^2$$
$$\text{(spontaneous strain)} \quad \text{(piezostriction)} \quad \text{(electrostriction)}$$

3. The piezoelectric constitutive equations:

$$\textit{Intensive parameter description} \quad x = -\frac{\partial G}{\partial X} = s^E X + dE$$
$$D = -\frac{\partial G}{\partial E} = dX + \varepsilon_0\varepsilon^X E.$$

s^E — elastic compliance under constant field, ε^X — dielectric constant under constant stress, d — piezoelectric charge coefficient

$$\text{Extensive parameter description} \qquad X = \tfrac{\partial A}{\partial x} = c^D x - hD$$

$$E = \tfrac{\partial A}{\partial D} = -hx + \kappa_0 \kappa^x D \quad [\kappa_0 = \tfrac{1}{\varepsilon_0}].$$

c^D — elastic stiffness under constant electric displacement, κ^x — inverse permittivity under constant strain, h — inverse piezoelectric charge coefficient

4. Electromechanical coupling factor (k):

$$k^2 = \frac{d^2}{s^E(\varepsilon^X \varepsilon_0)} = \frac{h^2}{c^D(\kappa^x/\kappa_0)}.$$

5. Constraint dependence of permittivity and elastic compliance:

$$\varepsilon^x/\varepsilon^X = (1-k^2), s^D/s^E = (1-k^2), \kappa^X/\kappa^x = (1-k^2), c^E/c^D = (1-k^2).$$

6. Three losses in piezoelectrics:

   a. Dielectric loss
   b. Mechanical/elastic loss
   c. Piezoelectric loss — This loss has not been focused on previously, but it is the key to understanding frequency dependence of the mechanical quality factor and/or the transducer efficiency.

7. *Intensive and Extensive Loss Definitions*:

$$\varepsilon^{X*} = \varepsilon^X(1 - j\tan\delta') \quad \kappa^{x*} = \kappa^x(1 + j\tan\delta)$$
$$s^{E*} = s^E(1 - j\tan\phi') \quad c^{D*} = c^D(1 + j\tan\varphi)$$
$$d^* = d(1 - j\tan\theta') \qquad h^* = h(1 + j\tan\theta)$$

Intensive and Extensive Loss Interrelation:

$$\begin{bmatrix} \tan\delta' \\ \tan\phi' \\ \tan\theta' \end{bmatrix} = [K] \begin{bmatrix} \tan\delta \\ \tan\phi \\ \tan\theta \end{bmatrix} \text{ or } \begin{bmatrix} \tan\delta \\ \tan\phi \\ \tan\theta \end{bmatrix} = [K] \begin{bmatrix} \tan\delta' \\ \tan\phi' \\ \tan\theta' \end{bmatrix}, \text{ where } [K] = \tfrac{1}{1-k^2}\begin{bmatrix} 1 & k^2 & -2k^2 \\ k^2 & 1 & -2k^2 \\ 1 & 1 & -1-k^2 \end{bmatrix}.$$

8. Mechanical Quality Factors at Resonance and Antiresonance Frequencies:

   (a) k_{31} mode: intensive elastic loss

$$Q_{A,31} = \frac{1}{\tan\phi_{11}'}, \frac{1}{Q_{B,31}} = \frac{1}{Q_{A,31}} - \frac{2}{1 + (\frac{1}{k_{31}} - k_{31})^2 \Omega_{B,31}^2}(2\tan\theta_{31}' - \tan\delta_{33}' - \tan\phi_{11}')$$

   (b) k_t mode: extensive elastic loss

$$Q_{B,t} = \frac{1}{\tan\phi_{33}}, \frac{1}{Q_{A,t}} = \frac{1}{Q_{B,t}} - \frac{2}{k_t^2 - 1 + \Omega_A^2/k_t^2}(2\tan\theta_{33} - \tan\delta_{33} - \tan\phi_{33})$$

9. Equivalency between mechanical and electrical systems:

$$M(d^2u/dt^2) + \zeta(du/dt) + cu = F(t), \text{ or } M(dv/dt) + \zeta v + c\int_0^t v\,dt = F(t)$$

$$L(d^2q/dt^2) + R(dq/dt) + (1/C)q = V(t), \text{ or } L(dI/dt) + RI + (1/C)\int_0^t I\,dt = V(t)$$

Mechanical	Electrical (F – V)
Force F(t)	Voltage V(t)
Velocity v / ú	Current I
Displacement u	Charge q
Mass M	Inductance L
Spring Compliance 1/c	Capacitance C
Damping ζ	Resistance R

10. *LCR circuit*: $L\left(\frac{d^2Q}{dt^2}\right) + R(\frac{dQ}{dt}) + \frac{Q}{C} = V(t)$ or $L\left(\frac{dI}{dt}\right) + RI + \frac{1}{C}\int I\,dt = V$

$$I(t) = \frac{V_0}{Z}\sin(\omega t\,\phi); \quad Z = \sqrt{R^2 + \left(L\omega - \frac{1}{C\omega}\right)^2},$$

$$\tan\phi = \frac{\left(L\omega - \frac{1}{C\omega}\right)}{R}$$

11. Steady state oscillation for a *mass-spring- dashpot model*: $m\ddot{u} + \xi \cdot \dot{u} + cu = f(t)$. $\omega_0 = \sqrt{c/m}$ (ω_0: resonance angular frequency for zero damping) $\xi = \zeta/2m\omega_0$ (ξ, ζ: damping factor, ratio)

$$u(t) = \frac{f_0\sin(\omega t + \phi)}{\sqrt{(c - m\omega^2)^2 + (2m\zeta\omega_0\omega)^2}}$$

$$u_0 = \frac{f_0}{\sqrt{(c - m\omega^2)^2 + (2m\zeta\omega_0\omega)^2}}$$

$$\tan\phi = -\frac{2m\zeta\omega_0\omega}{c - m\omega^2}$$

Bode plot: asymptotic curves –
0 dB/decade, – 40 dB/decade.
Resonance peak height $= 20log_{10}(\frac{1}{2\zeta})$
Mechanical quality factor:

$$Q_m = \frac{\omega_0}{\Delta\omega} = 1/2\zeta$$

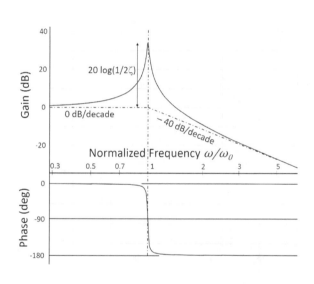

Bode plot for a standard second-order system.

12. Admittance of a piezoelectric specimen:
k_{31} Mode

$$Y = j\omega C_d \left[1 + \frac{k_{31}^2}{1 - k_{31}^2}\frac{\tan\left(\frac{\omega L}{2v_{11}^E}\right)}{\left(\frac{\omega L}{2v_{11}^E}\right)}\right]$$

k_{33} Mode

$$Y = \frac{j\omega C_d}{\left[1 - k_{33}^2\frac{\tan\left(\frac{\omega L}{2v_{33}^D}\right)}{\left(\frac{\omega L}{2v_{33}^D}\right)}\right]} = j\omega C_d + \frac{j\omega C_d}{\left[-1 + 1/k_{33}^2\left\{\frac{\tan(\Omega_{33})}{(\Omega_{33})}\right\}\right]} = j\omega C_d + \frac{1}{-\frac{1}{j\omega C_d} + \frac{1}{j\tan\left(\frac{\omega L}{2v_{33}^D}\right)\frac{2bwd_{33}^2}{\rho v_{33}^D L^2 s_{33}^E\,^2}}}$$

13. 2-terminal and 4-terminal equivalent circuits (k_{31} case)

2-terminal EC: (a)

$$L_n = (\rho/8)(Lb/w)(s_{11}^{E2}/d_{31}^2)$$

$$C_n = (8/n^2\pi^2)(Lw/b)(d_{31}^2/s_{11}^{E2})s_{11}^E$$

$$R_n = \sqrt{L_n/C_n}/Q$$

4-terminal EC: (b)

$$\Phi = \frac{2wd_{31}}{s_{11}^E}(force\,factor)$$

$$l_n = \Phi^2 L_n = (\rho/2)(Lbw)$$

$$c_n = C_n/\Phi^2 = (2/n^2\pi^2)(L/wb)s_{11}^E$$

$$r_n = \Phi^2 R_n = (l_n/c_n)^{1/2}/Q$$

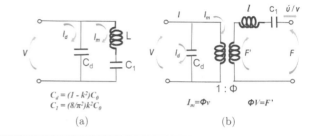

$C_d = (1 - k^2)C_0$
$C_1 = (8/\pi^2)k^2C_0$

$I_m = \Phi v$ $\Phi V = F'$

(a)   (b)

Check Point

1. (T/F) A polycrystalline piezoelectric PZT has three independent piezoelectric d matrix components, d_{33}, d_{13} and d_{15}. True or False?
2. What does the first suffix "1" stand for of the piezoelectric d matrix component, d_{15}, in a polycrystalline piezoelectric PZT?
3. (T/F) The following two force configurations are equivalent mathematically. True or False?

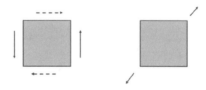

4. (T/F) There is a highly resistive (no electric carrier/impurity in a crystal) piezoelectric single crystal (spontaneous polarization P_S) with a mono-domain state without surface electrode. The "*depolarization electric field*" in the crystal is given by $E = -(\frac{P_S}{\varepsilon_0\varepsilon^X})$. True or False?
5. What is a typical number for k_{33} in Soft PZT ceramics? 1%, 10%, 50% or 70%?
6. There is a PZT multilayer actuator with a cross-section area $5 \times 5\,\text{mm}^2$. Provide a generative (blocking force) roughly. 1 N, 10 N, 100 N or 1 kN?
7. What is the fundamental longitudinal resonance frequency of a PZT multilayer actuator with a length 10 mm? 1 kHz, 10 kHz, 100 kHz or 1 MHz?
8. (T/F) The MPB composition of the PZT system exhibits the maximum electromechanical coupling k, piezoelectric coefficient d, and the minimum permittivity ε. True or False?
9. (T/F) Mechanical quality factor and the damping ratio ζ are related as $Q_m = \frac{\omega_0}{\Delta\omega} = 1/\zeta$. True or False?

10. (T/F) Complex spring constant is equivalent to the viscoelastic damping model. True or False?

11. (T/F) The high-frequency portion of the Bode plot for the second-order system is approximated with an asymptotic straight line having a negative slope of 20 dB/decade. True or False?

12. (T/F) Because the polarization is induced after the electric field is applied (time delay), the P vs. E hysteresis loop should show the clockwise rotation. True or False?

13. (T/F) The hysteresis area of the strain x vs. electric field E corresponds directly to the piezoelectric loss factor $\tan\theta$'. True or False?

14. (T/F) The permittivity under mechanically clamped condition is smaller than that under mechanically free condition. True or False?

15. (T/F) The elastic compliance under open-circuit condition is larger than that that under short-circuit condition. True or False?

16. (T/F) The piezoelectric resonance is only the mechanical resonance mode, and the antiresonance is not the mechanical resonance. True or False?

17. (T/F) The fundamental resonance mode of the k_{33} mode has an exact half-wave length vibration on the rod specimen. True or False?

18. Provide the relationship between the mechanical quality factor Q_M at the resonance frequency and the intensive elastic loss in the k_{31} type specimen.

19. (T/F) The strain distribution in a high k_{33} rod specimen is more uniform at the antiresonance mode than that at the resonance mode. True or False?

20. When $(\tan\delta'_{33} + \tan\phi'_{11})/2 < \tan\theta'_{31}$ is satisfied, which is larger Q_A or Q_B for the k_{31}-type specimen?

21. (T/F) When we consider an equivalent electric circuit of a mechanical system in terms of LCR series connection, the spring constant c corresponds directly to capacitance C. True or False?

22. (T/F) Because of the damped capacitance in the equivalent circuit of a piezoelectric oscillator, the antiresonance mode comes out, in addition to the resonance mode. True or False?

23. How can you describe the quality factor Q in a series connected LCR circuit?

24. The extensive losses are interrelated with the intensive losses in terms of the K-matrix in the 1-D model.

$$\begin{bmatrix} \tan\delta \\ \tan\phi \\ \tan\theta \end{bmatrix} = [K] \begin{bmatrix} \tan\delta' \\ \tan\phi' \\ \tan\theta' \end{bmatrix}, \text{ where } [K] = \frac{1}{1-k^2} \begin{bmatrix} 1 & k^2 & -2k^2 \\ k^2 & 1 & -2k^2 \\ 1 & 1 & -1-k^2 \end{bmatrix}.$$

Provide the inverse $[\boldsymbol{K}]$-matrix.

Chapter Problems

2.1 A "Hard" PZT shows the following performances:

$$s_{33}^E = 14.6 \times 10^{-12} \text{m}^2/N, k_t = 0.52, k_{33} = 0.64.$$

k_t and k_{33} modes generate 3D and 1D clamped conditions, respectively. When we calculate the D-constant elastic compliances for both modes, we obtain

$$s_{33}^D = s_{33}^E(1-k_t^2) = 10.6 \times 10^{-12} \text{m}^2/N, s_{33}^D = s_{33}^E(1-k_{33}^2) = 8.6 \times 10^{-12} \text{m}^2/N.$$

Consider the physical reason why the 1D clamped condition exhibits stiffer elasticity.

2.2 Knowing the mechanical system (mass M, spring c, and damper ζ) and the electric circuit (inductance L, capacitance C, and resistance R) equivalency, as shown in (a) Below, generate the electrical equivalent circuit corresponding to the mechanical system described in (b).

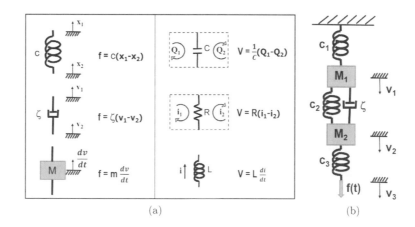

(a)   (b)

2.3 Using Mason's equivalent circuits for two length expander bars, surface and end electroded, as shown on the Right, calculate the maximum step-up voltage ratio for this Rosen-type transformer under an open-circuit condition. The Rosen-type transformer is a combination of the k_{31} (thin electrode gap) and k_{33} (large electrode gap) transducers.

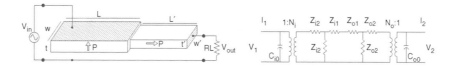

References

1. K. Uchino and S. Nomura, *Bull. Japan. Appl. Phys.*, 52, 575 (1983).
2. K. Uchino, *Bull. Amer. Ceram. Soc.*, 65(4), 647 (1986).
3. K. Uchino, S. Nomura, L. E. Cross, R. E. Newnham and S. J. Jang, *J. Mater. Sci.*, 16, 569 (1981).
4. Y. Ito and K. Uchino, Piezoelectricity, *Wiley Encyclopedia of Electrical and Electronics Engineering*, John Wiley & Sons, New York (1999).
5. W. A. Wallace, *Proc. SPIE — The Intl. Soc. Opt. Eng.*, 1733 (1992).
6. J. Kuwata, K. Uchino and S. Nomura, *Ferroelectrics* 37, 579 (1981).
7. J. Kuwata, K. Uchino and S. Nomura, *Japan. J. Appl. Phys.*, 21(9), 1298 (1982).
8. K. Yanagiwara, H. Kanai and Y. Yamashita, *Japan. J. Appl. Phys.*, 34, 536 (1995).
9. S. E. Park and T. R. Shrout, *Mater. Res. Innovt.*, 1, 20 (1997).
10. S. Wada, *Future Development of Lead-Free Piezoelectrics by Domain Wall Engineering*, Elsevier 2015/08/01 (2015).
11. E. Sawaguchi, G. Shirane and Y. Takagi, Y., *J. Phys. Soc. Japan*, 6, 333 (1951).
12. B. Jaffe, Piezoelectric transducers using lead titanate and lead zirconate, US Patent 2,708,244, May (1955).
13. H. Takeuchi, S. Jyomura, E. Yamamoto and Y. J. Ito, *Acoust. Soc. Am.*, 74, 1114 (1982).
14. Y. Yamashita, K. Yokoyama, H. Honda and T. Takahashi, *Japan. J. Appl. Phys.*, 20, Suppl. 20-4, 183 (1981).
15. Y. Ito, H. Takeuchi, S. Jyomura, K. Nagatsuma and S. Ashida, *Appl. Phys. Lett.*, 35, 595 (1979).
16. H. Takeuchi, H. Masuzawa, C. Nakaya and Y. Ito, *Proc. IEEE 1990 Ultrasonics Symp.*, 697 (1990).
17. H. Kawai, *Japan. J. Appl. Phys.*, 8, 975 (1969).
18. K. Uchino, *Sustainability in Environment*, 4(4) (2019). www.scholink.org/ojs/index.php/se.
19. T. Tou, Y. Hamaguchi, Y. Maida, H. Yamamori, K. Takahashi and Y. Terashima, *Jpn. J. Appl. Phys.*, 48, 07GM03, (2009).
20. Y. Saito, H. Takao, T. Tani, T. Nonoyama, K. Takatori, T. Homma, T. Nagaya and M. Nakamura, *Nature*, 432, 84 (2004).
21. Y. Doshida, *Proc. 81st Smart Actuators/Sensors Study Committee*, JTTAS, December 11, Tokyo (2009).
22. R. E. Newnham, D. P. Skinner and L. E. Cross, *Mater. Res. Bull.*, 13, 525 (1978).
23. W. A. Smith, *Proc. 1989 IEEE Ultrasonic Symp.*, 755 (1989).

24. M. Kitayama, *Ceramics*, 14, 209 (1979).
25. J. W. Hardy, J. E. Lefebre and C. L. Koliopoulos, *J. Opt. Soc. Amer.*, 67, 360 (1977).
26. B. Z. Janos and N. W. Hagood, *Proc. 6th Conf. New Actuators, Actuator '98*, Bremen, Germany (1998), p. 193.
27. S. Kalpat, X. Du, I. R. Abothu, A. Akiba, H. Goto and K. Uchino, *Japan. J. Appl. Phys.*, 40, 158 (2001).
28. Y. Sugawara, K. Onitsuka, S. Yoshikawa, Q. C. Xu, R. E. Newnham and K. Uchino, *J. Amer. Ceram. Soc.*, 75 (4), 996-998 (1992).
29. K. Nagai and T. Konno (eds.), *Electromechanical Vibrators and Their Applications*, Corona Pub., Tokyo, Japan (1974).
30. K. Abe, K. Uchino and S. Nomura, *Japan. J. Appl. Phys.*, 21, L408 (1982).
31. D. R. Tobergte and S. Curtis, IUPAC. Compendium of Chemical Terminology, (the "Gold Book") 53 (2013).
32. K. Uchino and S. Hirose, *IEEE Trans. Ultrasonics, Ferroelectrics, and Frequency Control*, 48, 307 (2001).
33. J. F. Nye, *Physical Properties of Crystals*, Oxford University Press, London (1972), p.123, 140.
34. N. Setter (Ed.), *Piezoelectric Materials in Devices*, Swiss Institute of Technology, Lausanne, Switzerland (2002).
35. G. Arlt and H. Dederichs, *Ferroelectrics*, 29, 47 (1980).
36. Y. Zhuang, S. O. Ural, A. Rajapurkar, S. Tuncdemir, A. Amin and K. Uchino, *Japan. J. Appl. Phys.*, 48, 041401 (2009).
37. K. Uchino, *Ferroelectric Devices*, 2nd Edition, CRC Press, New York, NY (2009).
38. W. P. Mason, *Physical Acoustics and the Properties of Solids*, Van Nostrand, New York (1958).
39. L. E. Kinsler, A. R. Frey, A. B Coppens and I. V. Sanders, *Fundamental of Acoustics*, John Wiley & Sons, New York (1982).
40. K. Uchino, *Micromechatronics*, 2nd Edition, CRC Press, Boca Raton, FL (2020), ISBN-13: 978-0-367-20231-6.
41. K. Uchino, J. Zheng, A. Joshi, Y. H. Chen, S. Yoshikawa, S. Hirose, S. Takahashi and J. W. C. de Vries, *J. Electroceramics*, 2, 33 (1998).
42. S. Hirose, M. Aoyagi, Y. Tomikawa, S. Takahashi and K. Uchino, *Ultrasonics*, 34, 213 (1996).
43. A. V. Mezheritsky, *IEEE Trans. Ultrason. Ferroelectr. Freq. Control*, 49, 484 (2002).
44. Y. Zhuang, S. O. Ural, S. Tuncdemir, A. Amin and K. Uchino, *Japan. J. Appl. Phys.*, 49, 021503 (2010).
45. ANSI/IEEE Std 176-1987, *IEEE Standard on Piezoelectricity*, The Institute of Electrical and Electronics Engineers, New York (1987), p. 56.
46. W. P. Mason, *Proc. I.R.E.*, 23 1252 (1935).
47. D. Damjanovic, *Ferroelectrics*, 110, 129 (1990).
48. W. Shi, H. N. Shekhani, H. Zhao, J. Ma, Y. Yao and K. Uchino, *J. Electroceram.*, 35, 1 (2015).
49. W. Shi, H. Zhao, J. Ma, Y. Yao and K. Uchino, *Japan. J. Appl. Phys.*, 54, 101501 (2015). http://dx.doi.org/10.7567/JJAP.54.101501
50. K. Uchino, *Micromechatronics*, 2nd Edition, CRC Press, Boca Raton, FL (2019). ISBN-13: 978-0-367-20231-6
51. W. P. Mason, *Electromechanical Transducers and Wave Filters*, D. Van Norstrand Co. Inc. (1948).
52. X. Dong, M. Majzoubi, M. Choi, Y. Ma, M. Hu, L. Jin, Z. Xu and K. Uchino, *Sensors & Actuators: A. Physical*, 256, 77 (2017). http://dx.doi.org/10.1016/j.sna.2016.12.026
53. X. Dong, T. Yuan, M. Hu, H. Shekhani, Y. Maida, T. Tou and K. Uchino, *Rev. Sci. Instruments*, 87, 105003 (2016). doi: 10.1063/1.4963920

Chapter 3

Principle of Piezoelectric Passive Dampers

Piezoelectric "passive dampers" are the original devices, which initiated "piezoelectric energy harvesting". Because the developing strategy is exactly the same for both, the reader will learn it in this chapter: how to dissipate (or cultivate) the electrical energy efficiently from two aspects: (1) electromechanical coupling factor and (2) electrical impedance matching.

3.1 Mechanisms of Piezoelectric Passive Dampers

3.1.1 *Piezo Passive Damper Principle*

The principle of the piezoelectric vibration damper is explained on the basis of a piezoelectric ceramic single plate in Fig. 3.1(a),[1,2] as it was a prior product to the energy harvesting devices. When an external impulse is applied to the piezo-plate, an electric charge is produced via the direct piezoelectricity (Fig. 3.1(b)), where a k_{31} piezo ceramic plate with the spontaneous polarization direction perpendicular to the force direction is illustrated. Accordingly, the vibration remaining after the removal of the external force induces alternating voltage, which corresponds to the intensity of that vibration, across the terminals of the single plate. The electric charge produced is allowed to flow and is dissipated as Joule heat, when a resistor is put between the terminals (see Fig. 3.1(c)). As we discuss below, when the external resistance is too large or too small, the vibration intensity is not readily reduced, and we need to tune the resistance to match exactly to the piezo-plate electric impedance, that is, $1/j\omega C$, where ω is the cyclic frequency (i.e., the fundamental mechanical resonance of the piezo-plate, and C is the piezo-plate capacitance).

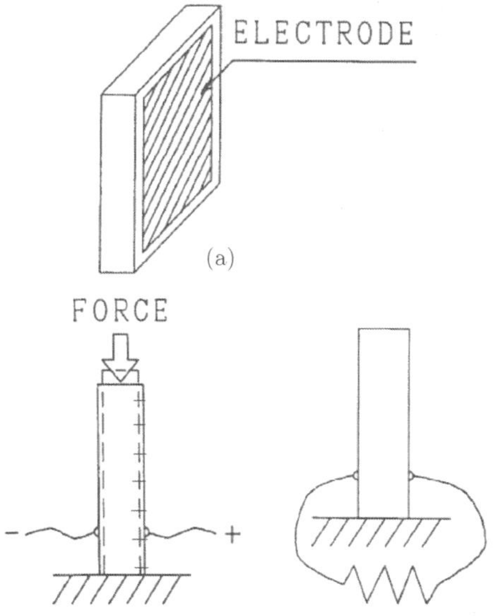

Fig. 3.1: Piezoelectric mechanical damper. (a) a piezoelectric sample; (b) direct piezoelectric effect; and (c) electric energy dissipation via a resistance.

3.1.2 *Electrical Impedance Matching*

Since the derivation process is detailed in Chapter 6, only the key points are described here on the *electrical impedance matching* of the energy harvesting system, using Fig. 3.2. An external electrical impedance Z is connected to a piezoelectric actuator. When we assume sinusoidal input stress $X = X_0 e^{j\omega t}$ and output electric displacement $D = dX_0 e^{j\omega t}$ (i.e., no time lag) under the vibration ringing-down process, we can derive the following current and voltage relationship, knowing the piezo actuator impedance $1/j\omega C$ (C: stress-free capacitance) under a off-resonance frequency. Note here that since the motional capacitance can be neglected in this scenario due to no vibration enhancement via the mechanical quality factor Q_m during the operating process.

$$i = \frac{\partial D}{\partial t} = i_{in} + i_{out} = j\omega dX_0;$$

$$Z_{in}i_{in} = Zi_{out}; \quad Z_{in} = 1/j(\omega)C;$$

$$i_{out}(1 + j\omega CZ) = j\omega dX_0$$

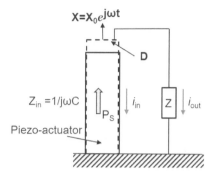

Fig. 3.2: Electric energy harvesting model under the external electrical impedance Z on a piezoelectric actuator.

Thus, we can obtain the output electric energy as

$$|P| = \frac{1}{2}Zi_{out}^2 = \frac{1}{2}Z\frac{(\omega dX_0)^2}{(1 + (\omega CZ)^2)}. \tag{3.1}$$

The maximum power energy $|P| = \frac{1}{4}\frac{\omega d^2 X_0^2}{C}$ can be obtained when the external impedance is adjusted to

$$Z = 1/\omega C \tag{3.2}$$

In other words, the "stored" electric energy can be spent maximum when the external load impedance matches exactly to the internal impedance. This is called *electrical impedance matching* for the mechanical damping. Usual electrical impedance matching learned in the university is to adjust the external impedance $Z = \left(\frac{1}{j\omega C}\right)^*$, that is, $j\omega L$ so that $\omega = 1/\sqrt{LC}$. However, in this case, the reader can easily understand that energy is not dissipated, nor is the vibration damping realized. The resistor connection is the key!

3.1.3 *Experimental Verification of Piezo Dampers*

The bimorph piezoelectric element shown in Fig. 3.3(a), an elastic beam sandwiched with two sheets of piezoelectric ceramic plates, is a typical example of a combination of a vibration object and piezoelectric

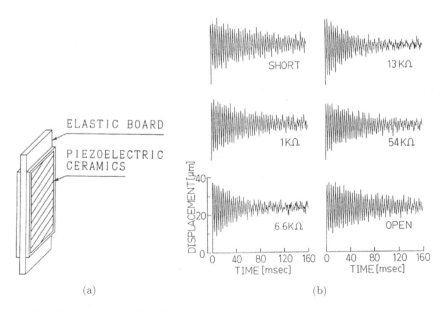

(a)   (b)

Fig. 3.3: Vibration damping change associated with external resistance change. (a) Bimorph transducer for this measurement. (b) Damped vibration with external resistor.[1]

ceramics. The bimorph edge was hit by an impulse force, and the transient vibration displacement decay was monitored by an eddy current-type non-contact displacement sensor (SDP-2300, Kaman, Middletown, CT). Figure 3.3(b) shows the measured displacement data, which vibrates at the bimorph fundamental resonance frequency (295 Hz), and Fig. 3.4 shows the relationship between the damping time constant and an external resistance. It can be seen in the figure that the damping time constant was minimized in the vicinity of 8 kΩ, which is close to the impedance $1/\omega C$.

3.1.4 *Damping Constant via k*

Let us evaluate the damping constant theoretically. The electric energy U_E generated can be expressed by using the electromechanical coupling factor k and the mechanical energy U_M:

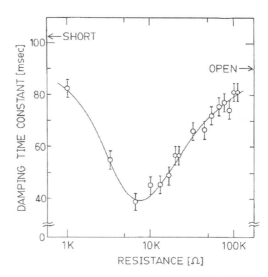

Fig. 3.4: Relation between the damping time constant and the external resistance.[1]

$$U_E = U_M \times k^2. \tag{3.3}$$

The piezoelectric damper transforms electric energy into heat energy when the resistor R is connected, and transforming rate of the damper can be raised to a level of up to 50% when the electrical resistive impedance is matched. Accordingly, the vibration energy is transformed at a rate of $(1 - k^2/2)$ times with energy vibration repeated, since $k^2/2$ multiplied by the amount of mechanical vibration energy is dissipated as heat energy. As the square of the amplitude is equivalent to the amount of energy, the amplitude decreases at a rate of $(1 - k^2/2)^{1/2}$ times with every vibration repeated. If the resonance period is taken to be T_0, the number of vibrations for t sec is $2t/T_0$ (taking ± twice per cycle). Consequently, the amplitude in t sec is $(1 - k^2/2)^{t/T_0}$. If the residual vibration period is taken to be T_0, the damping in the amplitude of vibration is t sec can be expressed as follows:

$$(1 - k^2/2)^{t/T_0} = e^{-t/\tau}. \tag{3.4}$$

Thus, the following relationship for the time constant of the vibration damping is obtained:

$$\tau = -\frac{T_0}{\ln(1 - k^2/2)} \tag{3.5}$$

Now, let us examine the time constant of the damping using the results for the bimorph in our study. Substitution in Eq. (3.5) of $k = 0.28$ and $T_0 = 3.4$ ms produces $\tau = 85$ ms, which seems to be considerably larger than the value of approximately 40 msec obtained experimentally for τ. This is because the theoretical derivation Eq. (3.5) was conducted under the assumption of a loss-free (high Q_M) bimorph. In practice, however, it involved originally mechanical loss, the time constant of which can be obtained as the damping time constant under a short-circuited condition, i.e., $\tau_s = 102$ ms. The total vibration displacement can then be expressed as $e^{-t/\tau_{\text{total}}} = e^{-t/\tau_s} \times e^{-t/\tau}$. Accordingly,

$$\frac{1}{\tau_{\text{total}}} = \frac{1}{\tau_s} + \frac{1}{\tau} \tag{3.6}$$

Substitution in Eq. (3.6) of $\tau_s = 102$ ms and $\tau = 85$ ms produces $\tau_{\text{total}} = 46$ ms. This conforms to the result shown in Fig. 3.3(b) and agrees with the experiment.

Because we used a bimorph structure in this pioneering paper merely due to the simplest geometry, the author feels a sort of guilt on providing a strong influence on subsequent researchers on the bimorph design.

Because the previous researchers did so! The bimorph is NOT an ideal piezo device design for piezo damper applications at all, as you will learn through this textbook. The higher the electromechanical coupling factor k_{eff} is, the lower the damping constant τ is, leading to the better damper. Example Problem 3.1 indicates the multilayer piezoelectric actuator usage.

Example Problem 3.1.

In order to damp a single mode of vibration in an aluminum cantilever beam, a mechanical engineer used a shunted soft-piezoelectric composite (Macro Fiber Composite (MFC) by Smart Material Corp., FL) bonded on the beam, as shown in Fig. 3.5. Though this design has already been published as a peer-reviewed academic paper, it is NOT an ideal design from the two fundamental aspects: (1) *electro-mechanical coupling factor* and (2) *mechanical impedance matching*. Discuss these problems, and provide an alternative design.

Fig. 3.5: Vibration damping in a published paper.

Solution.

(1) *Vibration Damping and Electromechanical Coupling*

The unimorph piezoelectric element shown in Fig. 3.5 is a typical misconception example of a combination of a vibration object and a piezoelectric component.

Let us evaluate the damping constant theoretically. The electric energy U_E generated can be expressed by using the electromechanical coupling factor k (in this case, effective k_{eff} for a unimorph structure) and the input mechanical energy U_M from the vibration cantilever beam:

$$U_E = U_M \times k^2. \tag{P3.1.1}$$

The piezoelectric damper transforms electric energy into heat energy when the external impedance R is connected, and transforming rate of the damper can be raised to a level of up to 50% when the electrical impedance is matched. That is, the external resistance R should be adjusted to $1/\omega C$, where C is the capacitance of the piezo transducer. Accordingly, the vibration energy is transformed at a rate of $(1 - k^2/2)$ times with energy vibration repeated, since $k^2/2$ multiplied by the amount of mechanical vibration energy is dissipated as heat energy. As the square of the amplitude is equivalent to the amount of energy, the amplitude decreases at a rate of $(1 - k^2/2)^{1/2}$ times with every vibration repeated. If the resonance period is taken to be T_0, the number of vibrations for t sec is $2t/T_0$ (Note "twice" for + and − vibration directions per cycle). Consequently, the amplitude in t sec is $(1 - k^2/2)^{t/T_0}$. If the residual vibration period is taken to be T_0, the damping in the amplitude of vibration in t sec can be expressed as follows:

$$(1 - k^2/2)^{t/T_0} = e^{-t/\tau}. \tag{P3.1.2}$$

Thus, the following relationship for the time constant τ of the vibration damping is obtained.

$$\tau = -\frac{T_0}{\ln(1 - k^2/2)}. \tag{P3.1.3}$$

We can easily find that "the higher k design, the better for vibration damping". As we will discuss in Chapter 5, the unimorph/bimorph structure is one of the worst designs from the electromechanical coupling factor's viewpoint. We had better adopt much higher k design, such as the k_{33} structure.

(2) *Mechanical Impedance Matching*

The mechanical work transferred from one object to the other is evaluated by the product of the applied force F on the interface and the displacement ΔL of this interface:

$$W = F \cdot \Delta L \tag{P3.1.4}$$

Figure 3.6 shows a conceptual cartoon illustrating two extreme cases with a popular "Japanese proverb". If the object is very soft, the force F can be very small, leading to very small W (practically no work!). This corresponds to "Pushing a curtain", exemplified by the case when the acoustic wave is generated in water directly by a hard PZT transducer surface. Most of the acoustic energy generated in the PZT is reflected at the interface, and only a small portion of acoustic energy transfers into water. On the other hand, if the object is very hard, the displacement will be very small, again leading to very small W. This corresponds to "Pushing a wall". Polymer piezoelectric polyvinylidene di-fluoride (PVDF) or soft MFC in Fig. 3.5 cannot drive a hard metal part efficiently. Therefore, the *mechanical/ acoustic impedance* $Z = \sqrt{\rho c}$ must be adjusted to maximize the output mechanical power (details in Chapter 4):

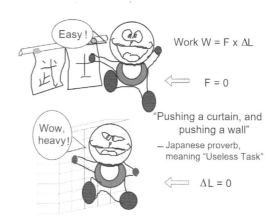

Fig. 3.6: Concept of mechanical impedance matching.

$$\sqrt{\rho_1 c_1} = \sqrt{\rho_2 c_2}, \tag{P3.1.5}$$

where ρ is the density and c is the elastic stiffness, and the subscripts 1 and 2 denote the two materials (i.e., actuator and load). This is one of the key factors for developing piezoelectric energy harvesting systems. In practice, in a medical array transducer, *acoustic impedance matching layers* (elastically intermediate materials between PZT and water, such as a polymer) are fabricated on the PZT transducer to optimize the transfer of mechanical energy to water. More precisely the matched acoustic impedance Z should be chosen as $\sqrt{Z_1 \cdot Z_2}$ (i.e., geometrical average), where Z_1 and Z_2 stand for the acoustic impedances of PZT and human tissue (close to water), respectively.

Because the researcher of Example Problem 3.1 used a unimorph structure in his study probably due to the simplest geometry, we cannot expect a sufficient damping performance because of a miserably small electromechanical coupling factor k ($< 10\%$). The multilayer (ML) PZT actuators with $k_{33} = 70\%$ will be a better alternative choice for this system. Refer to Fig. 3.7. Two ML actuators can sandwich the aluminum beam around the cantilever supporting nodal part with a fixed stiff vise with keeping the top and bottom sur-

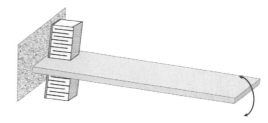

Fig. 3.7: Steel beam damping with ML actuators.

faces clamped. Since the ML actuator has a similar mechanical impedance to the vibrating metal beam, mechanical energy transfer from the beam to the ML devices is smooth. Furthermore, due to the PZT volume in the ML being much larger than the MFC composite, handling energy level will be sufficient to this large beam vibration in the level of mechanical energy 10–30 W. Note here that MFC composites are still useful against the "soft" systems, such as human, bio, polymer-related structure, as demonstrated in Chapter 4.

3.1.5 *Energy Loss Calculation*

Let us calculate the energy loss via the matched external impedance of the piezoelectric passive damping system, because this total loss should be accumulated actually in the piezo energy harvesting system. The "impulse" vibration energy U_M (represented by $\frac{1}{2}\frac{x_0^2}{s^E}$) is transformed into electric energy U_E by the factor of k^2 ($U_E = U_M \times k^2$), then a half of this energy can be consumed as Joule heat ($(1/2)k^2 \times U_M$) when we take the exact matching impedance condition. For the next half cycle, we start from the mechanical energy $(1 - k^2/2)U_M$, without taking into account elastic loss. If we consider the elastic loss (which corresponds to the τ_S in Eq. (3.6)), $(1 - k^2/2)e^{-T_0/2\tau_S}U_M$. Note that elastic loss or mechanical quality factor Q_M is related with τ_S as

$$\frac{T_0}{\tau_S} = \frac{\pi}{Q_M} \tag{3.7}$$

A similar process will be repeated every half cycle. Thus, we summarize the sequential energy ring-down process as follows:

Sequence n	Mechanical energy	Electrical energy	Lost energy
0	U_M	$k^2 U_M$	$\frac{1}{2}k^2 U_M$
1	$\left(1 - \frac{1}{2}k^2\right)e^{-\pi/2Q_M}U_M$	$k^2\left(1 - \frac{1}{2}k^2\right)e^{-\pi/2Q_M}U_M$	$\frac{1}{2}k^2\left(1 - \frac{1}{2}k^2\right)e^{-\pi/2Q_M}U_M$
2	$\left[\left(1 - \frac{1}{2}k^2\right)e^{-\frac{\pi}{2Q_M}}\right]^2 U_M$	$k^2\left[\left(1 - \frac{1}{2}k^2\right)e^{-\frac{\pi}{2Q_M}}\right]^2 U_M$	$\frac{1}{2}k^2\left[\left(1 - \frac{1}{2}k^2\right)e^{-\frac{\pi}{2Q_M}}\right]^2 U_M$
$\vdots$	$\vdots$	$\vdots$	$\vdots$
n	$\left[\left(1 - \frac{1}{2}k^2\right)e^{-\frac{\pi}{2Q_M}}\right]^2 U_M$	$k^2\left[\left(1 - \frac{1}{2}k^2\right)e^{-\frac{\pi}{2Q_M}}\right]^2 U_M$	$\frac{1}{2}k^2\left[\left(1 - \frac{1}{2}k^2\right)e^{-\frac{\pi}{2Q_M}}\right]^2 U_M$

The total lost energy can be calculated as

$$\frac{1}{2}k^2 U_M \sum_{n=0}^{n}\left[\left(1 - \frac{1}{2}k^2\right)e^{-\frac{\pi}{2Q_M}}\right]^n = \frac{1}{2}k^2 U_M \frac{1 - \left[\left(1 - \frac{1}{2}k^2\right)e^{-\frac{\pi}{2Q_M}}\right]^n}{1 - \left(1 - \frac{1}{2}k^2\right)e^{-\frac{\pi}{2Q_M}}} \tag{3.8}$$

This indicates the following results according to the mechanical quality factor:

- High Q_M ($\sim$10,000) $\rightarrow$ Since the heat generation is small, the total energy loss will reach the input mechanical energy U_M after more than 10,000 ring-down processes.
- Low Q_M ($\sim$0.5) $\rightarrow$ Since the original damping is large, the total energy loss will be $\frac{1}{2}k^2\frac{1}{0.96+0.02k^2}$, just slightly larger than $\frac{1}{2}k^2$ (only the first pulse electric energy).

Figure 3.3 is the vibration damping result after the impulse force for a bimorph with $Q_M \sim 20$. For this level of Q_M, the total loss energy is around $0.92k^2 U_M$ (summation roughly from the first 20 pulses) which is between $\frac{1}{2}k^2 U_M$ and U_M.

3.2 Passive Dampers to Piezo Energy Harvesting Systems

3.2.1 *Dawn of Piezoelectric Energy Harvesting Systems*

In the 1990s, Uchino's group decided to collect the electric energy without just dissipating the energy via Joule heat, which is the beginning of piezoelectric energy harvesting; the proverb "Chasing two hares to obtain BOTH", different from the original proverb "Chasing two hares to catch NEITHER", was our target because the condition for maximizing the vibration damping is exactly the same as the one for maximizing the energy harvesting rate.

Figures 3.8(a) and 3.8(b) illustrate two extreme examples we will treat in this book: (a) high (1 W) energy harvesting from a "hard" machine, such as an engine, and (b) low (1 mW) energy harvesting from a "soft" machine, such as human motion. Not only mechanical impedance matching, but also mechanical strength and damping factor (i.e., loss) of the device are important.

The "Cymbal" transducer is a preferable device for the high-power purposes.[3] A Cymbal transducer consists of a piezoelectric ceramic disk and a pair of metal endcaps (Fig. 3.8(a)). The metal endcaps play an important role as the displacement-direction convertor, displacement amplifier and mechanical impedance tuner. The Cymbal transducer has a relatively high coupling factor (k_{eff}) and a high stiffness, which makes it suitable for a high-force mechanical source, in comparison with popular unimorph/bimorph structures. The multilayer piezo-disk is adopted for "electrical impedance matching" purposes.

To the contrary, the flexible transducer is a preferable device for "soft" application. The Macro Fiber Composite (MFC) is an actuator that offers reasonably high performance and flexibility in a cost-competitive manner (Smart Material Corp.) (Fig. 3.8(b)).[4] The MFC consists of rectangular piezo ceramic rods sandwiched between layers of adhesive and electroded polyimide film. This film contains interdigitated electrodes that transfer the applied voltage directly to and from the ribbon-shaped rods. Uchino *et al.* developed Intelligent Clothing (IC) with piezoelectric energy harvesting system of flexible piezoelectric textiles[4] aiming at a power source for charging up portable equipment such as cellular phones and health monitoring units. Though the harvested energy density per unit area is small, if we increase this piezo textile area to all jacket (i.e., wearable device), a 10 mW level can be expected.

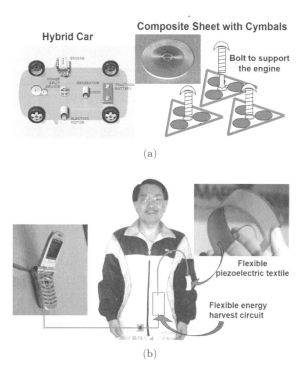

(a)

(b)

Fig. 3.8: (a) 1 W piezoelectric energy harvesting system for automobile with Cymbal devices. (b) 1 mW energy harvesting for portable electronic devices with flexible piezodevices.

3.2.2 *Three Phases in Energy Harvesting Process*

There are three major phases/steps associated with piezoelectric energy harvesting (see Fig. 3.9): (i) *mechanical–mechanical energy transfer*, (ii) *mechanical–electrical energy transduction* and (iii) *electrical–electrical energy transfer*, which are introduced successively in Chapters 4–6 in this textbook.

(i) *Mechanical–mechanical energy transfer* process includes (a) mechanical stability of the piezoelectric transducer under large stresses, and (b) mechanical impedance matching. Even if a large acoustic energy exists in water, it does not transmit effectively into a mechanical "hard" PZT ceramics directly.

(ii) *Mechanical–electrical energy transduction* process relates with the electromechanical coupling factor in the composite transducer structure. In order to increase the transduction rate from the initial mechanical energy to the output electric energy, we had better adopt the high k design, escaping from the most inefficient bimorph/unimorph designs.

(iii) *Electrical–electrical energy transfer* process includes electrical impedance matching. A suitable DC/DC converter is required to accumulate the electrical energy from a high-impedance piezo device into a rechargeable battery (low impedance).

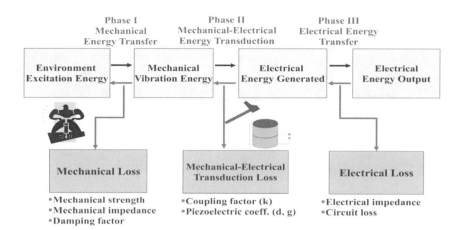

Fig. 3.9: Three major phases associated with piezoelectric energy harvesting): (i) mechanical-mechanical energy transfer, (ii) mechanical-electrical energy transduction and (iii) electrical-electrical energy transfer.

In the following chapters, we deal with detailed energy flow analysis in piezoelectric energy harvesting systems with stiff "Cymbals" (~100 mW) and flexible piezoelectric transducers (~1 mW) under cyclic mechanical load (off-resonance), in order to provide comprehensive strategies on how to improve the efficiency of the harvesting system. Energy transfer rates are practically evaluated for all three steps above. Our application target of the "Cymbal" was set to hybrid vehicles with both an engine and an electromagnetic motor, reducing the engine vibration and harvesting electric energy to car batteries to increase the mileage, while the target of the flexible piezocomposites was "wearable electric chargers" for portable electronic equipment.

We should also point out here that there is another development direction of piezo energy harvesting, that is, small energy harvesting (mW) for signal transfer applications, where the efficiency is not a primary objective. This research usually treats an impulse/snap action load to generate instantaneous electric energy for transmitting signals for a short period (100 ms–10 s), without accumulating the electricity in a rechargeable battery (but in a capacitor, or in the piezo device itself). Successful million-selling products in the commercial market belong mostly to this category at present, including the "Lightning Switch"[5] [remote switch for room lights, with using a unimorph piezoelectric component] by PulseSwitch Systems, VA, and the 25-mm-caliber "Programmable Ammunition"[6] (electricity generation with a multilayer piezo-actuator under shot impact) by ATK Integrated Weapon Systems, AZ, and Micromechatronics Inc., PA as introduced in Section 1.3. Impulse response analysis is detailed in Chapter 5.

Recall the minimum required usable electric energy higher than 1 mW level, in the components/devices such as in practice:

- Typical MOSFET
- Blue-tooth transmission device
- DC/DC converter
- Heart pacemaker
- Blood soaking syringe.

The present micro-, nanoenergy harvesting research, seeking "micro-" or "nano-" W level using the silicon MEMS technologies, should be redirected from the cascading or multi-connecting strategy viewpoints.

Chapter Essentials

1. Piezo energy harvesting research historically originates from the "piezoelectric damper", and the development strategy is basically the same.
2. Piezoelectric passive mechanical damping originates from the electric energy dissipation, which is converted from the mechanical noise vibration via the direct piezoelectric effect.
3. The damping time constant in a piezoelectric passive damper is given by

$$\tau = -\frac{T_0}{\ln(1 - k^2/2)}.$$

Thus, the higher the electromechanical coupling factor k, the better the damping rate.

4. The total loss energy on the piezo-damper after the pulse force can be calculated as

$$\frac{1}{2}k^2 U_M \sum_{n=0}^{n} \left[\left(1 - \frac{1}{2}k^2\right) e^{-\frac{\pi}{2Q_M}}\right]^n = \frac{1}{2}k^2 U_M \frac{1 - \left[\left(1 - \frac{1}{2}k^2\right) e^{-\frac{\pi}{2Q_M}}\right]^n}{1 - \left(1 - \frac{1}{2}k^2\right) e^{-\frac{\pi}{2Q_M}}}.$$

Depending on the mechanical quality factor Q_M, the total energy is between $\frac{1}{2}k^2 U_M$ and U_M.

5. The mechanical impedance matching is also an essential factor to increase the system damping factor.
6. The mechanical/acoustic impedance of a material is given by $\sqrt{\rho c}$, where ρ is the density and c is the elastic stiffness of the material.

Check Point

1. (T/F) Reducing the damping constant τ in a piezoelectric passive damping system is also a strategy to increase the energy harvesting rate in a piezoelectric energy harvesting system. True or False?
2. (T/F) The principle of the piezoelectric damper is to dissipate the electrical energy 100% per cycle via the externally connected resistor, which is converted from the mechanical noise vibration via direct piezoelectric effect. True or False?
3. (T/F) The unimorph piezoelectric structure is most popularly used, because it exhibits the highest vibration damping rate among various piezo component designs. True or False?
4. Calculate the electrical impedance $1/j\omega C$ of a piezoelectric component with capacitance 1 nF at the off-resonance frequency 100 Hz.
5. When we apply 1 Joule mechanical energy on a piezoelectric block with the electromechanical coupling factor $k = 50\%$, we can expect the electric energy $k^2 = 25\%$ converted. How much Joule can we expect to harvest in a rechargeable battery ideally through this pseudo-static process?
6. Is it possible to create a passive mechanical damping system by connecting an impedance $j\omega L$ to a piezoelectric component C, so as to satisfy the L value as $\omega = 1/\sqrt{LC}$ (ω: operating frequency)?
7. (T/F) The impedance matching in the piezoelectric passive damper system stands for the 50%:50% energy separation of the converted electrical energy into the external impedance and the internal piezoelectric component. True or False?
8. (T/F) The PZT fiber embedded polymer films (MFC) are recommended for controlling the steel beam structures, because of their simplest installation process. True or False?
9. (T/F) In the piezoelectric passive damper system, the damping constant τ becomes smaller with the electromechanical coupling factor k being higher. True or False?
10. In the piezoelectric passive damper system (with $Q_M > 200$) initiated by the impulse/snap action force (initial mechanical energy U_M), many ring-down vibrations are followed. Supposing that we use

the external resistor with the impedance matching condition, how much total energy loss is expected in this resistor (say, accumulated loss during the 200 cycles)? $(1/4)U_M$, $(1/2)U_M$, U_M, or else?

Chapter Problems

3.1 "Soft" PZT 5A shows the performances: $\rho = 7750 \, \text{kg/m}^3$, $s_{33}^E = 18.8 \times 10^{-12} \, \text{m}^2/\text{N}$, $d_{33} = 374 \times 10^{-12} \, \text{m/V}$.

Stainless steel shows the parameters: $\rho = 5800 \, \text{kg/m}^3$, $1/Y = 5.2 \times 10^{-12} \, \text{m}^2/\text{N}$ (Y: Young's modulus).

On the contrary, the polymer has $\rho = 1000 \, \text{kg/m}^3$, $1/Y = 770 \times 10^{-12} \, \text{m}^2/\text{N}$ (Y: Young's modulus). Calculate the mechanical/acoustic impedance for these materials.

3.2 Calculate the handling energy level of a multilayer (ML) piezo-actuator as used in Fig. 3.7. Suppose the ML actuator ($10 \, \text{mm} \times 10 \, \text{mm} \times 10 \, \text{mm}$) of soft PZT with $s_{33}^E = 20 \times 10^{-12} \, \text{m}^2/\text{N}$ and the blocking stress is 400 MPa. When the steel beam fundamental resonance frequency is 10 Hz, calculate the maximum handling energy of this ML piezo-actuator.

Hint:

When the maximum stress, i.e., blocking stress, is applied on this piezo ML actuator, the mechanical energy density (per m^3) can be calculated as

$$\frac{1}{2} s_{33}^E (X_{\max})^2 = \frac{1}{2} \times 20 \times 10^{-12} \text{m}^2/\text{N} \times (400 \times 10^6 \text{N/m}^2)^2$$

Multiplied by the ML actuator volume 10^{-6}m^3, and the drive frequency 10 Hz, the above energy level will be in the order of $16 \, \text{W/cm}^3$.

References

1. K. Uchino and T. Ishii, *J. Ceram. Soc. Japan.*, 96, 863 (1988).
2. K. Uchino and K. Ohnishi, Shock Preventing Apparatus, US Patent No. 4,883,248.
3. K. Uchino, *J. Japan. Soc. Appl. Electromag.*, 15(4) (20071210), 399 (2008).
4. http://www.smart-material.com/Smart-choice.php?from=MFC.
5. http://www.lightningswitch.com/.
6. http://www.atk.com/MediaCenter/mediacenter_videogallery.asp.

Chapter 4

Mechanical-to-Mechanical Energy Transfer

The higher the mechanical vibration energy transferred to the piezoelectric device is, the better the energy harvesting rate is! Though this is an obvious strategy, most of the current publications on the piezo energy harvesting systems have not discussed this issue seriously. Similar to the electromagnetic and light waves, elastic waves reflect at the material's interface, and lose the transmitting mechanical energy into the piezoelectric transducer. Refer to Fig. 4.1, the same as Fig. 3.5 in the previous chapter, which shows piezo energy harvesting systems found in multiple publications. Though many control engineers use elastically soft piezoelectric composites (Macro Fiber Composite, MFC, by Smart Material Corp., FL) bonded on the beam struc-

Fig. 4.1: Piezo energy harvesting system in a published paper.

tures to cultivate electrical energy, it is NOT an ideal design. Why? This chapter is devoted to the reader who cannot understand the reason at present. We consider the improvement of mechanical-to-mechanical energy transfer in terms of the concept of "mechanical impedance matching" in this chapter. Then, we also discuss the mechanical toughness from the device fracture's viewpoint.

4.1 Transmission/Reflection of Elastic Waves

4.1.1 *Mechanical Wave Equations in an Isotropic Material*

The dynamic equation in continuum elastic media is given by

$$\begin{cases} \rho\left(\frac{\partial^2 u}{\partial t^2}\right) = \left(\frac{\partial X_{11}}{\partial x}\right) + \left(\frac{\partial X_{12}}{\partial y}\right) + \left(\frac{\partial X_{13}}{\partial z}\right) \\ \rho\left(\frac{\partial^2 v}{\partial t^2}\right) = \left(\frac{\partial X_{21}}{\partial x}\right) + \left(\frac{\partial X_{22}}{\partial y}\right) + \left(\frac{\partial X_{23}}{\partial z}\right), \\ \rho\left(\frac{\partial^2 w}{\partial t^2}\right) = \left(\frac{\partial X_{31}}{\partial x}\right) + \left(\frac{\partial X_{32}}{\partial y}\right) + \left(\frac{\partial X_{33}}{\partial z}\right) \end{cases} \tag{4.1}$$

where ρ is the density of the elastic material, and u, v and w are the displacements of a small volume element in the material in the x-, y- and z-directions, respectively. In the pure mechanical wave discussion, we neglect the electromechanical coupling (i.e., piezoelectric effect) in order to simply the analysis.

If we consider the "elastically isotropic" assumption in this chapter because of its simplicity, we can adopt the simple Hooke's law using only two elastic stiffnesses, c_{11} and c_{12}, as in Eq. (4.2):

$$
\begin{bmatrix} X_1 \\ X_2 \\ X_3 \\ X_4 \\ X_5 \\ X_6 \end{bmatrix} = \begin{bmatrix} c_{11} & c_{12} & c_{12} & 0 & 0 & 0 \\ c_{12} & c_{11} & c_{12} & 0 & 0 & 0 \\ c_{12} & c_{12} & c_{11} & 0 & 0 & 0 \\ 0 & 0 & 0 & \frac{1}{2}(c_{11}-c_{12}) & 0 & 0 \\ 0 & 0 & 0 & 0 & \frac{1}{2}(c_{11}-c_{12}) & 0 \\ 0 & 0 & 0 & 0 & 0 & \frac{1}{2}(c_{11}-c_{12}) \end{bmatrix} \begin{pmatrix} x_1 \\ x_2 \\ x_3 \\ x_4 \\ x_5 \\ x_6 \end{pmatrix}. \tag{4.2}
$$

In an isotropic elastic material, *Lamé parameters* (first parameter λ and *shear modulus* μ) are often utilized conventionally. The first and second parameters, λ and μ, are defined by both shear parameters as

$$
\lambda = c_{12}, \text{ and } \mu = c_{66} = \frac{1}{2}(c_{11}-c_{12}). \tag{4.3}
$$

The elastic stiffness matrix can also be represented as follows in the isotropic symmetry:

$$
(c_{ij}) = \begin{pmatrix} (\lambda+2\mu) & \lambda & \lambda & 0 & 0 & 0 \\ \lambda & (\lambda+2\mu) & \lambda & 0 & 0 & 0 \\ \lambda & \lambda & (\lambda+2\mu) & 0 & 0 & 0 \\ 0 & 0 & 0 & \mu & 0 & 0 \\ 0 & 0 & 0 & 0 & \mu & 0 \\ 0 & 0 & 0 & 0 & 0 & \mu \end{pmatrix}. \tag{4.4}
$$

The above Lamé parameters' definitions derive Poisson's ratio

$$
\sigma = \frac{c_{12}}{c_{11}+c_{12}} = \frac{\lambda}{2(\lambda+\mu)} \tag{4.5}
$$

and the sound velocity for the longitudinal and transverse waves in an isotropic material, $c_l^2 = (\lambda+2\mu)/\rho$ and $c_t^2 = \mu/\rho$, taking the material's mass density ρ, as we derive them in the next section.

On the contrary, when we use the inverse notations with elastic compliances, we denote

$$
\begin{bmatrix} x_1 \\ x_2 \\ x_3 \\ x_4 \\ x_5 \\ x_6 \end{bmatrix} = \begin{bmatrix} s_{11} & s_{12} & s_{12} & 0 & 0 & 0 \\ s_{12} & s_{11} & s_{12} & 0 & 0 & 0 \\ s_{12} & s_{12} & s_{11} & 0 & 0 & 0 \\ 0 & 0 & 0 & 2(s_{11}-s_{12}) & 0 & 0 \\ 0 & 0 & 0 & 0 & 2(s_{11}-s_{12}) & 0 \\ 0 & 0 & 0 & 0 & 0 & 2(s_{11}-s_{12}) \end{bmatrix} \begin{pmatrix} X_1 \\ X_2 \\ X_3 \\ X_4 \\ X_5 \\ X_6 \end{pmatrix}. \tag{4.6}
$$

An alternative set of elastic parameters in an isotropic material is composed of Young's modulus E and Poisson's ratio σ, where $E = 1/s_{11}$ and $\sigma = -s_{12}/s_{11}$:

$$(s_{ij}) = \frac{1}{E} \begin{pmatrix} 1 & -\sigma & -\sigma & 0 & 0 & 0 \\ -\sigma & 1 & -\sigma & 0 & 0 & 0 \\ -\sigma & -\sigma & 1 & 0 & 0 & 0 \\ 0 & 0 & 0 & 2(1+\sigma) & 0 & 0 \\ 0 & 0 & 0 & 0 & 2(1+\sigma) & 0 \\ 0 & 0 & 0 & 0 & 0 & 2(1+\sigma) \end{pmatrix}. \tag{4.7}$$

We have also the following relations:

$$c_{11} = \frac{1-\sigma}{(1+\sigma)(1-2\sigma)}E, \quad c_{12} = \frac{\sigma}{(1+\sigma)(1-2\sigma)}E.$$

Knowing the strain and displacement relations, $x_1 = \frac{\partial u}{\partial x}$, $x_2 = \frac{\partial v}{\partial y}$ and $x_3 = \frac{\partial w}{\partial z}$, the following stress vs. displacement relations are obtained using the Lamé parameters:

$$\begin{cases} X_1 = \lambda\Delta + 2\mu\frac{\partial u}{\partial x} \\ X_2 = \lambda\Delta + 2\mu\frac{\partial v}{\partial y} \\ X_3 = \lambda\Delta + 2\mu\frac{\partial w}{\partial z} \end{cases}, \quad \begin{cases} X_4 = \mu\left(\frac{\partial v}{\partial z} + \frac{\partial w}{\partial y}\right) \\ X_5 = \mu\left(\frac{\partial w}{\partial x} + \frac{\partial u}{\partial z}\right) \\ X_6 = \mu\left(\frac{\partial u}{\partial y} + \frac{\partial v}{\partial x}\right) \end{cases} \tag{4.8}$$

Here, $\Delta = \frac{\partial u}{\partial x} + \frac{\partial v}{\partial y} + \frac{\partial w}{\partial z}$. Thus, Eq. (4.1) can be written as

$$\rho\frac{\partial^2}{\partial t^2}\begin{pmatrix} u \\ v \\ w \end{pmatrix} = (\lambda + \mu)\begin{pmatrix} \frac{\partial}{\partial x} \\ \frac{\partial}{\partial y} \\ \frac{\partial}{\partial z} \end{pmatrix}\Delta + \mu\nabla^2\begin{pmatrix} u \\ v \\ w \end{pmatrix}. \tag{4.9}$$

Here, $\nabla^2 = \frac{\partial^2}{\partial x^2} + \frac{\partial^2}{\partial y^2} + \frac{\partial^2}{\partial z^2}$.

4.1.2 *Plane Wave Propagation*

The propagating plane wave in an isotropic elastic material keeps the *equi-phase plane* always in parallel plane. Thus, if we assume a sinusoidal harmonic wave with time dependence of $e^{-j\omega t}$, and denote the *direction cosine* of the normal axis to the equi-phase plane as (l, m, n), this equi-phase plane can be described as

$$lx + my + nz = p, \tag{4.10}$$

where $l^2 + m^2 + n^2 = 1$. The displacement (u, v, w) can be described as

$$\begin{pmatrix} u \\ v \\ w \end{pmatrix} = \begin{pmatrix} u_0 \\ v_0 \\ w_0 \end{pmatrix} e^{-j\{\omega t - k(lx+my+nz)\}}. \tag{4.11}$$

Here, ω is the angular frequency and k is the wave vector with the following relations:

$$k_1^2 + k_2^2 + k_3^2 = k^2(l^2 + m^2 + n^2) = k^2. \tag{4.12}$$

Since the dynamic equation can be rewritten as

$$\frac{\partial X_{ij}}{\partial x_j} = -\frac{1}{2}k^2 c_{ijmn}(l_m l_j u_n + l_n l_j u_m) = -\rho\omega^2 u_i \tag{4.13}$$

by putting the above equation as

$$k^2 \begin{pmatrix} A_{11} & A_{12} & A_{13} \\ A_{21} & A_{22} & A_{23} \\ A_{31} & A_{32} & A_{33} \end{pmatrix} \begin{pmatrix} u \\ v \\ w \end{pmatrix} = \rho\omega^2 \begin{pmatrix} u \\ v \\ w \end{pmatrix}, \tag{4.14}$$

let us solve this eigenvalue problem. In an isotropic material expressed in Eq. (4.2), the A_{ij} can be given by

$$\begin{cases} A_{11} = c_{11}l^2 + c_{66}m^2 + c_{66}n^2 \\ A_{22} = c_{66}l^2 + c_{11}m^2 + c_{66}n^2 \\ A_{33} = c_{66}l^2 + c_{66}m^2 + c_{11}n^2 \end{cases}, \quad \begin{cases} A_{12} = A_{21} = (c_{12} + c_{66})lm \\ A_{23} = A_{32} = (c_{12} + c_{66})mn \\ A_{31} = A_{13} = (c_{12} + c_{66})nl \end{cases}. \tag{4.15}$$

Note $c_{66} = \frac{1}{2}(c_{11} - c_{12})$. Thus, denoting the eigenvalue as C, we solve the following equation:

$$\begin{vmatrix} c_{11}l^2 + c_{66}m^2 + c_{66}n^2 - C & (c_{12} + c_{66})lm & (c_{12} + c_{66})nl \\ (c_{12} + c_{66})lm & c_{66}l^2 + c_{11}m^2 + c_{66}n^2 - C & (c_{12} + c_{66})mn \\ (c_{12} + c_{66})nl & (c_{12} + c_{66})mn & c_{66}l^2 + c_{66}m^2 + c_{11}n^2 - C \end{vmatrix} = 0.$$

Three eigenvalues of the above equation, C_1, C_2 and C_3, can be obtained, irrelevant to (l, m, n) as

$$C_1 = c_{11}, \quad \text{and} \quad C_2 = C_3 = c_{66}. \tag{4.16}$$

From $C = \frac{\rho\omega^2}{k^2}$ and the phase velocity $v_p = \frac{\omega}{k} = \sqrt{\frac{C}{\rho}}$, we obtain two characteristic sound (phase) velocities, that is, longitudinal (P (primary)-wave) and transversal (two S (secondary)-waves, orthogonal to each other) types:

$$v_{p1} = \sqrt{\frac{c_{11}}{\rho}} = \sqrt{\frac{\lambda + 2\mu}{\rho}}, \quad \text{and} \quad v_{p2} = v_{p3} = \sqrt{\frac{c_{66}}{\rho}} = \sqrt{\frac{\mu}{\rho}}. \tag{4.17}$$

You recognize that $v_{p1} > v_{p2} = v_{p3}$; the P-wave propagates faster than the S-wave. Because the longitudinal P-wave reaches to the target point first in the "earthquake", this wave was named "primary", in fact historically.

4.1.3 *Elastic Wave Transmission/Reflection at the Interface*

There exists only the transverse electromagnetic wave in an isotropic material, while both longitudinal and transverse elastic waves exist in an elastic material, which makes the situation more complicated in the analysis. We consider here the transmission/reflection of three different waves: (1) SH (secondary horizontal), (2) P (primary) and (3) SV (secondary vertical) waves between two phases.[1] In an infinite elastic medium, these three waves exist independently, but in a finite size medium, P and SV are usually coupled at the medium interface, depending on the beam angle.

4.1.3.1 *SH waves*

The SH wave has the displacement direction perpendicular to the wave propagation direction and in parallel to the interface plane (sound velocity $\sqrt{\frac{c_{66}}{\rho}}$), which exhibits an equivalent "Snell's law" in optics. The model of this analysis is illustrated in Fig. 4.2, where two material phases 1 and 2 (with density, elastic shear stiffness (*Lamé* shear modulus) and wave numbers, ρ_1, μ_1, k_1, ρ_2, μ_2, k_2) create an interface, and incident, reflection and transmission waves are illustrated with cant angles θ_i, θ_t^1, and θ_t^2, respectively. Since the SH wave is the transverse type, only the shear elastic stiffness is sufficient for the analysis. We may assume the displacements v (only along the y-direction, normal direction of this page; $u = w = 0$) of the incident, reflection and transmission waves as

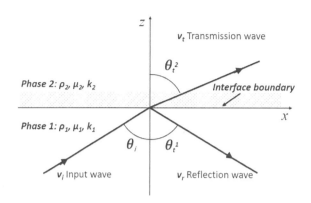

Fig. 4.2: Transmission/reflection of SH acoustic wave.

$$v_i = v_{i0}e^{-i\{\omega t - k_1(\sin\theta_i \cdot x + \cos\theta_i \cdot z)\}} \tag{4.18}$$

$$v_r = v_{r0}e^{-i\{\omega t - k_1(\sin\theta_t^1 \cdot x - \cos\theta_t^1 \cdot z)\}}, \tag{4.19}$$

$$v_t = v_{t0}e^{-i\{\omega t - k_2(\sin\theta_t^2 \cdot x + \cos\theta_t^2 \cdot z)\}}. \tag{4.20}$$

The stress boundary conditions at $z = 0$ should satisfy

$$\begin{cases} X_3^{(i)} + X_3^{(r)} = X_3^{(t)} \\ X_4^{(i)} + X_4^{(r)} = X_4^{(t)} \\ X_5^{(i)} + X_5^{(r)} = X_5^{(t)} \end{cases}, \quad v_i + v_r = v_t. \tag{4.21}$$

Here, the boundary conditions for X_3 and X_5 are automatically satisfied because all stress components are zero. Regarding the X_4 boundary condition, the displacement continuity should be maintained at $z = 0$, in order that displacement v should satisfy the above condition on a plane $z = 0$; that is, the relation

$$v_{i0}e^{-i\{\omega t - k_1(\sin\theta_i \cdot x)\}} + v_{r0}e^{-i\{\omega t - k_1(\sin\theta_t^1 \cdot x)\}} = v_{t0}e^{-i\{\omega t - k_2(\sin\theta_t^2 \cdot x)\}} \tag{4.22}$$

should be satisfied for any time t and position x, leading to the relations

$$\theta_i = \theta_t^1, \tag{4.23}$$

$$v_{i0} + v_{r0} = v_{t0}, \tag{4.24}$$

$$k_1(\sin\theta_t^1) = k_2(\sin\theta_t^2). \tag{4.25}$$

Here, the wave numbers k_1 and k_2 are given by

$$k_1 = \frac{\omega}{c_{t1}} = \omega\sqrt{\frac{\rho_1}{\mu_1}}, \quad k_2 = \frac{\omega}{c_{t2}} = \omega\sqrt{\frac{\rho_2}{\mu_2}}. \tag{4.26}$$

In order that stress X_4 should satisfy the continuity condition on a plane $z = 0$,

$$\mu_1 v_{i0}k_1\cos\theta_t^1 - \mu_1 v_{r0}k_1\cos\theta_t^1 = \mu_2 v_{t0}k_2\cos\theta_t^2 \tag{4.27}$$

Then, we can obtain the following reflectance and transmittance relations:

$$\frac{v_{r0}}{v_{i0}} = \frac{\left(1 - \sqrt{\frac{\rho_2\mu_2}{\rho_1\mu_1}} \cdot \frac{\cos\theta_t^2}{\cos\theta_t^1}\right)}{\left(1 + \sqrt{\frac{\rho_2\mu_2}{\rho_1\mu_1}} \cdot \frac{\cos\theta_t^2}{\cos\theta_t^1}\right)}, \quad \frac{v_{t0}}{v_{i0}} = \frac{2}{\left(1 + \sqrt{\frac{\rho_2\mu_2}{\rho_1\mu_1}} \cdot \frac{\cos\theta_t^2}{\cos\theta_t^1}\right)}. \tag{4.28}$$

Note that Eq. (4.28) indicates the energy conservation; that is, the incident wave energy is equal to the sum of the reflected and transmitted wave energy. In order to minimize the reflectance (or 100% transmittance), the following relation must be maintained:

$$\sqrt{\frac{\rho_2\mu_2}{\rho_1\mu_1}} \cdot \frac{\cos\theta_t^2}{\cos\theta_t^1} = 1. \tag{4.29}$$

When the incident wave is normal to the phase boundary plane ($\theta_t^1 = 0$) in particular,

$$\sqrt{\rho_1\mu_1} = \sqrt{\rho_2\mu_2} \quad \text{or} \quad \sqrt{\rho_1 c_{66,1}} = \sqrt{\rho_2 c_{66,2}}. \tag{4.30}$$

Since we define the mechanical impedance by the form of $\sqrt{\rho c}$, the reader can understand that the mechanical energy (SH wave) efficient transfer requires the mechanical impedance matching between two phases.

4.1.3.2 *P and SV waves*

We consider now the P wave with the displacement in parallel to the wave propagation direction (sound velocity $\sqrt{c_{11}/\rho}$) and the SV wave with the displacement direction perpendicular to the wave propagation direction and in perpendicular to the interface plane (the same sound velocity $\sqrt{c_{66}/\rho}$ as the SH wave). The displacements u and w of the incident, reflection and transmission waves can be described as

(Incident)

$$\text{P-Wave}: \begin{cases} u_i^{(P)} = A_i^{(P)}\sin\theta_p^1 e^{-i\left\{\omega t - k_p^1\left(\sin\theta_p^1 \cdot x + \cos\theta_p^1 \cdot z\right)\right\}} \\ w_i^{(P)} = A_i^{(P)}\cos\theta_p^1 e^{-i\left\{\omega t - k_p^1\left(\sin\theta_p^1 \cdot x + \cos\theta_p^1 \cdot z\right)\right\}}, \end{cases} \tag{4.31}$$

$$\text{SV-Wave}: \begin{cases} u_i^{(S)} = A_i^{(S)}\cos\theta_t^1 e^{-i\{\omega t - k_t^1(\sin\theta_t^1 \cdot x + \cos\theta_t^1 \cdot z)\}} \\ w_i^{(S)} = -A_i^{(S)}\sin\theta_t^1 e^{-i\{\omega t - k_t^1(\sin\theta_t^1 \cdot x + \cos\theta_t^1 \cdot z)\}}. \end{cases} \tag{4.32}$$

(Reflection)

$$\text{P-Wave}: \begin{cases} u_r^{(P)} = A_r^{(P)}\sin\theta_p^1 e^{-i\{\omega t - k_p^1(\sin\theta_p^1 \cdot x - \cos\theta_p^1 \cdot z)\}} \\ w_r^{(P)} = -A_r^{(P)}\cos\theta_p^1 e^{-i\{\omega t - k_p^1(\sin\theta_p^1 \cdot x - \cos\theta_p^1 \cdot z)\}}, \end{cases} \tag{4.33}$$

$$\text{SV-Wave}: \begin{cases} u_r^{(S)} = A_r^{(S)}\cos\theta_t^1 e^{-i\{\omega t - k_t^1(\sin\theta_t^1 \cdot x - \cos\theta_t^1 \cdot z)\}} \\ w_r^{(S)} = A_r^{(S)}\sin\theta_t^1 e^{-i\{\omega t - k_t^1(\sin\theta_t^1 \cdot x - \cos\theta_t^1 \cdot z)\}}. \end{cases} \tag{4.34}$$

(Transmission)

$$\text{P-Wave}: \begin{cases} u_t^{(P)} = A_t^{(P)}\sin\theta_p^2 e^{-i\{\omega t - k_p^2(\sin\theta_p^2 \cdot x + \cos\theta_p^2 \cdot z)\}} \\ w_t^{(P)} = A_t^{(P)}\cos\theta_p^2 e^{-i\{\omega t - k_p^2(\sin\theta_p^2 \cdot x + \cos\theta_p^2 \cdot z)\}}, \end{cases} \tag{4.35}$$

$$\text{SV-Wave}: \begin{cases} u_t^{(S)} = A_t^{(S)}\cos\theta_t^2 e^{-i\{\omega t - k_t^2(\sin\theta_t^2 \cdot x + \cos\theta_t^2 \cdot z)\}} \\ w_t^{(S)} = -A_t^{(S)}\sin\theta_t^2 e^{-i\{\omega t - k_t^2(\sin\theta_t^2 \cdot x + \cos\theta_t^2 \cdot z)\}}. \end{cases} \tag{4.36}$$

We can solve the wave forms based on the boundary conditions at $z = 0$:

$$\begin{cases} u_i^{(P)} + u_i^{(S)} + u_r^{(P)} + u_r^{(S)} = u_t^{(P)} + u_t^{(S)} \\ w_i^{(P)} + w_i^{(S)} + w_r^{(P)} + w_r^{(S)} = w_t^{(P)} + w_t^{(S)}, \end{cases} \tag{4.37}$$

$$X_3^{(i)} + X_3^{(r)} = X_3^{(t)}, \quad X_5^{(i)} + X_5^{(r)} = X_5^{(t)}. \tag{4.38}$$

Because further analyses are lengthy, we skip the full derivation, and leave it for the reader's homework by referring to Ref. 1. Instead, we introduce the simplest transmission/reflection model of the P acoustic wave under the normal incident wave situation illustrated in **Fig. 4.3**. We describe the incident, reflected and transmitted stress waves with normalized w displacement notations (which are proportional to the stress) as

$$\begin{cases} w_i = e^{ik_p^1 \cdot z} \\ w_r = Re^{-ik_p^1 \cdot z} \\ w_t = Te^{ik_p^2 \cdot z}, \end{cases} \qquad (4.39)$$

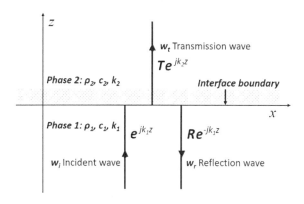

Fig. 4.3: Transmission/reflection of P acoustic wave.

where R and T denote the reflectance and transmittance coefficients, respectively. The total normalized stress in Phase 1 and Phase 2 (the equi-stress wave plane is parallel to the phase boundary) are expressed as

$$\begin{cases} X^1(z) = e^{ik_p^1 \cdot z} + Re^{-ik_p^1 \cdot z} \\ X^2(z) = Te^{ik_p^1 \cdot z} \end{cases} \qquad (4.40)$$

From the stress continuation boundary condition at $z = 0$, we obtain first the relation

$$1 + R = T. \qquad (4.41)$$

Second, taking into account $\rho \left(\frac{\partial^2 w}{\partial t^2} \right) = \left(\frac{\partial X_3}{\partial z} \right)$, and imposing continuity of particle velocity (along the z direction), we need also the relation

$$(k_p^1/\rho_1)(1 - R) = (k_p^2/\rho_2)T. \qquad (4.42)$$

Knowing the phase velocity (i.e., sound velocity) $v_p = \frac{\omega}{k}$, Eqs. (4.41) and (4.42) result in

$$\begin{cases} R = \dfrac{\rho_2 v_{p2} - \rho_1 v_{p1}}{\rho_2 v_{p2} + \rho_1 v_{p1}} \\ T = \dfrac{2\rho_2 v_{p2}}{\rho_2 v_{p2} + \rho_1 v_{p1}} \end{cases}. \qquad (4.43)$$

Thus, minimizing the reflectance (or 100% transmittance), the following relation must be maintained:

$$\rho_1 v_{p1} = \rho_2 v_{p2}. \qquad (4.44)$$

We may define the mechanical impedance by the product of density and phase sound velocity $\rho \cdot v_p$. Because

$$v_{p1} = \sqrt{\frac{c_{11}^1}{\rho_1}} \quad \text{and} \quad v_{p2} = \sqrt{\frac{c_{11}^2}{\rho_2}},$$

the above condition is equivalent to

$$\sqrt{\rho_1 c_{11}^1} = \sqrt{\rho_2 c_{11}^2}. \qquad (4.45)$$

We can also define the mechanical/acoustic impedance by the form of $\sqrt{\rho c}$ (c: elastic stiffness). The reader can again understand that the mechanical energy efficient transfer requires the mechanical impedance matching between two phases, irrelevant to the longitudinal (P wave) or transverse (SH, SV) waves.

4.2 Acoustic Impedance Matching

Though the acoustic impedance parameter is not unique for piezoelectric energy harvesting systems, it is closely associated with the piezoelectric device designing. This parameter is a measure of the impedance that a system presents to the acoustic flow against acoustic pressure applied to the system. The acoustic impedance Z is used for evaluating the acoustic energy transfer between two materials, which is defined, in general, by

$$Z = (\text{pressure/volume velocity}). \tag{4.46}$$

4.2.1 *Acoustic/Mechanical Impedance Derivation*

In a solid material, from the following dynamic equation,

$$\rho\left(\frac{\partial^2 u}{\partial t^2}\right) = \left(\frac{\partial X_{11}}{\partial x}\right) + \left(\frac{\partial X_{12}}{\partial y}\right) + \left(\frac{\partial X_{13}}{\partial z}\right) = -\left(\frac{\partial p}{\partial x}\right). \tag{4.47}$$

From $p = X_{11}$, $X_{11} = c_{11}x_{11}$, we obtain

$$\rho\left(\frac{\partial^2 u}{\partial t^2}\right) = c_{11}\left(\frac{\partial^2 u}{\partial x^2}\right) \text{ or } \left(\frac{\partial^2 u}{\partial t^2}\right) = v_{11}^2\left(\frac{\partial^2 u}{\partial x^2}\right)\left[\text{sound velocity } v_{11} = \sqrt{\frac{c_{11}}{\rho}}\right]. \tag{4.48}$$

The constitutive law of non-dispersive linear acoustics in one dimension gives a relation between stress (pressure) and strain:

$$p = -\rho v^2\left(\frac{\partial u}{\partial x}\right), \tag{4.49}$$

where p is the acoustic pressure, ρ mass density and v the sound wave speed in the medium. This equation is valid both for fluids and solids:

- Fluids $-\rho v^2 = K$ (K: bulk modulus);
- Solids $-\rho v^2 = K + (4/3)G$ (G: shear modulus) for longitudinal waves and $\rho v^2 = G$ for transverse waves.

From Eq. (4.48), we assume a general displacement solution $u(x,t)$ for a traveling plane wave as

$$u(x,t) = f(x - vt). \tag{4.50}$$

Then, we obtain pressure $p(x,t)$ and each volume/particle velocity $v(x,t) = \frac{\partial u}{\partial t}$ is represented by

$$\begin{cases} p(x,t) = -\rho v^2 f'(x - vt) \\ v(x,t) = -v f'(x - vt). \end{cases} \tag{4.51}$$

From the definition in Eq. (4.46), and Eq. (4.51), the acoustic/mechanical impedance is obtained:

$$Z = \frac{p(x,t)}{v(x,t)} = \rho v. \tag{4.52}$$

Since $v = \sqrt{c/\rho}$ in a solid material, Z can be translated to

$$Z = \sqrt{\rho c}, \tag{4.53}$$

where ρ is the density and c is the elastic stiffness of the material.

4.2.2 Acoustic Impedance Matching

4.2.2.1 Concept of acoustic impedance matching

Acoustic (or mechanical) impedance matching is necessary for transferring mechanical energy from one material to the other efficiently. Figure 4.4 illustrates a conceptual cartoon for two extreme cases. The mechanical work done by one material on the other is evaluated by the product of the applied force F and the displacement ΔL: $W = F \times \Delta L$. If the material is very soft, the force F can be very small, leading to very small W (practically no work!). This corresponds to "Pushing a curtain", exemplified by the case when the acoustic wave is generated in water directly by an elastically hard PZT transducer. Most of the acoustic energy generated in the PZT (large force and small displacement) is reflected at the interface, and only a small portion of acoustic energy

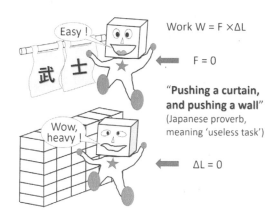

Fig. 4.4: Mechanical impedance matching concept.

transfers into water (large displacement is required). On the other hand, if the material is very hard, the displacement ΔL will be very small, again leading to very small W. This corresponds to "Pushing a wall". Polymer piezoelectric PVDF (polyvinylidene di-fluoride) (large displacement & small force) cannot drive a hard steel part effectively (large force is required). Therefore, the *acoustic impedance* must be adjusted to maximize the output mechanical power:

$$\sqrt{\rho_1 c_1} = \sqrt{\rho_2 c_2}, \tag{4.54}$$

where ρ and c are the density and elastic stiffness, respectively, and the subscripts 1 and 2 denote the two materials. In practice, an acoustic impedance matching layer (elastically intermediate material between PZT and water, such as a polymer) is inserted between two phases (1 and 2). More precisely, the acoustic impedance Z should be chosen as the geometrical average $\sqrt{Z_1 \cdot Z_2}$ of Z_1 in phase 1 and Z_2 in phase 2, so that the transfer of mechanical energy in the PZT to water will be optimized.

In more advanced discussions, there are three kinds of impedances: specific acoustic impedance (pressure/particle speed), acoustic impedance (pressure/volume speed) and radiation impedance (force/speed). See Ref. 2 for the details.

4.2.2.2 Acoustic impedance of piezoelectric materials

(a) PZT Ceramics

The acoustic wave transmitting object in medical diagnostic and underwater sonar applications is basically water. First, let us calculate the acoustic impedance of water. Taking into account $\rho = 1,000$ kg/m^3, and the sound velocity $v = 1.5$ km/s, we obtain $Z = \rho v = 1.5 \times 10^6$ kg/m$^2 \cdot$ s = 1.5 MRayls. The unit, [Rayl] = [kg/m$^2 \cdot$ s], is used for the acoustic impedance, named after John William Strutt, 3rd Baron Rayleigh. On the other hand, the acoustic impedance of the PZT is calculated from $\rho = 7750$ kg/m^3, $s_{33}^E = 18.8 \times 10^{-12}$ m^2/N for "Soft" PZT 5A, and $\rho = 7600$ kg/m^3, $s_{33}^E = 13.9 \times 10^{-12}$ m^2/N for "Hard" PZT 8, for example, $Z = \sqrt{\rho c} \approx \sqrt{\rho/s_{33}^E} = 20$ MRayls, and 24 MRayls, respectively. Because of this large difference of the acoustic impedance in the PZT ceramics and water, an energy transmission problem happens, if we set the PZT transducer directly in water medium.

(b) Piezoelectric Polymers

Polymer piezo transducer materials such as *polyvinylidene difluoride* (PVDF) are more suitable from the acoustic impedance matching viewpoint, because $\rho = 1780$ kg/m^3, Young's modulus $Y = 8.3 \times 10^9$ N/m^2

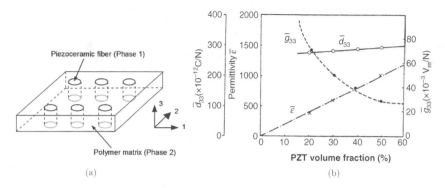

Fig. 4.5: (a) 1-3 composite of PZT rods and polymer. The top and bottom planes are rigid electrodes. (b) Volume fraction dependence of the permittivity ε and the piezoelectric constants d_{33} and g_{33} in the 1-3 PZT:polymer composite.

and $Z = \sqrt{\rho Y} = 3.8$ MRayls, much closer to water. However, due to small piezoelectric actuator figures of merit, that is, piezoelectric d constants, their application is limited to sensors such as heart-rate/pulse measurements or hydrophones.

(c) PZT: Polymer Composites

Piezo composites comprised of a piezoelectric ceramic and a polymer phase are promising materials because of their excellent and readily tailored properties. The geometry for two-phase composites can be classified according to the dimensional connectivity of each phase into 10 structures: 0-0, 0-1, 0-2, 0-3, 1-1, 1-2, 1-3, 2-2, 2-3 and 3-3.[3] A 1-3 piezo composite, such as the PZT-rod/polymer composite, is the most promising candidate, which is composed of PZT fibers embedded in a polymer matrix, as shown in Fig. 4.5(a). The original fabrication process involves the injection of epoxy resin into an array of PZT fibers assembled with a special rack.[4] After the epoxy is cured, the sample is cut, polished, electroded on the top and bottom, and finally electrically poled. The die casting technique has also been employed to make rod arrays from a PZT slurry.[5]

The effective piezoelectric coefficients d^* and g^* of the composite can be interpreted as follows: When an electric field E_3 is applied to this composite, the piezo ceramic rods extend easily because the polymer is elastically very soft (assuming the electrode plates which are bonded to its top and bottom are rigid enough). Thus, d^*_{33} is almost the same as $^1d_{33}$ of the PZT itself,

$$d^*_{33} = {}^1d_{33}. \tag{4.55}$$

Similarly,

$$d^*_{33} = {}^1V{}^1d_{33}, \tag{4.56}$$

where ^1V is the volume fraction of phase 1 (piezoelectric). On the other hand, when an external stress is applied to the composite, the elastically stiff piezo ceramic rods will support most of the load, and the effective stress is drastically enhanced and inversely proportional to the volume fraction. Thus, larger induced electric fields and larger g^* constants are expected:

$$g^*_{33} = d^*_{33}/\varepsilon_0\varepsilon^*_3 = {}^1d_{33}/{}^1V\varepsilon_0^1\varepsilon_3$$
$$= {}^1g_{33}/{}^1V. \tag{4.57}$$

Figure 4.5(b) shows the piezoelectric coefficients for a PZT-Spurrs epoxy composite with 1-3 connectivity, measured with a Berlincourt d_{33} meter.[4] As predicted by the model for this composite, the measured d^*_{33} values are almost independent of volume fraction, but are only about 75% of the d_{33} value of the PZT 501A ceramic. This discrepancy may be due to incomplete poling of the rods. A linear relation between the permittivity and the volume fraction ^1V is satisfied, resulting in a significant increase in g_{33}^* with decreasing fraction of PZT. Therefore, the 1-3 composites can enhance the piezoelectric g coefficient by an order of magnitude with decreasing volume fraction of PZT, while the d coefficient remains constant.

The advantages of this composite are high coupling factors, low acoustic impedance, good matching to water or human tissue (more than 70% of a human body is water!), mechanical flexibility, broad bandwidth in combination with a low mechanical quality factor and the possibility of making undiced arrays by simply patterning the electrodes. The thickness-mode electromechanical coupling of the composite can exceed the k_t (0.40–0.50) of the constituent ceramic, approaching almost the value of the rod-mode electromechanical coupling, k_{33} (0.70–0.80), of that ceramic.[6] The acoustic impedance matching to tissue or water (1.5 Mrayls) of the typical piezo ceramics (20–30 Mrayls) is significantly improved when they are incorporated in forming a composite structure, that is, by replacing the dense, stiff ceramic with a low-density, soft polymer. Piezoelectric composite materials are especially useful for underwater sonar and medical diagnostic ultrasonic transducer applications. This composite design is suitable for the energy harvesting under plate-perpendicular force (cyclic or impact) because of a very high effective thickness electromechanical coupling factor, close to k_{33} value of the PZT ceramic with much lower effective acoustic impedance.

Although the PZT composites are very useful for acoustic transducer applications, care must be taken when using them in continuously operating transducer applications. Under an applied DC field, the field-induced strain exhibits large hysteresis and creep due to the "viscoelastic property" of the polymer matrix. More serious problems are found when they are driven under a high AC field, related to the generation of heat. The heat generated by ferroelectric hysteresis in the piezo ceramics cannot be dissipated easily due to the very low thermal conductivity of the polymer matrix, which results in rapid degradation of piezoelectricity.

Several flexible PZT composites such as "active fiber composite" (AFC) and "macro fiber composite" (MFC) (Fig. 4.6) are commercially available, and used for a high bending displacement actuator and transducer even for energy harvesting systems. The major advantages of those fiber composites over conventional piezoelectric elements are their flexibility and toughness, both far superior compared to monolithic PZT bulk ceramics, although fiber composites fail to drive elastically hard metal or ceramic structures. Furthermore, due to their thin, planar geometry, the fiber composites can easily be integrated into composite laminates.[7] Previously, a PVDF film was used for implantable physiological power supply,[8] and inserted for recovering some of the mechanical power in the process of human walking.[9] For an alternative, the advanced piezo fiber composite (MFC) was experimentally demonstrated and confirmed for the energy conversion component by the Penn State group, which is introduced in this book.

The MFC is an actuator design which was developed at the NASA Langley Research Center. The piezoelectric fibers manufactured by computer-controlled dicing saw (rectangular cross-section) and embedded in the epoxy matrix were sandwiched between two layers of polyimide film that had a conductive interdigitated electrode pattern printed on the inner surface. There are two types of MFC depending on the

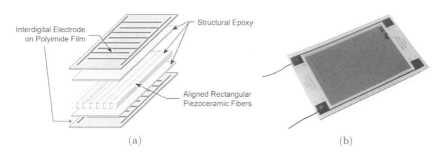

(a)   (b)

Fig. 4.6: MFC by Smart Material Corporation. (a) Composite structure, (b) photo of MFC.

poling direction. The poling direction of d_{33} type is parallel along the fiber length and each segment has opposite poling directions by interdigitated electrodes as shown in Fig. 4.6(a). For the d_{31} type, the poling direction is from the top to the bottom along the fiber thickness. The MFC is extremely flexible, durable and has the advantage of higher electromechanical coupling coefficients granted through the interdigitated electrodes (Fig. 4.6(b)).

Previously, the d_{33} mode type of the MFC was tested to charge batteries, but Sodano *et al.*[10] claimed that the MFC did not produce high current because of the construction of the MFC. The interdigitated electrodes of the d_{33} type made the small segments connect in series. High voltage but low current is obtained. To the contrary, the d_{31} mode type of MFC (M8528 P2), fabricated by Smart Material Corp., is composed of the piezo ceramic fibers in the MFC cut at 350 μm width and 170 μm thickness from a piezoelectric wafer by a computer-controlled dicing saw. The total dimensions of MFC are 85 mm length, 28 mm width and 0.3 mm thickness. Uchino *et al.* tested the d_{31} mode type of MFC under a small mechanical vibration source to generate enough current for battery charging.[11]

4.2.3 *Acoustic Impedance Matching Layer in Ultrasonic Transducers*

Ultrasonic waves are now used in various fields. The sound source is made from piezoelectric ceramics with high acoustic impedance. In particular, hard piezoelectric materials with a high Q_M are preferable because of high-power generation without heat generation. A liquid medium (water) is usually used for sound energy transfer. Ultrasonic washers, ultrasonic microphones and sonars for short-distance remote control, underwater detection and fish finding, and non-destructive testers are typical applications. Ultrasonic scanning detectors are useful in medical electronics for clinical applications ranging from diagnosis to therapy and surgery.

4.2.3.1 *Medical ultrasonic probe*

One of the most important applications is based on the ultrasonic echo field.[12,13] Ultrasonic transducers convert electrical energy into mechanical form when generating an acoustic pulse and convert mechanical energy into an electrical signal when detecting its echo. The transmitted waves propagate into a body and echoes are generated which travel back to be received by the same transducer. These echoes vary in intensity according to the type of tissue or body structure, thereby creating images. An ultrasonic image represents the mechanical properties of the tissue, such as *density* and *elasticity*. We can recognize anatomical structures in an ultrasonic image since the organ boundaries and fluid-to-tissue interfaces are easily discerned. The ultrasonic imaging process can also be done in real time. This means we can follow rapidly moving structures such as the heart without motion distortion.

In addition, ultrasound is one of the safest diagnostic imaging techniques. It does not use ionizing radiation like X-rays and thus is routinely used for fetal and obstetrical imaging. Useful areas for ultrasonic imaging include cardiac structures, the vascular systems, the fetus and abdominal organs such as the liver and kidneys.

Figure 4.7 shows the basic ultrasonic transducer geometry. The transducer is mainly composed of (1) *matching*, (2) *piezoelectric material* and (3) *backing layers*.[14] One or more matching layers are used to increase sound transmissions into tissues. The backing is added to the rear of the transducer in order to damp the acoustic backwave and to reduce the pulse duration (i.e., quick decay of the ring-down vibra-tion). Piezoelectric materials are used to generate and detect ultrasound

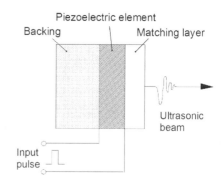

Fig. 4.7: Basic transducer geometry for acoustic imaging applications.

($f = 2$–4 MHz, $\lambda = 0.4$–0.7 mm). In general, broadband transducers should be used for medical ultrasonic imaging. The broad bandwidth response corresponds to a short pulse length (due to low Q_m or large elastic

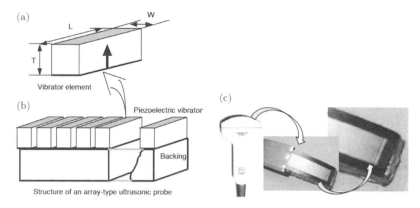

Fig. 4.8: (a) Basic transducer geometry for acoustic imaging applications. (b) Linear array type ultrasonic probe. (c) Photos of convex linear array ultrasonic probes.

loss), resulting in better axial resolution. Three factors are important in designing broad bandwidth transducers: *acoustic impedance matching*, a *high electromechanical coupling coefficient* of the transducer and *electrical impedance matching*. These pulse echo transducers operate based on thickness k_t mode resonance of the piezoelectric thin plate. Further, a low planar mode coupling coefficient, k_p (or k_{31}), is beneficial for limiting energies being expended in a non-productive lateral mode, which generates so-called "Ghost" images.[15] A large dielectric constant is necessary to enable a good electrical impedance match to the system, especially with tiny piezoelectric sizes.

There are various types of transducers used in ultrasonic imaging. Multiple element array transducers permit discrete elements to be individually accessed by the imaging system and enable electronic focusing in the scanning plane to various adjustable penetration depths through the use of phase delays. A linear array is a collection of elements (Fig. 4.8(a)) arranged in one direction, producing a rectangular display. As illustrated in Fig. 4.8(b), a thin PZT (thickness 0.3 mm for 24-MHz ultrasonic wave) wafer is segmented into 128 elements with the separation pitch 0.3–0.5 mm. In a phased array transducer, the acoustic beam is steered by signals that are applied to the elements with delays, creating a sector display. This segmentation also enhances the electromechanical coupling much higher from the k_t to k_{33} mode. The acoustic impedance of the "matching layer" is to be chosen as a geometric average value of the piezoelectric element impedance Z_1 ($\sim$20 Mrayls) and the water impedance Z_2 ($\sim$1.5 Mrayls), that is, $\sqrt{Z_1 Z_2}$ ($\sim$5.5 Mrayls). In practice, the PZT thin wafer is initially coated by Epoxy-based polymer (Glass-filled type to tune the Young's modulus) with the acoustic impedance 3–7 Mrayls. Then, the rubber covers as the second impedance matching layer (see Center photo of Fig. 4.8(c)). A curved linear (or convex) array is a modified linear array whose elements are arranged along an arc to permit an enlarged trapezoidal field of view (Fig. 4.8(c)).

4.2.3.2 *Cymbal underwater sonar*

In contrast to the flexural bimorph designs, flextensional "Cymbal" transducers use the flexural motion of only the metal shells. The use of metal endcap motion offers one great advantage over the flexural disks, that is, no bending deformation on the ceramic material. The Cymbal transducers consist of a piezoelectric disk (poled in the thickness direction) sandwiched between two metal endcaps (Fig. 4.9(a)). The caps contain a

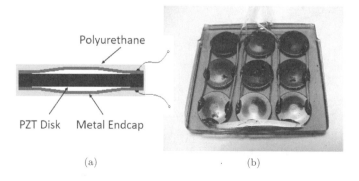

Fig. 4.9: (a) Single cymbal potted in polyurethane; (b) 3×3 array, potted in polyurethane.[17]

shallow cavity on their inner surface. When acting as a sensor or energy harvesting device, the cavities allows the incident axial stress to be converted into large radial and tangential stresses of opposite sign, causing d_{31} and d_{33} contributions of the piezoelectric to add in the effective sensitivity of the device (i.e., force amplification mechanism). Conversely, the presence of the cavities allows the caps to convert and amplify the small radial displacement of the disk into a much larger axial displacement normal to the surface of the caps when used as an actuator (i.e., displacement amplifier).[16]

For underwater sonar applications, in the low-frequency range, the flextensional transducer is one of the competing technologies with 1-3 composites and Langevin/Tonpilz transducers. The Cymbal, which is a miniaturized Class V flextensional transducer, has attracted interest because of its small size, low cost and weight, and thin profile. There are no other technologies that can match these characteristics, which are key to some of the Navy applications, including underwater automated vehicles (UAVs). As discussed previously, single Cymbal transducers are not well suited for most underwater applications because of their high Q_m and low efficiency. Thus, Cymbal transducer elements were built into 3×3 arrays to enhance their underwater characteristics in terms of the power level and acoustic beam directivity. Figure 4.9(b) shows a photo of the 3×3 array. We used "polyurethane" for the potting material of the Cymbal transducers for the following three reasons: (1) fixing structure of the 3×3 arrays, (2) electric insulation and (3) acoustic impedance matching to water.[17] The acoustic impedance of polyurethane is rather tunable from 1.6 to 6.9 Mrayls depending on the foam rate prepared in the polymeric material.

4.3 Mechanical Fracture of Piezo Ceramics

Piezoelectric actuators and energy harvesting devices are used in mechanical systems, occasionally under certain prestress conditions, in order to optimize efficiency as well as to improve mechanical reliability. We will consider in this section the stress dependence of piezoelectricity and the mechanical strength of materials.

4.3.1 *Uniaxial Stress Dependence of Piezoelectric Strains*

Even elastically stiff ceramic actuators will deform under an applied stress. The electric field-induced strain is affected by a bias stress. The uniaxial compressive stress dependence of the longitudinal field-induced strain in $BaTiO_3$ (BT)-based, PZT and lead magnesium niobate-lead titanate (0.65PMN-0.35PT) ceramics is shown in Fig. 4.10(a).[18] The PZT- and PMN-based ceramics exhibit *maximum field-induced strains*, which are indicated on the plot where the individual lines intersect with the strain axis (around 0.1% under the standard electric field 1 kV/mm applied), and are doubled larger than the BT-based material (0.04%). On the other hand, the *maximum generative stress* (indicated where the lines intersect the stress axis, and is sometimes referred to as the *blocking force*) is nearly the same (about 3.5×10^7 N/m^2) for all the samples. This is because the elastic compliance of lead-based ceramics tends to be relatively large (that is, they are elastically soft). Note this "blocking force" level, 40 MPa, which is almost the same for any piezo ceramic materials, and 20 MPa is a suitable compressive DC bias stress for obtaining the maximum energy transmission coefficient or the maximum output (i.e., usable) mechanical energy (or electrical for energy harvesting, see Section 2.6.2.2).

The strain as a function of compressive stress is plotted also for electrostrictive and piezoelectric PMN-PT actuators in Fig. 4.10(b).[19] We see from these data that the 0.9PMN-0.1PT devices exhibit the larger maximum field-induced strains than the 0.65PMN-0.35PT composition, but the maximum generative forces (i.e., blocking force) for these two compositions under a similar field strength are nearly the same ($\approx$40 MPa). It is worth noting here that (1) the electrostriction exhibits significant nonlinear behavior

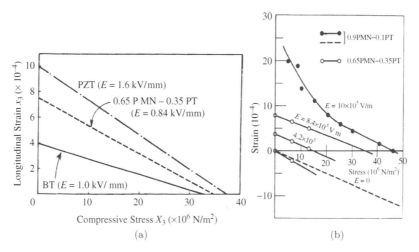

Fig. 4.10: (a) Uniaxial (compressive) stress dependence of the electric field-induced strain in $BaTiO_3$ (BT)-, $Pb(Mg_{1/3}Nb_{2/3})O_3$ (PMN)- and $Pb(Zr,Ti)O_3$ (PZT)-based piezoelectric ceramics. (b) Uniaxial stress dependence of the longitudinal strain for PMN-PT actuators. 0.9PMN-0.1PT: electrostrictor, 0.65PMN-0.35PT: piezoelectric.

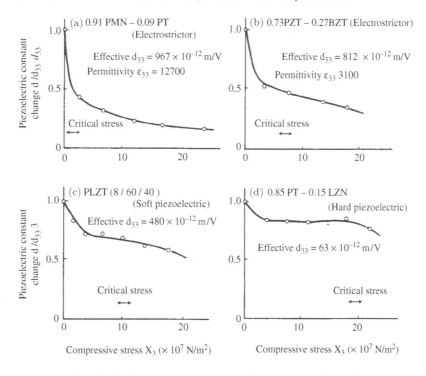

Fig. 4.11: Variation of the large field "effective" piezoelectric strain coefficient, d_{33}, with applied compressive stress for various electrostrictive, soft piezoelectric and hard piezoelectric ceramics.[20]

against the stress (NB: electrostriction has a quadratic relation with the electric field), (2) the elastic compliance of the 0.9PMN-0.1PT electrostrictor decreases abruptly with increasing bias stress, while (3) the piezoelectric 0.65PMN-0.35PT specimens exhibit practically linear/straight at each applied electric field strength. Precisely speaking, note also that (4) with increasing the applied electric field, the stress-strain curve slope decreases, that is, the elastic compliance decreases (i.e., elastically hardening).

The variation of the large field "effective" piezoelectric strain coefficient, d_{33} (which corresponds to the maximum value of $(\partial x/\partial E)$ for the electrostrictive materials), with applied compressive stress for various electrostrictive, soft piezoelectric and hard piezoelectric ceramics is shown in Fig. 4.11.[20] Note that the horizontal stress scale in Fig. 4.11 is 10 MPa, 10 times higher than that in Fig. 4.10. All the samples show a dramatic decrease in the piezoelectric coefficient above a certain critical level of applied stress. This change is reversible later in the stress cycle for electrostrictive materials, but in soft and hard piezoelectric ceramics it is not, and the degraded d_{33} is not recovered even after the compressive stress is removed once the applied stress exceeds the critical stress (marked by the arrow range). This degradation has been related to the reorientation of domains induced by the applied stress.[21] Note that this critical level of stress tends to be highest (200 MPa) for the "hard" piezoelectrics, somewhat less (100 MPa) for soft piezoelectrics and lowest for electrostrictors. We can deduce from these data that for conditions where the maximum stress is less than $100\,\mathrm{MPa} = 10^{8}\,\mathrm{N/m^{2}}$ ($1\,\mathrm{ton/cm^{2}}$), piezoelectric ceramics are useful, while under larger stresses electrostrictive ceramics seem to be more reliable as long as the ceramic can endure such high stress levels (200 MPa).

The so-called "bolt-clamped" Langevin transducers[22] exhibit very intriguing performance improvement with the compressive stress. Figure 4.12 shows the performance change with the compressive stress observed in the bolt-clamped transducer.[23] You can notice that with an increase in the compressive stress, the resonance/antiresonance frequencies shift to a higher frequency range (i.e., elastically stiffening), and the peaks become sharp (i.e., higher mechanical quality factor Q_m). Daneshpajooh *et al.*[24] explained the origin of this performance improvement. Figure 4.13 summarizes the dependence on the compressive

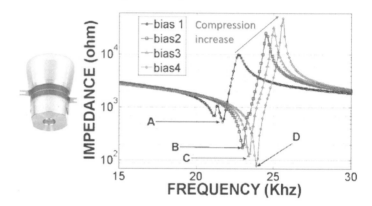

Fig. 4.12: Performance improvement with the compressive stress in a bolt-clamped Langevin transducer.

stress along the polarization direction of (a) intensive dielectric constant ε_{33}^{E} and dielectric loss $\tan\delta_{33}'$; (b) intensive elastic compliance (s_{33}^{E}) and elastic loss ($\tan\phi_{33}'$); (c) piezoelectric constant (d_{33}) and intensive piezoelectric loss ($\tan\theta_{33}'$); and (d) piezoelectric coupling factor (k_{33}) and its corresponding loss. The permittivity shows a monotonic increase under compressive stress (+3.67% increase at maximum stress level 42 MPa), while the dielectric loss decreases slightly with an increase of compressive stress along the polarization direction (−3.6% decrease at maximum stress level 42 MPa). The elastic constant showed a monotonous stiffening behavior (elastic compliance decrease) under compressive stress (+2% increase at maximum stress level 42 MPa). Also, as can be seen in Fig. 4.13, similar to the dielectric loss, the elastic loss decreases remarkably under compressive stress (−12% at maximum stress level). While the piezoelectric constant showed a small increase under compressive stress (1% at highest stress level), the intensive piezoelectric loss showed considerable increase (∼12% at highest stress level). Also, as can be seen in Fig. 4.13(d), the piezoelectric coupling factor (k_{33}) was almost constant over a small compressive stress range, while the electromechanical coupling loss factor ($2\tan\theta_{33}' - \tan\phi_{33}' - \tan\delta_{33}'$) is increasing with an increase of compressive stress.

While the piezoelectric constant and coupling factor of soft PZT do not show a major change, the loss parameters, especially elastic loss, show a considerable improvement. The decrease in elastic loss under compressive stress (which indicates lower heat generation), in conjunction with stable piezoelectric constant over low stress levels, can be beneficial for piezoelectric resonance/high-power applications.

In a brief summary, the compressive (negative) stress along the spontaneous polarization P_S direction exhibits a similar decreasing trend in real physical parameters such as permittivity, elastic compliance (except for piezoelectric d constant), while imaginary elastic and dielectric losses decrease with the

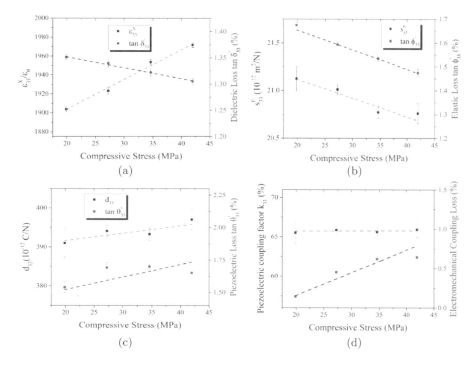

Fig. 4.13: Dependence on the compressive stress along the polarization direction of (a) intensive dielectric constant ε_{33}^E and dielectric loss tan δ'; (b) intensive elastic compliance (s_{33}^E) and elastic loss (tan ϕ_{33}'); (c) piezoelectric constant ($d_3 3$) and intensive piezoelectric loss (tan ϕ_{33}'); (d) piezoelectric coupling factor (k_{33}) and its corresponding k_{33}^2 loss.[24]

compressive stress. This performance enhancement is the "motivation of the bias stress usage" for piezo energy harvesting systems.

4.3.2 *Mechanical Strength*

The mechanical strength of the piezoelectric ceramics used in the actuator and energy harvesting devices is as important as the electrical and electromechanical properties. Mechanical strength limits the mechanical excitation level of the piezo devices, in practice. Because the piezo ceramic is much weaker for the tensile stress than for the compressive stress, we set the mechanical excitation (or vibration) level below the fracture strength (typically less than the blocking stress, ~40 MPa). We now consider some general principles of fracture mechanics, and the mechanical strength of the material as it is affected by field-induced strains.

4.3.2.1 *Fracture mechanics*

Ceramics are generally *brittle*, and fracture of these bodies tends to occur suddenly and catastrophically.[25] Improvement of the *fracture toughness* (which is essentially the material's resistance to the initiation and propagation of cracks within it) is thus a key issue in the design of ceramics for actuator applications. The brittle nature of ceramic materials is directly related to their crystal structure. The atoms in the ceramic crystallites are typically *ionically or covalently bonded* and their displacement under external influences, such as an applied stress, is limited. The mechanisms for relieving stress in such structures are few and, therefore, even a small stress can cause fracture. In polycrystalline samples, there are two types of crack propagation: (1) *transgranular*, in which the cracks pass through grains, and (2) *intergranular*, in which the cracks propagate along grain boundaries.

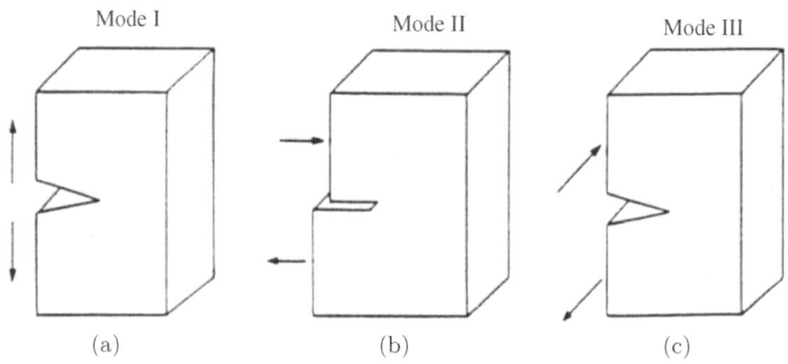

Three fundamental stress modes, which may act on a propagating crack, are depicted in Fig. 4.14. Mode I involves a tensile stress as shown in Fig. 4.14(a). Mode II leads to a sliding displacement of the material on either side of the crack in opposite directions as shown in Fig. 4.14(b). The stress distribution associated with Mode III-type produces a tearing of the medium. The fracture of most brittle ceramics corresponds primarily to Mode I; thus, this is the mode of principle concern to designers of new ceramic materials for actuator and energy harvesting devices under high stress conditions.[26]

Fig. 4.14: Three fundamental stress modes for propagating a crack: (a) Mode I: tensile mode in which a tensile stress acts normal to the crack plane, (b) Mode II: sliding mode in which a shear stress acts normal to the crack edge plane, (c) Mode III: tearing mode in which a shear stress acts parallel to crack edge plane.

Theoretically, the fracture strength of a material can be evaluated in terms of the cohesive strength between its constituent atoms, and is approximately one tenth of Young's modulus, Y. The experimentally determined values do not support this premise, however, as they are typically three orders of magnitude smaller than the theoretically predicted values. This discrepancy has been associated with the presence of microscopic cracks in the material before any stress is applied to it, which are sometimes called *stress raisers*.[27] It has been shown that for a crack with an elliptical shape, oriented with its major axis perpendicular to the applied stress, σ_0 (Mode I), the local stress concentration, σ_m, at the crack tip increases significantly as the ellipticity of the defect increases. The relationship between the applied and local stresses is given by

$$\sigma_m = K_t \sigma_o, \qquad (4.58)$$

where K_t is the stress concentration factor. Fracture is expected to occur when the applied stress level exceeds some critical value, σ_c. The *fracture toughness*, K_{Ic}, is defined as follows:

$$K_{Ic} = F\sigma_c\sqrt{\pi a}, \qquad (4.59)$$

where F is a dimensionless parameter that depends on both the sample and crack geometries, σ_c is the critical stress and a is the microcrack size. The fracture toughness is a fundamental material property that depends on temperature, strain rate and microstructure.

4.3.2.2 *Measurement of fracture toughness*

There are four common methods for measuring the fracture toughness of a ceramic: (1) *Indentation Microfracture* (IM), (2) *Controlled Surface Flow* (CSF), (3) *Chevron Notch* (CN) and (4) *Single Edge Notched Beam* (SENB). We focus on the first, as it is perhaps one of the most commonly employed among these four. The test is initiated by artificially generating cracks on a polished surface of the sample with a Vickers pyramidal diamond indenter. The fracture toughness, K_{Ic}, is then determined from the indentation size and the crack length.

Three types of cracks are produced by indentation: (1) the *Palmqvist crack*, (2) the *median crack* and (3) the *lateral crack* as illustrated in Fig. 4.15. A *Palmqvist crack* is generated at the initial stage of loading. It has the shape of a half ellipse, and occurs around the very shallow region of plastic deformation near the surface. As the load is increased, a crack in the shape of a half circle starts to form at the boundary between the plastic and elastic deformation regions. This is called a *Median crack*. Above a certain critical indenter load, the median crack will reach the surface. The indentation also produces a residual stress

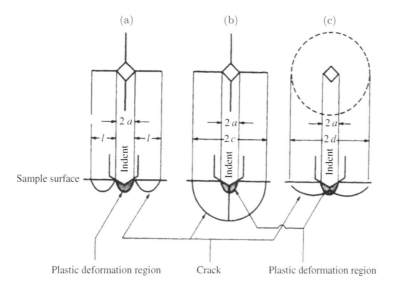

Fig. 4.15: Crack shapes generated by the Vickers indentation: (a) the Palmqvist crack, (b) the median crack and (c) the lateral crack. [Top and sectional views are shown.]

around the indented portion of the sample, in which are generated more Palmqvist and median cracks, as well as *lateral cracks* in the plastic deformation region which will not reach the surface. The Indentation Microfracture (IM) method is used to characterize Palmqvist and/or median cracks in order to determine K_{Ic}. A theoretical equation descriptive of median cracks has been derived, which has the following form:

$$\left[\frac{K_{Ic}\phi}{H\sqrt{a}}\right]\left(\frac{Y\phi}{H}\right)^{2/5} = (0.055)\log\left(\frac{8.4\,a}{c}\right), \tag{4.60}$$

where ϕ is a restriction constant, H is the hardness, Y is Young's modulus, and a and c are half of the indentation diagonal and the crack length, respectively.[28] The theory that produced this equation was further refined and modified expressions,[29] which individually characterize the Palmqvist and median cracks, were derived as follows:

$$\left[\frac{K_{Ic}\phi}{H\sqrt{a}}\right]\left[\frac{H}{Y}\right]^{2/5} = (0.018)\left[\frac{(c-a)}{a}\right]^{-1/2} \text{[Palmqvist crack]}, \tag{4.61}$$

$$\left[\frac{K_{Ic}\,\phi}{H\,\sqrt{a}}\right] = (0.203)\left[\frac{c}{a}\right]^{-3/2} \text{[Median crack]}. \tag{4.62}$$

The fracture toughness, K_{Ic}, can be calculated using the crack length determined from the IM method. Other methods require pre-cracking a sample prior to failure testing, and then K_{Ic} is calculated using the fracture stress determined from the test. The CSF method, for example, uses the median crack generated by a Vickers indenter as a pre-crack to determine the relationship between the fracture stress, σ_f, and the crack length, $2c$, under three- or four-point bending tests. The fracture toughness, K_{Ic}, is then calculated from

$$K_{Ic} = (1.03)\sigma_f\sqrt{\frac{\pi\,b}{q}}. \tag{4.63}$$

The parameter q is defined to be

$$q = \Phi^2 - 0.212\left[\frac{\sigma_f}{\sigma_y}\right]^2, \tag{4.64}$$

where σ_y is the tensile yield stress and Φ is the second perfect elliptic integration given by

$$\Phi = \int_0^{\pi/2} [\cos^2\theta + (b/c)^2 \sin^2\theta]^{1/2} d\theta. \tag{4.65}$$

Here, b is half of the minor axis of the elliptically shaped crack (referred to as the crack depth) and c is half of the major axis of the elliptical crack (referred to as the crack length).

The probability of fracture, $p(\sigma_f)$, for a sample subjected to a stress, σ_f, in a three- or four-point bend test is described by the following equation:

$$1 - p(\sigma_f) = \exp[-(\sigma_f/\sigma_o)^m], \tag{4.66}$$

where σ_0 is the average bending strength and m is referred to as the *Weibull coefficient*. The Weibull stress is a well-known means of predicting the likelihood of *weakest link* brittle fracture. Imagine a necklace chain which is composed of many small gold rings connected/linked to each other. When we pull this chain with a strong force, the chain will be torn off into two pieces. The torn part should be at the mechanically weakest gold ring. Weibull stress is based on its use of a two-parameter "Weibull distribution", a commonly used distribution in probabilistic engineering. The distribution is defined by a shape parameter, the Weibull modulus and a scaling parameter. The parameters σ_0 and m in Eq. (4.66) are the average fracture stress and distribution level of the fracture stress, respectively. When this relationship is rendered in terms of the natural logarithm (see Example Problem 4.1), we should find that the quantity $\ln(\ln[1/(1-p)])$ is linearly related to $\ln(\sigma_f)(\sigma_f$: measured fracture stress). A graph of this function is called a *Weibull plot*.

Example Problem 4.1

"Three-point bend tests" were carried out on a series of barium titanate-based multilayer actuators (laminated along l direction). The sample configuration is illustrated in Fig. 4.16. The results of the tests are summarized in Table 4.1. The samples were collapsed under the load 2.34–4.60 kgf on the rod center (shown by M). Create the "Weibull plot" curve on this set of bending tests, and calculate the average bending strength, σ_0, and the Weibull coefficient, m.

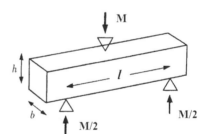

Fig. 4.16: Sample configuration for the three-point bend test.

Solution:

The fracture stress, σ_f, can be obtained from

$$\sigma_f = 3Ml/2bh^2, \tag{P4.1.1}$$

which yields for each of the samples tested the following table:

Table 4.1: Results of the three-point bend test on BT-based ceramic rods.

Sample	Length, l (cm)	Width, b (cm)	Height, h (cm)	Load, M (kg)
1	0.38	0.420	0.106	4.60
2	0.48	0.415	0.121	4.00
3	0.49	0.420	0.093	2.00
4	0.40	0.420	0.112	5.50
5	0.31	0.415	0.102	2.90
6	0.55	0.420	0.099	2.34

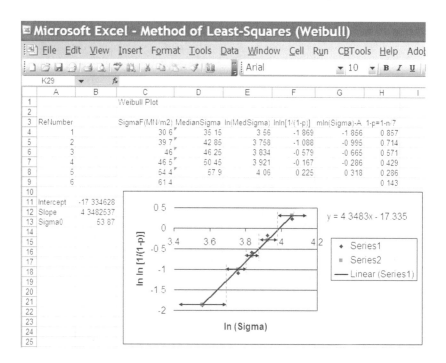

Fig. 4.17: Weibull plot for the data collected from the three-point bend tests conducted on a series of BT-based multilayer actuators.

Sample	1	2	3	4	5	6
σ_f (MN/m^2)	54.4	46.5	39.7	61.4	30.6	46.0

The sample sequence is changed according to the fracture stress (small to large) in Excel Table in Fig. 4.17 Top. The non-fracture probability, $1 - p(\sigma_f)$, of the sample under the applied stress, σ_f, of a three-point bend test is described by

$$1 - p(\sigma_f) = \exp[-(\sigma_f/\sigma_o)^m], \tag{P4.1.2}$$

where $p(\sigma_f)$ is the fracture probability, σ_0 is the average bending strength and m is the Weibull coefficient. This equation can be rewritten in the following form:

$$\ln(\ln[1/(1 - p(\sigma_f))]) = m[\ln(\sigma_0) - \ln(\sigma_f)]. \tag{P4.1.3}$$

Designating the total number of samples as N, the non-fracture probability can be evaluated by

$$[1 - p(\sigma)] = 1 - [n/(N + 1)], \tag{P4.1.4}$$

where $(\sigma_n < \sigma < \sigma_{n+1})$.

When the sample number is small (only 6 in this case), one certain non-fracture probability has a wide range, which is shown in the plot Fig. 4.17: from one fracture strength to the next fracture strength. Thus, in such a case, the intermediate fracture strength (median value, or average) may be used to enhance the calculation accuracy. Using just the obtained fracture stresses calculated directly from the experimental data without considering this wide range increases the calculation error in the Weibull constant and average sigma zero increase. The Weibull plot for this set of fracture tests appears in Fig. 4.17, with the regression line with Excel software. We are able to determine the Weibull coefficient to be $m = 4.3$ and the average bending strength to be $\sigma_0 = 54$ [MN/m^2] from this plot. The Weibull coefficient is a measure of the fracture strength distribution, and directly related with the product quality control: the smaller, the better in the production line!

4.3.2.3 *Electric poling and mechanical strength*

Piezoelectric ceramics require an "electric poling" process, which induces an anisotropy to the mechanical strength.[30] A schematic representation of the micro-indentation and cracks generated in a poled PLZT 2/50/50 (tetragonal) specimen appears in Fig. 4.18. This study confirmed that a crack that is oriented perpendicular to the poling direction propagates much faster than one oriented parallel to it. Figure 4.19 shows the bending test results in terms of the poling direction. Test setting is illustrated in Fig. 4.19(a), and Weibull plots for samples bending perpendicular and parallel to the poling direction in a 3-point bending test are shown in Fig. 4.19(b). These data indicate that cracks propagate more easily in the direction perpendicular to the poling direction.[31] The non-destruction probability $(1 - p(\sigma))$ for the unpoled sample is plotted in between the above two samples. This is attributed to the anisotropic internal stress caused by the strain induced during the poling process. We suppose that

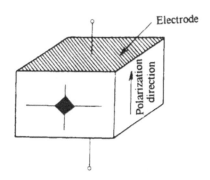

Fig. 4.18: Micro-indentation and the cracks generated in a poled piezo ceramic.

the extension of the piezo ceramic under the poling process still remains after the electric field removal, which generates the tensile stress along the poling or spontaneous polarization direction. This residual tensile stress enhances the crack propagation perpendicular to the P_S direction.

4.3.2.4 *Strain level versus life time*

Acoustic emission (AE) is utilized popularly for monitoring crack propagations. In a piezoelectric material, AE accompanies the reorientation of domains, a phase transformation, in addition to crack propagation in a ceramic. It is in general an inaudible (to human), high-frequency acoustic burst signal caused by mechanical vibration in the specimen.[32–34] Thus, we need to distinguish the two types of AE signals, crack-related and normal domain-reorientation AE, during monitoring.

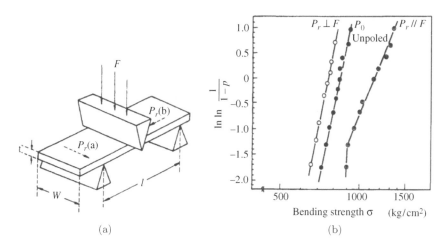

Fig. 4.19: (a) Three-point bend test configuration, and (b) Weibull plots for samples bending perpendicular and parallel to the poling direction.

The AE counts accumulated in a single drive cycle are plotted in Fig. 4.20 as a function of the total number of drive cycles for three different multilayer devices.[35] The field-induced strain generates a large stress concentration near the internal electrode edge, which can initiate a crack. All the samples exhibit a dramatic increase in the AE count while the cracks are propagating, and level off after the crack propagation ceases. The significant differences in the durability/lifetime (measured by the number of drive cycles before failure occurs) among these three samples are attributed mainly to the magnitude of the maximum strain attainable by each device: 0.1% for the electrostrictor, 0.2% for the

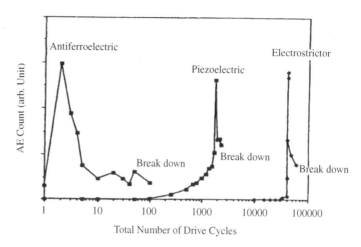

Fig. 4.20: AE counts as a function of total number of drive cycles for electrostrictive, piezoelectric, and phase-change materials.[35]

piezoelectric and 0.4% for the antiferroelectric sample. Assuming the grain–grain adhesion is similar among the samples, it is reasonable to expect that the larger strain will lead to larger stress, therefore a greater likelihood of the fracture.

In summary, mechanical strength of piezoelectric ceramics is up to 50 MPa for tensile stress, and 200 MPa for compressive stress. Piezo ceramics improve the performance under a DC compressive stress condition: 20 MPa is recommended bias compressive stress on the piezo actuator. In order to keep sufficient lifetime, the maximum strain of the PZT-based ceramics should be less than 0.1%.

4.4 Cymbal Transducer Case Study

4.4.1 *Pushing a Curtain & Pushing a Wall*

Wasted or unused mechanical energy (vibration source) should be transferred properly to the energy converter such as piezoelectric devices. Mechanical impedance matching is one of the important factors we have to take into account. The mechanical impedance of the material is defined by $Z = (\rho c)^{1/2}$, where ρ and c are the density and elastic stiffness, respectively, or "effective parameter" values in a composite structure. The receiving part of the mechanical energy in the piezo device should be designed to match the mechanical/acoustic impedance with the vibration source. Otherwise, most of the vibration energy will be reflected at the interface between the vibration source and the harvesting piezoelectric device. Remember a Japanese proverb, *pushing a curtain and pushing a wall* [useless task!], both cases of which will not transfer mechanical energy efficiently (Fig. 4.4). Figures 4.21(a) and 4.21(b) exhibit these two extreme examples we will treat in this article: (a) high-energy harvesting from a "hard" machine, such as an engine, and (b) low-energy harvesting from a "soft" machine, such as human motion. Not only mechanical impedance matching, but also mechanical strength and damping factor (i.e., loss) of the device are important.

The "Cymbal" transducer is a preferable device for the high-power purpose.[36,37] A Cymbal transducer consists of a piezoelectric ceramic disk and a pair of metal endcaps. The metal endcaps play an important role as displacement-direction convertor and displacement amplifier under electric field operation (Fig. 4.21(c)). For the energy harvesting application, the Cymbal structure behaves as a force amplifier and the effective "mechanical impedance transformer/tuner". The Cymbal transducer has a relatively high coupling factor (k_{eff}) and a high effective stiffness (but tunable!), in comparison with unimorphs/bimorphs, which makes it suitable for a high-force mechanical source. To the contrary, the flexible transducer is a preferable device for "soft" application.[38] The MFC is an actuator that offers reasonably high performance and flexibility in a cost-competitive manner (Smart Material Corp.) (Fig. 4.21(d)). The MFC consists of

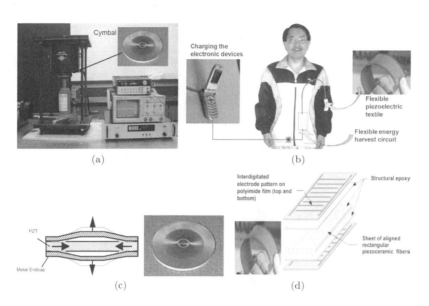

Fig. 4.21: Two extreme vibration source examples: (a) high-energy harvesting from a "hard" machine, such as an engine, and (b) low-energy harvesting from a "soft" machine, such as human motion. (c) Operation principle of the Cymbal and its photo. (d) MFC (Courtesy by Smart Materials Corp.).

rectangular piezo ceramic rods sandwiched between layers of adhesive and electroded polyimide film. This film contains interdigitated electrodes that transfer the applied voltage directly to and from the ribbon-shaped rods. This assembly enables in-plane poling, actuation and sensing in a sealed, durable, ready-to-use package. The MFC composites are useful for harvesting energy in flexible, large bending structures, though the effective electromechanical coupling factor k_{eff} is not large. However, usage with a metal or ceramic elastically stiff structure is not recommended due to the mechanical impedance mismatch. Uchino *et al.* developed Intelligent Clothing (IC, i.e., "wearable energy harvesting system") with a piezoelectric energy harvesting system of flexible piezoelectric textiles[39] aiming at a power source for charging up portable equipment such as cellular phones, health monitoring units or medical drug delivery devices.

4.4.2 *Energy Transfer Measurement for Cymbal's Case*

Energy transfer or reflection rate from the "stiff" electromagnetic shaker to a Cymbal was analyzed by changing the rigidity of the Cymbal endcap (i.e., by changing the endcap thickness, 0.3, 0.4, and 0.5 mm). The characterization system used for measuring the response of the Cymbal under controlled stress environment is shown in Fig. 4.21(a). The mounting assembly was designed to transfer the maximum mechanical energy on to the piezoelectric transducer. The vibration generated using the mechanical shaker resembles that of a car engine and was applied on to the Cymbal without any damping material. A large amplitude shaker (model type 4808, Bruel and Kjaer Instruments Inc., Norcross, GA) has the capability of applying a high-force level up to 112 N in a frequency range of 5 Hz to 10 kHz. The shaker was driven at various voltages and frequencies using the function generator (HP 33120 A, Agilent Technologies, Inc., Santa Clara, CA) and a high-power current amplifier (type 2719, Bruel and Kjaer Instruments Inc.) to produce a cyclic force of required magnitude and frequency. The output signal from the Cymbal was monitored using Tektronix digital oscilloscope (TDS 420 A, Tektronix, Richardson, TX). The output voltage generated from the Cymbal was passed through the rectifier and charged to a capacitor and successively discharged through a resistive load. The Cymbal transducers were tested under a high vibration source with prestressed conditions applied using a hydraulic press (Fred S. Carver Inc.). All experiments were performed on an isolated bench to avoid any interference from the mechanical noise in the surrounding environment.

Figure 4.22 shows the example output voltage waveforms from the Cymbal transducer under a cyclic force depending on prestress condition at 100 Hz frequency. Prestress is the pressure applied by the mass of vibration source to the Cymbal transducer. AC forces of 7.8 N and 40 N at 100 Hz were applied for zero prestress and prestress (67 N) conditions, respectively. Under the zero prestress condition, intermittent spiky high voltage was observed because the vibration source bottom block hit the Cymbal like a bang-bang mode, while under the prestress (larger than the AC amplitude), steady $\pm$ sinusoidal voltage was observed, which seems to be much more stable as the energy harvesting system.

Once we obtain the sinusoidal vibration on the Cymbal device, we can evaluate the transferred mechanical energy into the Cymbal transducer, as follows. Refer to Fig. 4.23. In order to obtain the source energy, we measure initially the free vibration

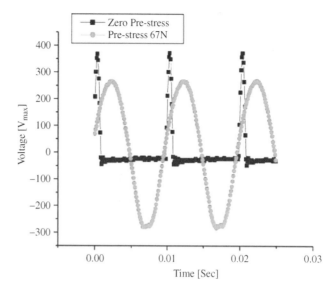

Fig. 4.22: Output voltage of the cymbal transducer under two prestress conditions. A force of 7.8 N and 40 N at 100 Hz was applied for zero prestress and prestress condition, respectively.

amplitude u of the load mass M at the angular frequency ω (here at 100 Hz). Then, the average kinetic energy (per second, i.e., "power") can be evaluated as

$$\int_0^1 \frac{1}{2} M \left(\frac{du}{dt} \right)^2 dt = \frac{1}{4} M \omega^2 u_0^2. \tag{4.67}$$

To obtain the transferred mechanical energy to the Cymbal, we now measure the Cymbal AC displacement w change under prestress condition at the angular frequency ω. Knowing the effective spring constant c^{eff} (i.e., effective elastic stiffness) of the Cymbal transducer by measuring the displacement w and force F beforehand, $F = c^{eff} \cdot w$, we can evaluate the mechanical energy in the Cymbal (per second) from

$$\int_0^1 \frac{1}{2} c^{eff} (w)^2 dt = \frac{\omega}{8\pi} c^{\text{eff}} (w_0)^2. \tag{4.68}$$

4.4.3 *Endcap Thickness Dependence of Mechanical Energy Transmission*

The force generated by the electromagnetic shaker is proportional to the payload mass M and the generated acceleration which is controlled by the applied voltage. The acceleration of the payload was computed by performing the real-time differentiation of measured vibration velocity. The vibration velocity was measured by using Polytec Vibrometer (Tustin, CA). Several payload masses (100–1000 g) were used in the experiment. The Cymbal transducer was bonded on the payload by using a silicone rubber sealer. A bias DC force was applied on the transducer using a hydraulic system (Fred S. Carver INC.) to avoid the separation problem, i.e., bang-bang shock (demonstrated in Fig. 4.22). The mechanical energy transferred to the

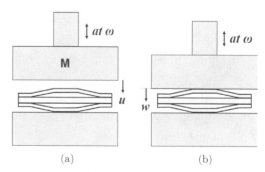

Fig. 4.23: Energy evaluation methods for the vibration source (a) and for the transferred energy to the Cymbal (b).

Cymbal was evaluated from the Cymbal deformation and its effective stiffness.

Figure 4.24 shows mechanical energy flow analysis from the vibration source to the Cymbal transducer, measured on three types of Cymbal transducers: endcap thickness of 0.3 and 0.4 mm with and without bias force under various cyclic vibration levels and drive durations. We used "burst" drive from 1- to 20-second time periods for accumulating the mechanical energy by using payload masses of 120 g (for low stress) or 820 g (for high stress). This limited time measurement was originated from the used capacitor's specification; the rectified voltage was used for charging the capacitor C_{rec} of 10 μF in open condition during the charging time (t), for which the capacitor was charged up to 200 V. Note that the measurement results in Fig. 4.24 include the energy Joule (not the power W, Joule per second) during a certain charging period.

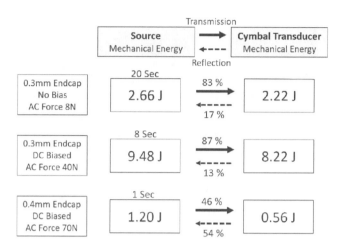

Fig. 4.24: Energy flow on three types of Cymbal transducers: endcap thickness of 0.3 and 0.4 mm with and without bias force under various vibration levels and drive durations.

(1) The energy transmission rate depends significantly on the endcap thickness (0.3 mm 83–87%, while 0.4 mm 46%), because of the effective elastic stiffness difference, that is, 0.4 mm may be too thick/rigid from the optimized *mechanical impedance matching* to the shaker's effective elasticity.

(2) With increasing the bias stress level, the handling mechanical energy level by changing the payload mass increases significantly by suppressing the bang-bang shock.

(3) However, when we increased the AC force level up to 70 N by using a payload mass of 820 g, the sample with 0.3-mm thick endcaps was damaged, because it is too compliant or fragile for this high stress.

Therefore, we need to take a compromised strategy among the followings: (1) keeping the mechanical impedance matching, reducing the applying stress level, or (2) increasing the applying stress level, sacrificing the impedance matching.

The above discussion is supported from the generated electrical energy level in the Cymbal transducers, as shown in Fig. 4.25.[37] A rectification circuit is composed of a full wave rectifier and a capacitor for storing generated electrical energy of the Cymbal transducer in the case of off-resonance (details are described in Chapter 6). Using this circuit, the output power was measured across the resistive load directly without any amplification circuit to characterize the performance of different Cymbal transducers. The maximum rectified voltage V_{rec} of a capacitor C_{rec} (10 μF) was charged up to 248 V after saturation. Figure 4.25 shows the output electrical power from various Cymbal transducers under AC and DC mechanical loads as a function of external load resistance. Endcap thickness made of steel was changed from 0.3 to 0.5 mm. Prestress (DC bias load) to the Cymbal was set constant

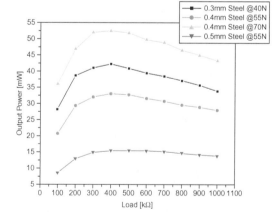

Fig. 4.25: Change in output electrical power from the various Cymbal transducers under different 100 Hz AC mechanical loads (shown as xx N under 66 N constant DC bias) with external electrical load resistance.

at 66 N, and applied AC (100 Hz) force was varied experimentally from 40 to 70 N by changing the load mass. For a small force drive (40 and 55 N), the power level increased with decreasing the endcap thickness (from 0.5, then 0.4, and finally 0.3 mm), which is due to the energy transmission rate from the vibration source via the mechanical impedance matching. With increasing the force level up to 70 N, the maximum power of 53 mW was obtained at electric load of 400 kΩ with a 0.4-mm steel endcap, because the Cymbal sample with a 0.3-mm-thick endcap could not endure under this high-force drive (i.e., the cavity depth is collapsed). The electric impedance matching is described in Chapter 6. Note that the maximum power was obtained due not only to large input mechanical vibration level (via mechanical impedance matching), but also to the effective electromechanical coupling factor k of the Cymbal design with different endcap thicknesses. Just from the energy efficiency viewpoint, the 0.3-mm thick sample is the best because of the best mechanical impedance matching, though it is too fragile for the practical high-power-level applications.

Chapter Essentials

1. Improvement of mechanical-to-mechanical energy transfer between two phases requires the concept of *mechanical impedance matching*.
2. The mechanical/acoustic impedance is expressed by

$$Z = \rho v_p \text{ [including liquid, } \rho\text{: mass density, } v_p\text{: sound phase velocity], or}$$

$$Z = \sqrt{\rho c} \text{ [solid material, c: elastic stiffness]}$$

3. Example of acoustic impedance:
   Water: $Z = 1.5 \times 10^6$ kg/m$^2 \cdot$ s = 1.5 MRayls
   Polymer: 3.8 MRayls
   PZT: 20–24 MRayls
   Steel: 45 MRayls
4. An acoustic piezoelectric medical and underwater transducer is mainly composed of three layers: (1) *matching*, (2) *piezoelectric material* and (3) *backing layers*. One or more matching layers are used to increase sound transmissions from the piezo material to the medium, or vice versa. The backing is added to the rear of the transducer in order to damp the acoustic backwave and to reduce the pulse duration. Piezoelectric materials are used to generate and detect ultrasound (20–40 kHz for SONAR, 2–4 MHz for medical diagnosis).
5. Compressive DC bias stress effect on piezo ceramic materials:

   (1) Maximize the output electric (or electrical) work under a certain input mechanical (or electric) energy.
   (2) Minimize the fracture probability under tensile stress.
   (3) Increase the elastic stiffness, and mechanical quality factor Q_m (or elastic loss decrease).
   (4) Optimized stress (~20 MPa) is half of the blocking stress/force (~40 MPa).

6. Mechanical fracture strength of PZT ceramics: tensile stress 50–100 MPa; compressive stress ~200 MPa
   Bending strength is lower when the force line is perpendicular to the spontaneous polarization direction.
7. When the strain level exceeds 0.1% in piezo ceramics, the lifetime decreases significantly.

Check Point

1. (T/F) Mechanical impedance in a solid material is given by $Z = \sqrt{\rho s}$ (solid material, s: elastic compliance). True or False?
2. Knowing $\rho = 7750$ kg/m^3, $s_{33}^E = 18.8 \times 10^{-12}$ m^2/N for "Soft" PZT 5A, calculate its acoustic impedance.

3. (T/F) When we consider a mechanical wave transmission from one phase to another, as long as the incident wave is normal to the boundary interface, 100% energy transmittance is expected. True or False?

4. What is the unit of mechanical impedance Z, which is equivalent to the MKS unit of $(kg/m^2 \cdot s)$?

5. Acoustic impedance definition has two ways in solid materials, $Z = \rho v_p$ and $Z = \sqrt{\rho c}$. Derive the sound velocity formula v_p in terms of density ρ and elastic stiffness c.

6. Fill in the blank: Acoustic piezoelectric medical and underwater transducer is mainly composed of three layers: (1) _____, (2) piezoelectric material and (3) backing layers.

7. (T/F) Ceramic materials are mechanically weaker for the compression than for the tension. True or False?

8. Provide the mechanical fracture strength of PZT ceramics for the tensile stress. 1 MPa, 10 MPa, 100 MPa, or 1 GPa?

9. (T/F) We found one piezo ceramic specimen which exhibits 0.4% strain under electric field application. We can say that this specimen is much superior to the piezo actuator applications. True or False?

10. (T/F) In the poled PZT ceramics, cracks propagate more easily in the direction parallel to the poling direction. True or False?

Chapter Problems

4.1 Right-hand-side figure shows a piezo energy harvesting system with an elastically soft (polyimide-base) piezoelectric composite (MFC) bonded on the steel beam structure, a part of machinery vibrating at 50 Hz. Discuss why this system is not recommended in terms of the "mechanical impedance matching" step by step (Figure P4.1).

(1) Calculate the acoustic impedance of steel from the data: $\rho = 8{,}000$ kg/m^3, $Y(\sim c_{11}) = 200$ GPa.

(2) Calculate the acoustic impedance of polyimide from the data: $\rho = 1{,}450$ kg/m^3, Y $(\sim c_{11}) = 2.5$ GPa.

(3) Supposing the SH wave (transversal wave) from the steel, and using the transmittance ratio equation for the normal elastic wave, $\frac{v_{t0}}{v_{i0}} = \frac{2}{(1+\sqrt{\frac{\rho_2 c_{66,2}}{\rho_1 c_{66,1}}})}$. calculate the transmittance ratio taking into account the acoustic impedances in (1) and (2). You can find very small mechanical energy is transmitted into the piezoelectric component due to large acoustic impedance mismatch.

Fig. P4.1: Piezo energy harvesting system.

4.2 A *monolithic hinge lever mechanism* is applied to amplify small displacement generated in a piezoelectric multilayer (ML) actuator. Figure P4.2 illustrates a structure of the hinge lever used in a dot-matrix printer in the 1980s. Discuss the design principle of the mechanism in terms of the concept of the mechanical impedance matching.

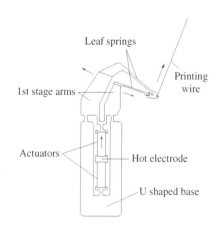

Fig. P4.2: Hinge lever mechanism with multilayer actuators.

Hint:

The displacement 20 μm generated in a ML with the blocking force 200 N (with the elastic stiffness constant $200/20 \times 10^{-6}$ (N/m)) is amplified to 500 μm (amplification $\times$ 25) at the tip of the printing wire, by sacrificing the force level a little less than 8 N. The first stage arms generate $\times$ 5 displacement via the geometric distance ratio, and the push–pull leaf springs provide additional $\times$ 5 amplification. Thus, the effective elastic stiffness constant can be estimated by $8/500 \times 10^{-6}$ [N/m]. Using this mechanical amplifier (or "mechanical transformer") with amplification ratio 25, we can generate the effective elastic compliance by a factor of $(25)^2$. Supposing that the mass density of the materials is the same order, we can expect the effective *mechanical impedance reduction* by a factor of 25 in order to transfer mechanical energy of the printing wire on to the soft paper. The "mechanical transformer" is recognized as a "mechanical impedance converter", analogous to the relation of an "electromagnetic transformer" with an "impedance/admittance converter.

References

1. M. Onoe (ed.), *Fundamentals of Solid State Vibrators for Electric & Electronic Engineers*, Ohm Publ. Company, Tokyo (1982).
2. L. E. Kinsler, A. R. Frey, A. B. Coppens and J. V. Sanders, *Fundamentals of Acoustics*, John Wiley & Sons, New York (1982).
3. R. E. Newnham, D. P. Skinner and L. E. Cross, *Mater. Res. Bull.*, 13, 525 (1978).
4. K. A. Klicker, J. V. Biggers and R. E. Newnham, *J. Amer. Ceram. Soc.*, 64, 5 (1981).
5. http://www.matsysinc.com/, U.S. Patent 5,340,510.
6. W. A. Smith, *Proc. IEEE Ultrasonic Symp. '89* (1989), p. 755.
7. L. J. Nelson, *Mater. Sci. Technol.*, 18, 1245 (2002).
8. E. Hausler, L. Stein, and G. Harbauer, *Ferroelectrics*, 60, 277 (1984).
9. J. Kymissis, C. Kendall, J. Paradiso and N. Gershenfeld, *Digest of Papers, Second IEEE International Symposium on Wearable Computers*, 132 (1998).
10. H. A. Sodano, G. Park, D. J. Leo, D. J. Inman, *Proc. of SPIE*, 5050, 101 (2003).
11. K. Uchino, *Proceedings of the 9th Japan Int'l SAMPE Symposium*, Tokyo, November 18–21, 2005, pp. 11–14.
12. B. A. Auld, *Acoustic Fields and Waves in Solids*, 2nd edn., Melbourne: Robert E. Krieger (1990).
13. G. S. Kino, *Acoustic Waves: Device Imaging and Analog Signal Processing*, Englewood Cliffs, N. J.: Prentice-Hall (1987).
14. C. S. Desilets, J. D. Fraser and G. S. Kino, *IEEE Trans. Sonics Ultrason.*, SU-25, 115 (1978).
15. S. Saitoh, T. Takeuchi, T Kobayashi, K. Harada, S. Shimanuki and Y. Yamashita, *Japan. J. Appl. Phys.*, 38(5B), 3380 (1999).
16. Y. Sugawara, K. Onitsuka, S. Yoshikawa, Q. C. Xu, R. E. Newnham and K. Uchino, *J. Amer. Ceram. Soc.*, 75(4), 996 (1992).
17. J. Zhang, Mater. Sci. Eng., Ph.D. Thesis, The Pennsylvania State University (2000).
18. K. Uchino, *Piezoelectric/Electrostrictive Actuators*, Morikita Pub. Co., Tokyo (1985).
19. Y. Nakajima, T. Hayashi, I. Hayashi and K. Uchino, *Jpn. J. Appl. Phys.*, 24(2), 235 (1985).
20. S. Nomura, O. Osawa, K. Uchino and I. Hayashi, *Abstract Jpn. Appl. Phys.*, Spring, 764 (1982).
21. H. Cao and A. G. Evans, *J. Amer. Ceram. Soc.*, 76, 890 (1993).
22. E. Mori, S. Ueha, Y. Tsuda, *Proc. Ultrasonics Int.*, 83, 154 (1983).
23. E. Moreno *et al.*, *2005 IEEE 2nd International Conference* (2005).
24. H. Daneshpajooh, M. Choi, Y. Park, T. Scholehwar, E. Hennig, and K. Uchino, *Rev. Sci. Instrum.*, 90, 075001 (2019), https://doi.org/10.1063/1.5096905.
25. W. D. Callister Jr., *Materials Science and Engineering*, Wiley (1984) p. 189.
26. T. Nishida and E. Yasuda, *Evaluation of Mechanical Characteristics in Ceramics*, Nikkan-Kogyo (1986).
27. A. A. Griffith, *Phil. Trans. Roy. Soc. (London)*, A221, 163 (1920).
28. A. G. Evans and E. A. Charles, *J. Amer. Ceram. Soc.*, 59, 371 (1976).
29. K. Niihara, R. Morena and D. P. H. Hasselman, *Commun. Amer. Ceram. Soc.*, C-116 (1982).
30. T. Yamamoto, H. Igarashi and K. Okazaki, *Ferroelectrics*, 50, 273 (1983).

31. H. Wang and R. N. Singh, *Ferroelectrics*, 168, 281 (1995).

32. H. Aburatani, S. Harada, K. Uchino, A. Furuta and Y. Fuda, *Jpn. J. Appl. Phys.*, 33, 3091 (1994).

33. H. Aburatani and K. Uchino, *Jpn. J. Appl. Phys.*, 35, L516 (1996).

34. H. Aburatani, J. P. Witham and K. Uchino, *Jpn. J. Appl. Phys.*, 37, 602 (1998).

35. H. Aburatani, S. Harada, K. Uchino, A. Furuta and Y. Fuda, *Japan. J. Appl. Phys.*, 33(5B), 3091 (1994).

36. H.-W. Kim, A. Batra, S. Priya, K. Uchino, D. Markley, R. E. Newnham and H. F. Hofmann, *Japan. J. Appl. Phys.*, 43(9A) 6178–6183 (2004).

37. H. W. Kim, S. Priya, K. Uchino and R. E. Newnham, *J. Electroceram.*, 15, 27 (2005).

38. http://www.smart-material.com/Smart-choice.php?from=MFC.

39. K. Uchino, *Proceedings of the 9th International Conference New Actuators*, A1.0, Bremen, Germany, June 14–16, 2004, pp. 38–48.

Chapter 5

Mechanical-to-Electrical Energy Transduction

This chapter describes the reason why the author has frustration on the following research trends:

> "Though the electromechanical coupling factor k is the smallest (i.e., the energy conversion rate from the input mechanical to electric energy is the lowest) among various device configurations, the majority of researchers primarily use the 'unimorph' design. Why?"

The mechanical to electrical energy transduction/conversion rate is dependent on three factors: (1) piezoelectric material's physical properties, (2) design/size and vibration mode of the piezo ceramic specimen and (3) piezo device design in a composite structure.

5.1 Figure of Merit (FOM) for Piezo Energy Harvesting

5.1.1 *FOM for Stress Input*

Mechanical energy transferred to the piezo transducer, such as a Cymbal and a flexible transducer, will be converted into electrical energy through the piezoelectric effect. Voltage induced in the transducer under a certain force/stress input can be explained by Eqs. (5.1) and (5.2) for a bulk piezoelectric:

$$V = g \times \frac{F(N) \times t(m)}{A(m^2)}, \tag{5.1}$$

$$g = d/\varepsilon_0 \varepsilon_r. \tag{5.2}$$

Here, g, F, t and A are *piezoelectric voltage constant*, applied force normal to the piezo material area, thickness and area of the piezo material, respectively. Since piezoelectric voltage constant g can be expressed by $g = d/\varepsilon_0 \varepsilon_r$, the output electric power (J/s) can be calculated (under off-resonance mode) as follows:

$$\begin{aligned} P &= \frac{1}{2}CV^2 \cdot f \\ &= \frac{1}{2} \cdot g_{33} \cdot d_{33} \cdot F^2 \cdot \frac{t}{A} \cdot f, \end{aligned} \tag{5.3}$$

$C \left(= \varepsilon_0 \varepsilon_r \frac{A}{t}\right)$ and f are capacitance of the transducer and frequency of the vibration, respectively. The suffixes "33" are used for a k_{33}-type rod specimen, for example. Hence, output power can be evaluated by the product of g and d as *Figure of Merit* of the transducer under a force constant condition:

$$P \propto g \cdot d = d^2/\varepsilon_0 \varepsilon_r \tag{5.4}$$

Note that when the piezo component is a composite or a hybrid structure, the above constants g and d should be replaced by the "*effective*" values. It is well known, for example, that the effective d coefficient for a cymbal (refer to Figs. 2.15 and 5.34] is expressed by[1]

$$d_{\text{eff}} = d_{33} + \alpha|d_{31}|, \tag{5.5}$$

where the amplification factor α (refer to Section 5.4.3.2.3 for the details) is given by

$$\alpha \propto (1/2) \text{ [Cavity diameter/Cavity depth]}. \tag{5.6}$$

5.1.2 *FOM for Mechanical Energy Input*

The same input force/stress does not mean the same input mechanical power. The "unit" input stress on a piezoelectric material, with piezo material's elastic compliance s^E, provides the mechanical energy input of

$$U_M = \frac{1}{2} s^E X^2 \text{ (per unit volume)}. \tag{5.7}$$

The higher the elastic compliance is, the higher the input mechanical energy is! Hence, in comparison with Eq. (5.4), the converted electric power can be evaluated by the *electromechanical coupling factor* k, which is the *Figure of Merit (FOM)* of the transducer under unit mechanical energy input:

$$U_{ME} = \frac{d^2}{s^E \varepsilon_0 \varepsilon_r} U_M = k^2 U_M. \tag{5.8}$$

If we rephrase the expression of the electromechanical coupling factor as

$$k^2 = \left(\frac{1}{s^E}\right) \cdot (d/\varepsilon_0 \varepsilon_r) \cdot (d),$$

the meaning of the FOM is much clearer: the second term $\left(\frac{d}{\varepsilon_0 \varepsilon_r}\right)$ is the output electric field E (or voltage V) under a certain stress X, equivalent to the piezoelectric g constant, the last term (d) is the output charge P (or current I, differentiated by time) under a certain stress X, and the first term $(\frac{1}{s^E})$ is to convert the "unit" from the input stress X to the input mechanical energy. Some of the recently submitted or published papers, entitled "Piezoelectric Energy Harvesting", include only the voltage generated from the input force (without measuring the current and power), which is a serious misconception of the authors: voltage monitoring verifies merely for "sensor" applications, not for "energy harvesting" devices. Because the electromechanical coupling factor k of a composite piezo device depends not only on the materials constant itself, but also on the specimen geometry, we discuss the k derivation process in detail in Sections 5.3 and 5.4, in particular, for bimorph and cymbal composite structures.

Mechanical input energy sometimes has a form of either "high force & low displacement" such as from a building, bridge and machinery, or "low force & high displacement" such as from environmental wind, water flow, tidal and human motion. Even under the same mechanical input energy, as learned in Chapter 4, harvesting energy is significantly different, depending on the mechanical/acoustic impedance matching of the input mechanical source and the transducer device. The concept of "mechanical transformer" introduced in Chapter 4 can also be understood as the mechanical impedance converter. The FOM of $g \cdot d$ is only useful, when we compare just the performance of piezo materials by keeping the device structure almost the same (i.e., keeping the effective elastic compliance constant).

5.2 Piezoelectric Materials Selection

5.2.1 *Piezoelectric Ceramics*

Popularly used piezoelectric materials are perovskite-type ceramics. Lead zirconate titanate ($Pb(Zr_x, Ti_{1-x})O_3$, PZT) based ceramics have dominated in the market in these 60 years. In particular, the morphotropic phase boundary (MPB) composition (52PZ-48PT) between the rhombohedral and tetragonal phases is the base, because the MPB composition exhibits the maximum piezoelectric performances, as explained in Section 2.3.2. We summarized electromechanical properties of three commercially available piezoelectric PZT-based ceramics in Table 5.1.[1] Hard APC 841, Soft APC 850 (American Piezo Ceramics, Mackeyville, PA), and High g D210 (Dong Il Technology, Korea). Among these three, we can conclude that D210 is the best material for the energy harvesting application by comparing $g \cdot d$ product values, primarily due to the smallest permittivity in this material. Note, however, that there is no guarantee for the material with the maximum $g \cdot d$ exhibits the largest electromechanical coupling factor k (depending on the elastic compliance value).

5.2.2 *Piezoelectric Single Crystals*

How high *electromechanical coupling factor* k can we practically find in the piezoelectric materials? 100% may be impossible, but our research group firstly reported an enormously large electromechanical coupling factor $k_{33} = 92$–95% and piezoelectric constant $d_{33} = 1500\,pC/N$ in solid solution single crystals between relaxor and normal ferroelectrics, $Pb(Zn_{1/3}Nb_{2/3})O_3$–$PbTiO_3$.[2,3] Figure 5.1(a) shows the piezoelectric

Table 5.1: Electromechanical properties of the commercial PZT ceramics.

Parameter	Hard (APC 841)	Soft (APC 850)	High g (D210)
ε_r	1350	1750	681
$k_p(\%)$	60	63	58
$d_{31}(10^{-12}\,\mathrm{C/N})$	109	175	120
$g_{31}(10^{-3}\,\mathrm{Vm/N})$	10.5	12.4	20
Q_m	1400	80	89.7
T_c (deg.)	320	360	340
$g_{31} \cdot d_{31}$	1.14×10^{-12}	2.17×10^{-12}	2.40×10^{-12}

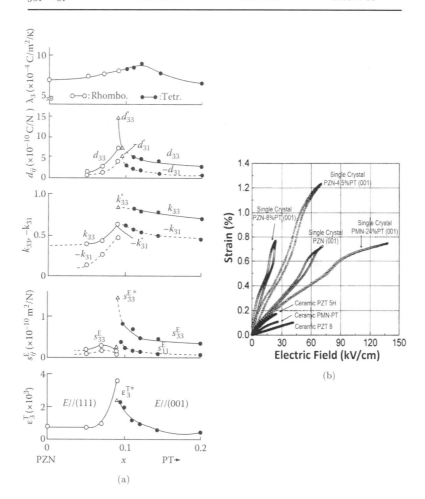

Fig. 5.1: (a) Piezoelectric properties in the $Pb(Zn_{1/3}Nb_{2/3})O_3$–$PbTiO_3$ system,[2] and (b) the electric field induced strain curve.[5]

properties in the PZN-PT system, which has the morphotropic phase boundary (MPB) around 9 mol% of PT between the rhombohedral and tetragonal phases. The MPB composition exhibited the giant piezoelectric d_{33} constant and electromechanical coupling factor k_{33}^{eff} only when the sample was prepared along the perovskite [001] direction. These data have been reconfirmed by Yamashita *et al.*, and Park *et al.*, and improved data were obtained recently, aiming at medical acoustic applications.[4,5] The strains as large as 1.7% can be induced practically for the PZN-PT solid solution single crystals, as shown in Fig. 5.1(b).[5] It is notable that the highest val-

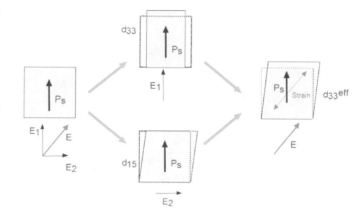

Fig. 5.2: Basic concept of the piezoelectric enhancement in a provskite ferroelectric. Note the fact $d_{15} \gg d_{33} > |d_{31}|$.

ues are observed for a rhombohedral composition only when the single crystal is poled along the perovskite [001] axis, not along the [111] spontaneous polarization axis. An intuitive explanation for this enhancement of piezoelectricity is schematically shown in Fig. 5.2. Supposing that an electric field is applied on a single crystal with a cant angle from the spontaneous polarization direction, the field component parallel to the P_S generates the extension and contraction strains via d_{33} and d_{31}, and the component perpendicular to the P_S generates the shear strain via d_{15}. Because d_{15} is much larger than d_{33} or d_{31} in perovskite ferroelectrics, this large shear strain exhibits a drastic enhancement in the effective strain magnitude, as illustrated in Fig. 5.2.

Another epoch-making paper was published in 1998; a series of theoretical calculations made on perovskite-type ferroelectric crystals suggested that large d and k values in similar magnitudes to PZN-PT can also be expected in PZT. Crystal orientation dependence of piezoelectric properties was phenomenologically calculated for compositions around the morphotropic phase boundary of PZT (see Fig. 5.3).[6] The maximum longitudinal piezoelectric constant d_{33}^{eff} (4–5 times enhancement) and electro-mechanical coupling factor k_{33}^{eff} (more than 90%) in the rhombohedral composition were found to be at around 57° angle, canted from the spontaneous polarization direction [111], which corresponds roughly to the perovskite [100] axis. Figure 5.4 illustrates the domain configuration difference between the [111] oriented film (mono-domain-like with P_S along the 1 axis) and the [100] oriented film (uniformly distributed domain pattern with P_S along the 1, 2, 3 and 4 axes). This is a trigger report on *domain engineering*.[7] This paper also predicted the superior design of the PZT epitaxially grown thick/thin films (in addition to single crystal forms); that is, the rhombohedral composition film of the [100] orientation shows lower E_C, sharper polarization reversal and more linear induction of strain, in comparison with the film of the [111] orientation, though the remnant polarization is lower by a factor of $1/\sqrt{3}$ (Fig. 5.5).

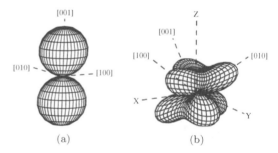

Fig. 5.3: Crystal orientation dependence of the effective piezoelectric constant d_{33}^{eff} in PZT single crystal-like samples. (a) PZT 40/60 tetragonal, and (b) PZT 60/40 rhombohedral composition.[6]

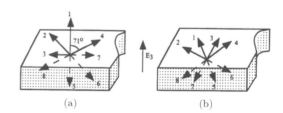

Fig. 5.4: Domain configuration models for rhombohedral PZT films with (a) [111] plate and (b) [100] plate.

Figure 5.6 shows the comparison between the theoretical and experimental results for the piezoelectric d_{31} and e_{31} constants in epitaxially grown PZT thin films.[8] Note that the maximum e_{31} constant can be obtained in the rhombohedral phase near the morphotropic phase boundary, and in the [100] specimen (rather than in the [111] specimen). Damjanovic *et al.* reported similar results.[9] Note also that even in polycrystalline poled specimen PZTs the piezoelectric constant enhancement is observed by changing cutting angle, depending on the composition around the MPB region.[10]

5.2.3 *Piezoelectric Polymers and Composites*

Flexible polymer piezoelectric materials are utilized for harvesting energy from flexible and large vibration amplitude source, such as wearable harvesting devices and medical body-

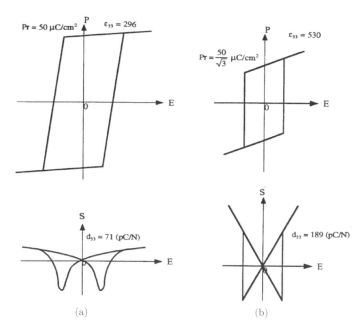

Fig. 5.5: Polarization and strain curves predicted for rhombohedral PZT films/single crystals with (a) [111] plate and (b) [100] plate forms.[7]

embeddable devices. One of the popular materials is polyvinylidene difluoride (PVDF or PVF_2). As seen in Table 5.2, because pure PVDF does not exhibit a large FOM, $g \cdot d$ product, PZT: polymer composites seem to be better alternative materials.[11] Table 5.2 summarizes Electromechanical properties of PZT: polymer composites (3-1 connectivity with PZT rod, and 3-0 connectivity with PZT powder) in comparison with pure PZT ceramic and PVDF. It is noteworthy that PZT 501A and PZT: PVDF 3-0 show almost the same FOM ($g \cdot d$), while PZT: Epoxy 3-1 exhibits a significantly large FOM. This fact suggests that the composite geometrical/structural designing is important to enhance the piezoelectric energy harvesting capability.

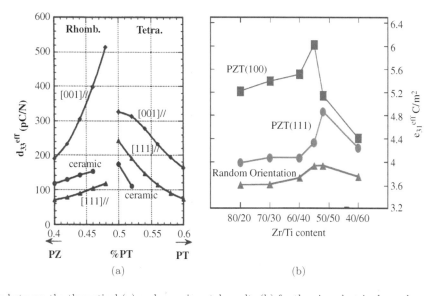

Fig. 5.6: Comparison between the theoretical (a) and experimental results (b) for the piezoelectric d_{31} and e_{31} constants in epitaxially grown PZT thin films.[8]

Table 5.2: Electromechanical property comparisons among PZT, PZT: polymer composities and PVDF.

Parameter	PZT (501A)	PZT:Epoxy (3-1)	PZT:PVDF 3-0	Extended PVDF
$\rho(10^3\text{kg/m}^3)$	7.9	3.0	5.5	1.8
c_{33} (GPa)	81	19	2.6	3
ε_{tau}	2000	400	120	13
$d_{33}(10^{-12}\,\text{C/N})$	400	300	90	20
$g_{33}(10^{-3}\,\text{Vm/N})$	20	75	85	160
$g_{33}\cdot d_{33}$	8×10^{-12}	23×10^{-12}	7.7×10^{-12}	3.2×10^{-12}
Z (Mrayl)	25	7.5	3.8	2.3

The reader is also requested to notice that the mechanical impedance $Z=\sqrt{\rho c}$ [ρ: mass density, c: elastic stiffness] can be significantly tuned in composite structures. Since bimorphs and cymbals are also a sort of composite design with two material phases, we consider the design optimization principle in Section 5.4.

5.3 Operation Method Consideration

We considered and analyzed mostly the sinusoidal stress application on the piezoelectric energy harvesting devices so far. However, since a majority of the present best-selling devices are based on the impulse stress application type, such as programmable air-burst munition (PABM for 25-mm Φ caliber) and Lightning-Switch (remote control relay switch), we consider the electric energy process on a piezoelectric specimen under an impulse stress application theoretically in this section first. Then, we discuss the sinusoidal stress application with a similar mathematical tool to enhance deeper understanding.

5.3.1 *Pulse Drive*

When an impulse force is applied to a piezoelectric actuator, an electric voltage/current vibration is excited via a mechanical vibration, the characteristics of which depend on the pulse profile. Voltage overshoot and ringing are frequently observed in piezoelectric "igniters", even if we start from a compressive stress input. Because the pulse drive may lead to the destruction of the piezo actuator due to large tensile stress and high voltage associated with the vibration overshoot, we need to examine more precisely the transient response of a piezoelectric device driven by a pulse stress in this section. Since we need the basics of the piezoelectric constitutive equations and dynamic vibration equations in a continuum medium, reviewing Section 2.7 is required prior to studying this section. The reader is reminded that an "equivalent circuit" (EC) cannot be adopted in the transient response analysis under impulse drive.

5.3.1.1 *Longitudinal vibration mode via transverse piezoelectric effect — k_{31} mode*

We consider a longitudinal mechanical vibration in a simple piezoelectric ceramic plate via the transverse piezoelectric effect d_{31} with thickness b, width w and length $L(b\ll w\ll L)$, pictured in Fig. 5.7. When the polarization is in the z direction and the $x-y$ planes are the planes of the electrodes, the extensional vibration along the x direction is represented by the following dynamic equation:[11]

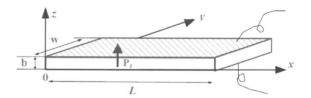

Fig. 5.7: Longitudinal vibration k_{31} mode of a rectangular piezoelectric plate.

$$\rho\frac{\partial^2 u}{\partial t^2}=F=\frac{\partial X_{11}}{\partial x}+\frac{\partial X_{12}}{\partial y}+\frac{\partial X_{13}}{\partial z} \tag{5.9}$$

where u is the displacement in the x direction of a small volume element $(wb \cdot dx)$ in the piezo ceramic plate, ρ is mass density of the piezoelectric material, X_{ij}'s are stresses (only the force along the x-direction is our target, because the stress along y and z directions can be neglected owing to this specimen geometry). Based on the assumptions, (1) poly-crystalline specimen (limited number of the tensor components) and (2) only z-direction components of electric field, the general piezoelectric constitutive equations

$$x_i = s_{ij}^E X_j + d_{mi} E_m \tag{5.10a}$$

$$D_m = d_{mi} X_i + \varepsilon_0 \varepsilon_{mk}^X E_k \quad (i, j = 1, 2, \ldots, 6; \; m, k = 1, 2, 3). \tag{5.10b}$$

can be transformed as follows:

$$x_1 = s_{11}^E X_1 + s_{12}^E X_2 + s_{13}^E X_3 + d_{31} E_z$$

$$x_2 = s_{12}^E X_1 + s_{11}^E X_2 + s_{13}^E X_3 + d_{31} E_z,$$

$$x_3 = s_{13}^E X_1 + s_{13}^E X_2 + s_{33}^E X_3 + d_{33} E_z,$$

$$x_4 = s_{44}^E X_4,$$

$$x_5 = s_{44}^E X_5,$$

$$x_6 = 2(s_{11}^E - s_{12}^E) X_6, \tag{5.11a}$$

and

$$D_1 = d_{15} X_5,$$

$$D_2 = d_{15} X_4,$$

$$D_3 = d_{31} X_1 + d_{31} X_2 + d_{33} X_3 + \varepsilon_0 \varepsilon_{33}^X E_z. \tag{5.11b}$$

If we summarize again the assumptions

(1) Only E_z exists, because $E_x = E_y = 0$ due to the electrodes on the top and bottom, and
(2) Only X_1 exists, because X_2 and X_3 may be set equal to zero through the plate because the plate is very long with thin thickness and width under low to the fundamental resonance frequency range, only the following two equations are essential to solve the dynamic equation around the resonance/antiresonance frequencies:

$$x_1 = s_{11}^E X_1 + d_{31} E_z, \tag{5.12}$$

$$D_3 = d_{31} X_1 + \varepsilon_0 \varepsilon_{33}^X E_z. \tag{5.13}$$

Let us refresh the basic idea on the mechanical resonance that when an AC stress is applied along the length 1-axis direction to this piezoelectric plate, length, width and thickness extensional resonance vibrations are excited. If we consider a typical PZT plate with dimensions $100\,\text{mm} \times 10\,\text{mm} \times 1\,\text{mm}$, these resonance frequencies correspond roughly to $10\,\text{kHz}$, $100\,\text{kHz}$ and $1\,\text{MHz}$, respectively. Recall the resonance frequency is given by $v/2L$ (v: sound velocity). We consider here the fundamental mode for this configuration, the length extensional mode. When the frequency of the applied stress is well below $10\,\text{kHz}$, the induced displacement follows the AC stress cycle, and the displacement magnitude is given by $s_{11}^E X_1 L$ (when the top and bottom electrodes are "short-circuited"). As we approach the fundamental resonance frequency, a delay in the length displacement with respect to the applied stress begins to develop, and the amplitude of the displacement becomes enhanced by the factor of Q_m (mechanical quality factor) [precisely speaking, $\frac{8}{\pi^2} Q_m$]. Though the displacements along 2- and 3-axes are also developed via s_{12}^E and s_{13}^E, since the length resonance frequency is much lower than the width or thickness resonance frequency, we neglect the displacements

x_2 and x_3. At frequencies above $10\,\mathrm{kHz}$, the length displacement no longer follows the applied field and the amplitude of the displacement is significantly reduced. This status is called "longitudinally clamped" condition, which provides the "damped" capacitance. As described in the following sections, the mechanical resonance is slightly higher when the top and bottom electrodes are "open-circuited" (antiresonance state), than the above short-circuit condition (piezo-resonance state).

Transforming Eq. (5.12) to $X_1 = x_1/s_{11}^E - (d_{31}/s_{11}^E)E_z$, following relation comes out: $\frac{\partial X_1}{\partial x} = \frac{1}{s_{11}^E}\frac{\partial x_1}{\partial x} - \frac{d_{31}}{s_{11}^E}\frac{\partial E_z}{\partial x}$.

When the top and bottom surfaces have electrodes (the k_{31} configuration), by keeping the potential/voltage constant along the x direction, that is, $\frac{\partial E_z}{\partial x} = 0$, we obtain a simple relation: $\frac{\partial X_1}{\partial x} = \frac{1}{s_{11}^E}\frac{\partial x_1}{\partial x}$. Now, when a very long, thin plate is driven by the length direction force F_1 at a frequency lower than the fundamental resonance, X_2 and X_3 may be neglected throughout the plate. Since shear stress will not be generated by the applied stress X_1, and the strain definition, $x_1 = \partial u/\partial x$, the dynamic equation (5.9) is reduced to

$$\rho\frac{\partial^2 u}{\partial t^2} = \frac{1}{s_{11}^E}\frac{\partial^2 u}{\partial x^2} \tag{5.14}$$

5.3.1.2   *General solution for longitudinal vibration k_{31} mode*

We solve Eq. (5.14) for a piezo-plate shown in Fig. 5.7, using the "*Laplace transform*". The Laplace transform is a useful mathematical tool, generally employed for treating the *transient response* to a pulse input. On the contrary, the *Fourier transform* is preferred for cases where a continuous sinusoidal input is applied, such as for resonance-type actuators, which are handled in Section 5.4.2.

• *Laplace Transform Definition*

We consider a function $u(t)$ (1D x-direction displacement of a small volume element $(S \cdot dx)$ with area S) which is defined for $t \geq 0$ ($u(t) = 0$ for $t < 0$), and satisfies $|u(t)| \leq ke^{\delta t}$ for all δ not less than a certain positive real number δ_o. When these conditions are satisfied, $e^{-st}u(t)$ is absolutely integrable for $\mathrm{Re}(s) \geq \delta_o$. We define the Laplace transform as

$$U(s) = L[u(t)] = \int_0^\infty e^{-st}u(t)dt \tag{5.15}$$

The Laplace formula transforms the time-dependent function $u(t)$ to the s (occasionally frequency $j\omega$ as in Fourier Transform) domain function $U(s)$. The inverse Laplace transform is represented as $L^{-1}[U(s)]$. The simplest example for a step function $u(t)[u(t) = 0(t < 0); 1(t \geq 0)]$ is $U(s) = \int_0^\infty e^{-st} \cdot 1 dt = \left(-\frac{1}{s}\right)e^{-st}|_0^\infty = 1/s$ (see Table 5.3 Item 1).

Application of the useful theorems for the Laplace transform (listed below) reduces the work of solving certain differential equations by reducing them to simpler algebraic forms. The procedure is applied as follows:

1. Transform the differential equation to the s-domain by means of the appropriate Laplace transform.
2. Manipulate the transformed algebraic equation and solve for the output variable.
3. Obtain the inverse Laplace transform from Table 5.3.

See Ref.[1] for learning the Laplace transform in details.

Table 5.3: Some common forms of the Laplace transform.

	H(t)	G(s)
1	$1(t)$: Heaviside Step function $1(t) = 1$, $t > 0$; $1(t) = 0$, $t < 0$	$1/s$
2	$\delta(t)$: Dirac Impulse function $\delta(t) = \infty$, $t = 0$; $\delta(t) = 0$	1
3	$t^n/n!$ (n: positive integer)	$1/s^{n+1}$
4	e^{-at} (a: complex)	$1/(s+a)$
5	$\cos(at)$	$s/(s^2 + a^2)$
6	$\sin(at)$	$a/(s^2 + a^2)$
7	$\cosh(at)$	$s/(s^2 - a^2)$
8	$\sinh(at)$	$a/(s^2 - a^2)$
9	$e^{-bt}\cos(at)a^2 > 0$	$\frac{s+b}{(s+b)^2+a^2}$
10	$e^{bt}\sin(at)a^2 > 0$	$\frac{a}{(s+b)^2+a^2}$
11		$\frac{1}{s}(e^{-as} - e^{-bs})$
12		$\frac{m}{s^2}(1 - e^{-as})$

- *Useful Theorems for the Laplace Transform*

(a) *Linearity*:

$$L[au_1(t) + bu_2(t)] = aU_1(s) + bU_2(s) \tag{5.16a}$$

$$L^{-1}[aU_1(s) + bU_2(s)] = au_1(t) + bu_2(t) \tag{5.16b}$$

(b) *Differentiation with respect to t*:

$$L\left[\frac{du(t)}{dt}\right] = sU(s) - u(0) \tag{5.17a}$$

$$L\left[\frac{d^n u(t)}{dt^n}\right] = s^n U(s) - \sum s^{n-k} u^{k-1}(0) \tag{5.17b}$$

(c) *Integration*:

$$L\left[\int u(t)dt\right] = U(s)/s + (1/s)\left[\int u(t)dt\right]_{t=0} \tag{5.18}$$

(d) *Scaling formula*:

$$L[u(t/a)] = aU(sa)(a > 0) \tag{5.19}$$

(e) *Shift formula with respect to t*: $u(t - k) = 0$ *for* $t > k$ [k: positive real number]. The $u(t)$ curve shifts by k along the positive t axis.

$$L[u(t - k)] = e^{-ks}U(s) \tag{5.20}$$

(f) *Differentiation with respect to an independent parameter*:

$$L\left[\frac{\partial u(t, x)}{\partial x}\right] = \frac{\partial U(s, x)}{\partial x} \tag{5.21}$$

(g) *Initial and final values*:

$$\lim_{t \to 0}[u(t)] = \lim_{|s| \to \infty}[sU(s)] \tag{5.22a}$$

$$\lim_{t \to \infty}[u(t)] = \lim_{|s| \to 0}[sU(s)] \tag{5.22b}$$

Equation (5.14) can be transformed by Theorem (b), Eq. (5.17b) by denoting the Laplace transforms of $u(t, x)$ and $E_z(t)$ as $U(s, x)$ and $\widetilde{E(s)}$, respectively (x: coordinate along plate length):

$$\rho s_{11}^E s^2 U(s, x) = \frac{\partial^2 U(s, x)}{\partial x^2} \tag{5.23}$$

We assume the following *initial* ($t = 0$) *conditions*:

$$\text{Displacement: } u(0, x) = 0 \text{ and velocity}: \frac{\partial[u(0, x)]}{\partial t} = 0 \tag{5.24}$$

We also make use of the sound velocity v along the x-direction that

$$\rho s_{11}^E = 1/v^2. \tag{5.25}$$

To obtain a general solution of Eq. (5.23) in terms of space coordinate x, we assume

$$U(s, x) = Ae^{(sx/v)} + Be^{-(sx/v)} \tag{5.26}$$

Now, we consider the input force (pressure) $F(t)$ on the both ends of the piezo-plate. Since the stress is defined as positive for the tensile, we obtain

$$F(t) = -bwX_1(t) \tag{5.27}$$

and the geometrical *boundary condition* $X_1 = X(t)$ at $x = 0$ and L:

$$X(t) = \frac{(x_1 - d_{31}E_z)}{s_{11}^E} \tag{5.28}$$

We denote the Laplace transform of external stress $X(t)$ as $\widetilde{X(s)}$. We also utilize the fact on the strain that

$$L[x_1] = (\partial U / \partial x) = A(s/v)e^{(sx/v)} - B(s/v)e^{-(sx/v)} \tag{5.29}$$

5.3.1.2.1 Short-circuit condition.

Equation (5.12) is reduced to $x_1 = s_{11}^E X(t)$ at $x = 0$ and L, because of $E_z = 0$. Thus, using Eq. (5.29) yields the following two equations to solve the parameters A and B at $x = 0$ and L:

$$A(s/v) - B(s/v) = s_{11}^E \tilde{X} \tag{5.30a}$$

$$A(s/v)e^{(sL/v)} - B(s/v)e^{-(sL/v)} = s_{11}^E \tilde{X} \tag{5.30b}$$

Thus, we obtain

$$A = \frac{s_{11}^E \tilde{X}(1 - e^{-sL/v})}{(s/v)(e^{sL/v} - e^{-sL/v})}, \tag{5.31a}$$

$$B = \frac{s_{11}^E \tilde{X}(1 - e^{sL/v})}{(s/v)(e^{sL/v} - e^{-sL/v})}. \tag{5.31b}$$

and, consequently, Eqs. (5.26) and (5.29) become

$$U(s,x) = s_{11}^E \tilde{X}\left(\frac{v}{s}\right) \frac{\cosh\left[\frac{s(2x-L)}{2v}\right]}{\cosh\left(\frac{sL}{2v}\right)}, \tag{5.32a}$$

$$L[x_1(s,x)] = s_{11}^E \tilde{X} \frac{\sinh\left[\frac{s(2x-L)}{2v}\right]}{\cosh\left(\frac{sL}{2v}\right)}. \tag{5.33a}$$

By transforming the above equations into

$$U(s,x) = \frac{s_{11}^E \tilde{X}(v/s)[e^{-s(L-x)/v} + e^{-s(L+x)/v} - e^{-sx/v} - e^{-s(2L-x)/v}]}{(1 - e^{-2sL/v})}, \tag{5.32b}$$

$$L[x_1] = \frac{s_{11}^E \tilde{X}[e^{-s(L-x)/v} - e^{-s(L+x)/v} + e^{-sx/v} - e^{-s(2L-x)/v}]}{(1 - e^{-2sL/v})}, \tag{5.33b}$$

and making use of the expansion series

$$1/(1 - e^{-2sl/v}) = 1 + e^{-2sl/v} + e^{-4sl/v} + e^{-6sl/v.....} \tag{5.34}$$

the strain, $x_1(t,x)$, can now be obtained by shifting the $s_{11}^E X(t)$ curve with respect to t according to Theorem (e). That is to say, the *strain profile* $x_1(t,x)$ should be exactly the same as the *stress profile* $X(t)$, and the displacement profile is the inverse Laplace of $s_{11}^E \tilde{X}\left(\frac{v}{s}\right)$. We may also consider that since $u(t,L/2) = 0$ (from $U(s,L/2) = 0$) and $u(t,0) = -u(t,L)$ (from $U(s,0) = -U(s,L)$) (i.e., symmetrical vibration), the total displacement of the plate device ΔL becomes equal to $2u(t,L)$. We finally arrive at the following relation:

$$\Delta L = 2U(s,L) = \frac{2s_{11}^E \tilde{X}(v/s)(1 - e^{-sL/v})}{(1 + e^{-sL/v})} = 2s_{11}^E \tilde{X}(v/s)[\tanh(sL/2v)] \tag{5.35}$$

• We consider first a particular input of *Heaviside step stress* $X(t) = X_0 \cdot H(t)$. Since the Laplace transform of the step function is given by $(1/s)$ (Item 1 of Table 5.3), the total displacement Eq. (5.35) can be expressed by

$$2U(s, x = L) = 2s_{11}^E X_0(v/s^2)(1 - e^{-(sL/v)})/(1 + e^{-(sL/v)})$$
$$= 2s_{11}^E X_0(v/s^2)(1 - 2e^{-(sL/v)} + 2e^{-(2sL/v)} - 2e^{-3sL/v} + 2e^{-4sL/v} \ldots). \tag{5.36}$$

Note that the base function of $U(s,L)$, $1/s^2$, gives the base function of $u(t,L)$ in terms of t (i.e., $\propto t$, Item 3 in Table 5.3). The inverse Laplace transform of Eq. (5.36) yields (by superposing the e^{-sk} terms via Laplace transformation Theorem (e)):

$$2u(t,L) = 2s_{11}^E X_0 v \cdot t \qquad\qquad\qquad\qquad 0 < t < L/v$$
$$2u(t,L) = 2s_{11}^E X_0 v[t - 2(t - L/v)] \qquad\qquad L/v < t < 2L/v$$
$$2u(t,L) = 2s_{11}^E X_0 v[t - 2(t - L/v) + 2(t - 2L/v)] \qquad 2L/v < t < 3L/v$$
$$2u(t,L) = 2s_{11}^E X_0 v[t - 2(t - L/v) + 2(t - 2L/v) - 2(t - 3L/v)] \quad 3L/v < t < 4L/v$$

$$\ldots\ldots\ldots\ldots \tag{5.37}$$

The transient displacement, $\Delta L (= 2 \cdot u(t,L))$, produced by the step stress is pictured in Fig. 5.8(a) (here, X_0 is positive for the tensile, Fig. 5.8(a) is for $X_0 > 0$). The fundamental resonance period of this piezoelectric plate corresponds to $(2L/v)$, and the time interval in Eq. (5.37) is every $(T/2)$. It is worth noting that the displacement changes "linearly", not sinusoidally. Note also that the vibration "ringing" will continue permanently without considering any loss. The piezo damper measurement in Fig. 3.3(b) shows a similar

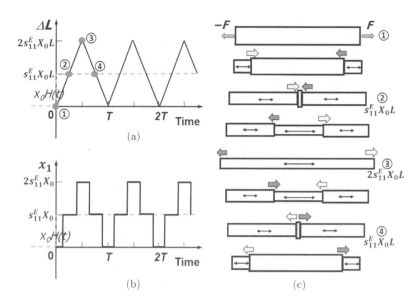

Fig. 5.8: (a) Total displacement ΔL, (b) strain x_1 at the point $x = L/4$ and (c) strain wave dynamic profile, responding to a Heaviside
step input stress in a continuum piezoelectric plate (the k_{31} mode).

vibration "ringing" from the initial abrupt force reduction $F \to 0$ (i.e., negative step stress). We discussed
on the loss factor increase to enhance the vibration "ring-down" period.

The strain distribution on a piezoelectric plate is also intriguing for this step excitation case. From
Eq. (5.33a), $L(x_1)$ is directly proportional to $\tilde{X}$; that is, the strain distribution $x_1(x)$ follows exactly to
$X(t)$, the Heaviside step function. The strain x_1 at a certain point x becomes "$s_{11}^E X_0$" suddenly from "zero"
with a certain time lag depending on its coordinate x. Figure 5.8(b) shows the stress change at the point
$x = L/4$, which indicates two discrete strain levels $s_{11}^E X_0$, and $2s_{11}^E X_0$ for different time intervals. Strain
wave dynamic modes are illustrated in Fig. 5.8(c); the strained portion starts from both ends ($x = 0$ and
L) of the piezo-plates at the time step force is applied. These two symmetrical boundaries/walls between
the strained and strain-free portion (analogous to a shock-wave) propagate with a piezo material's sound
velocity, v, opposite to each other, crossing over at the plate center, then generating the doubly strained
part in the center area (i.e., $2s_{11}^E X_0$). Thus, when the wall reaches the plate end, the plate length becomes
the maximum ($\Delta L = 2s_{11}^E X_0 L$), and we can understand the reason why 100% overshoot occurs under a
step force applied. Note also that at a minimum a half of the resonance period $T/2$ is required to reach
to the maximum total displacement. Because the ceramic is weaker for the tensile stress than for the
compressive stress, the ceramic plate may collapse $T/2$ later after the initial pressure is applied. After
$T/2$, the wall bounces back in the opposite direction, and the plate starts to shrink linearly according
to the shrinkage of the strained portion. The linear displacement change is originated from the constant
wall velocity (which is equal to the piezo material's sound velocity). This triangular vibration ringing
will continue long, if the loss is small. The reader is reminded that the step-like force/stress application
generates a distinct step-like strain discontinuity in the specimen, and this wave front propagates in the
specimen with a sound velocity. The vibration ringing will continue for a long period with the average
bias displacement at $\Delta L = s_{11}^E X_0 L$, as seen from Fig. 5.8(a). When you use an EC analysis, you will
obtain only the sinusoidal vibration ringing even under a step stress, because it cannot handle the step-
like strain discontinuity inside the ceramic specimen (the EC will handle a specimen just as a discrete
spring).

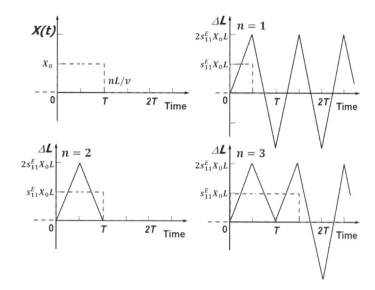

Fig. 5.9: Transient displacement ΔL produced by a rectangular pulse stress with the pulse width of 1, 2, 3 of $n(T/2)$. Note that the time interval, $T = (2l/v)$, corresponds to the resonance period of the piezoelectric plate.

- Let us consider next the response to a *rectangular pulse stress* such as pictured in the top left-hand corner of Fig. 5.9. This model corresponds to practical clicking, kicking the piezo-plate. We begin by substituting

$$\tilde{X} = (X_0/s)(1 - e^{-(nsL)/v}) \tag{5.38}$$

into Eq. (5.35), which allows us to obtain the displacement ΔL for $n = 1, 2$ and 3. The quantity n is a time scale based on half of the resonance period $(= T/2)$ of the piezoelectric plate.

For $n = 1$,

$$U(s, x = L) = s_{11}^E X_0 (v/s^2)(1 - e^{-(sL/v)})^2/(1 + e^{-(sL/v)})$$
$$= s_{11}^E X_0 (v/s^2)(1 - 3e^{-sL/v} + 4e^{-2sL/v} - 4e^{-3sL/v} + \cdots). \tag{5.39}$$

Similar to the step case, the base function of $U(s, L)$, $1/s^2$, gives the base function of $u(t, l)$ in terms of t. The inverse Laplace transform of Eq. (5.39) yields

$$
\begin{aligned}
2u(t, L) &= 2s_{11}^E X_0 vt & 0 < t < L/v \\
2u(t, L) &= 2s_{11}^E X_0 v[t - 3(t - L/v)] & L/v < t < 2L/v \\
2u(t, L) &= 2s_{11}^E X_0 v[t - 3(t - L/v) + 4(t - 2L/v)] & 2L/v < t < 3L/v
\end{aligned}
\tag{5.40}
$$

$\cdots\cdots\cdots\cdots$

The transient displacement, ΔL, produced by the rectangular pulse stress is pictured in Fig. 5.9 for top-right $n = 1$. The resonance period of this piezoelectric plate corresponds to $(2L/v)$. Notice how continuous ringing occurs under this condition.

For $n = 2$, since $\tilde{X} = (X_0/s)(1 - e^{-(2Ls)/v})$ includes the denominator of Eq. (5.35),

$$U(s, L) = s_{11}^E X_0 (v/s^2)(1 - 2e^{-sL/v} + e^{-2sL/v}). \tag{5.41}$$

Thus,

$$
\begin{aligned}
2u(t, L) &= 2s_{11}^E X_0 vt & 0 < t < L/v \\
2u(t, L) &= 2s_{11}^E X_0 v[t - 2(t - L/v)] & L/v < t < 2L/v \\
2u(t, L) &= 2s_{11}^E X_0 v[t - 2(t - L/v) + (t - 2L/v)] = 0 & 2L/v < t < 3L/v
\end{aligned}
\tag{5.42}
$$

In this case, the displacement, ΔL, occurs in a single pulse and does not exhibit ringing as depicted in Fig. 5.9 bottom-left. Remember that the applied field $\tilde{X}$ should include the denominator term $(1 + e^{-sL/v})$ to realize finite expansion terms, leading to a complete suppression of vibrational ringing.

For $n = 3$, $U(s, L)$ is again expanded as an infinite series:

$$2U(s, x = L) = 2s_{11}^{E} X_0 (v/s^2)(1 - e^{-(3sL/v)})(1 - e^{-(sL/v)})/(1 + e^{-(sL/v)})$$

$$= 2s_{11}^{E} X_0 (v/s^2)(1 - 2e^{-sL/v} + 2e^{-2sL/v} - 3e^{-3sL/v} + 4e^{-4sL/v} - 4e^{-5sL/v} \ldots). \qquad (5.43)$$

The displacement response for this case is pictured in Fig. 5.9 bottom-right. Note the displacement slope (plate edge vibration velocity) has twice the difference among the stress applied period and zero stress.

The short-circuit condition, $E_z = 0$, yields also the current response. The constitutive equation of the electric displacement, $D_3 = d_{31} X_1 + \varepsilon_0 \varepsilon_{33}^{X} E_z$, becomes simply $D_3 = d_{31} X_1$. Integrating this equation with respect to the electrode area

$$Q = w \int_0^L D_3 dx,$$

and from $x_1 = s_{11}^{E} X_1$, we obtain

$$Q = w d_{31} \int_0^L X_1 dx = w \frac{d_{31}}{s_{11}^{E}} \int_0^L x_1 dx = w \frac{d_{31}}{s_{11}^{E}} \Delta L \qquad (5.44)$$

Since the total current is provided by $I = \frac{\partial Q}{\partial t}$, and $\Delta L = 2u(t, L)$,

$$I = \frac{\partial Q}{\partial t} = w \frac{d_{31}}{s_{11}^{E}} \frac{\partial}{\partial t}(2u(t, L)). \qquad (5.45)$$

The total current response profile with time is given by the slope of the total displacement ΔL. The proportionality constant $\left(\frac{2w d_{31}}{s_{11}^{E}}\right)$ against the vibration velocity $\dot{u}(t, L)$ at the plate edge is called "*force factor*". Or, since the Laplace transform of the total displacement is expressed by

$$2U(s, L) = \frac{2s_{11}^{E} \tilde{X}(v/s)(1 - e^{-sL/v})}{(1 + e^{-sL/v})} = 2s_{11}^{E} \tilde{X}(v/s)[\tanh(sL/2v)],$$

we may express the Laplace transform of the total current $\tilde{I}$ as

$$\tilde{I} = 2w d_{31} \tilde{X} v \tanh\left(\frac{sL}{2v}\right) \qquad (5.46)$$

Figure 5.10 illustrates the transient displacement ΔL (top) and current I (bottom) produced by a rectangular pulse stress with the pulse width of 1, 2 of $n(T/2)$. Note that the time interval, $T = (2l/v)$, corresponds to the resonance period of the piezoelectric plate.

In conclusion, when we use the impulse stress input, the pulse width is important: since the pulse width is exactly equal to the resonance period of the piezo device, no vibration ringing is followed, in order to generate "large displacement continuous vibration ringing" for the energy harvesting purpose, the pulse width should be adjusted to

(1) Exactly to $T/2$ to obtain 100% overshoot (never take exactly to the resonance period T);
(2) Longer than 100 T to realize practically the "negative Step-force" (typically $T \approx 0.1$ ms).

In order to properly account for loss effects, the Laplace transform of the displacement $U(s, L)$ can be obtained by making the following substitution in Eq. (5.35):

$$s \to \frac{s}{\sqrt{1 + \zeta s}} \qquad (5.47)$$

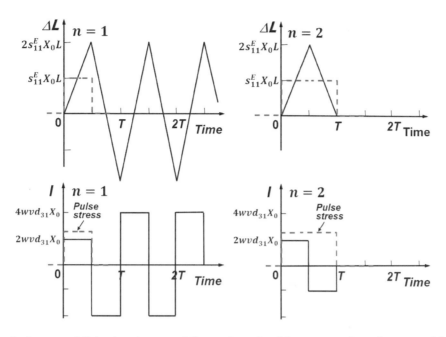

Fig. 5.10: Transient displacement ΔL (top) and current I (bottom) produced by a rectangular pulse stress with the pulse width of 1, 2 of $n(T/2)$. Note that the time interval, $T = (2l/v)$, corresponds to the resonance period of the piezoelectric plate.

which is practically equivalent to the dynamic equation with the viscous damping term:

$$\rho\frac{\partial^2 u}{\partial t^2} + \zeta\frac{\partial u}{\partial t} + \frac{1}{s^E}u = 0.$$

5.3.1.2.2 Open-circuit condition

The open-circuit condition means the total current $I = 0$, which yields the voltage response generated on the electrode. Remember that the mechanical resonance under the short- or open-circuit corresponds to the piezoelectric "resonance" or "antiresonance" frequency, respectively, which indicates that the vibration ringing time period under the open-circuit condition should be "shorter" (i.e., elastically stiffer!) than that under the short-circuit condition. When electromechanical coupling factor is not large, the antiresonance frequency is given by $f_A \approx (v^E/2L)\left(1 + \left(\frac{4}{\pi^2}\right)k_{31}^2\right)$, higher than the resonance frequency $f_R = (v^E/2L)$. Integrating the constitutive equation of the electric displacement, $D_3 = d_{31}X_1 + \varepsilon_0\varepsilon_{33}^X E_z$, with respect to the electrode area

$$Q = w\int_0^L D_3 dx = w\int_0^L [d_{31}X_1 + \varepsilon_0\varepsilon_{33}^X E_z]dx. \tag{5.48}$$

On the contrary, from $X_1 = x_1/s_{11}^E - (d_{31}/s_{11}^E)E_z$, we obtain

$$\int_0^L X_1 dx = \frac{1}{s_{11}^E}\int_0^L (x_1 - d_{31}E_z)dx \tag{5.49}$$

Knowing that $\int_0^L x_1 dx = 2u(tL)$ and $E_z(t) = $ constant in terms of the coordinate x, inserting Eq. (5.49) into Eq. (5.48), we obtain

$$Q = w\left\{\left(\frac{d_{31}}{s_{11}^E}\right)[2u(t,L) - d_{31}E_z L] + \varepsilon_0\varepsilon_{33}^X E_z L\right\}$$

$$= w\left\{\left(\frac{d_{31}}{s_{11}^E}\right)2u(t,L) + \varepsilon_0\varepsilon_{33}^X(1 - k_{31}^2)E_z(t)L\right\} \tag{5.50}$$

Open-circuit condition, $I = \frac{\partial Q}{\partial t} = 0$, results in the relation between the electric field E_z and the total displacement $2u(t, L)$ as

$$\frac{\partial E_z}{\partial t} = -\frac{k_{31}^2}{(1 - k_{31}^2)}\frac{1}{Ld_{31}}\frac{\partial}{\partial t}2u(t, L) \tag{5.51}$$

Thus, the open-circuit boundary condition indicates that time dependence of E_z is negatively proportional to the time dependence of average strain $\Delta L/L$. Note also that $u(t, L) = -u(t, 0)$, symmetric for the displacement profile. In other words, the Laplace transform describes

$$\tilde{E} = -\frac{k_{31}^2}{(1 - k_{31}^2)}\frac{2}{Ld_{31}}U(s, L) \tag{5.52}$$

This electric field is the *depolarization field* along the thickness direction, originated from the D-constant condition merely along the z-direction. If the D-constant condition is held for all x, y and z-directions, the sound velocity will be v^D. However, in this k_{31} case, the sound velocity along the x-direction is still v^E, because of the top-electrode covering all length directions.

Thus, we solve the same dynamic equation as under the short-circuit condition,

$$\rho\frac{\partial^2 u}{\partial t^2} = \frac{\partial X_{11}}{\partial x}. \tag{5.53}$$

Since $\frac{\partial X_1}{\partial x} = \frac{1}{s_{11}^E}\frac{\partial x_1}{\partial x} - \frac{d_{31}}{s_{11}^E}\frac{\partial E_z}{\partial x}$, and the top and bottom electrodes keep the potential/voltage constant along the x direction, $\frac{\partial E_z}{\partial x} = 0$, the dynamic equation (5.53) is reduced to

$$\rho\frac{\partial^2 u}{\partial t^2} = \frac{1}{s_{11}^E}\frac{\partial^2 u}{\partial x^2}. \tag{5.54}$$

Note here again $x_1 = \frac{\partial u}{\partial x}$. To obtain a general solution of Eq. (5.54) in terms of space coordinate x, we assume

$$U(s, x) = Ae^{(s^{x/v})} + Be^{-(s^{x/v})} \tag{5.55}$$

Here, we denote the sound velocity $v = 1/\sqrt{\rho s_{11}^E}$. Since the surface is the electrode, the sound velocity is the same for both short- and open-circuit conditions. Now, we consider the input force (pressure) $F(t)$ on both ends of the piezo-plate. Since the stress is defined as positive for the tensile, we obtain

$$F(t) = -bwX_1(t) \tag{5.56}$$

and the geometrical *boundary condition* $X_1 = X(t)$ at $x = 0$ and L:

$$X(t) = \frac{(x_1 - d_{31}E_z)}{s_{11}^E} \tag{5.57}$$

We denote the Laplace transform of external stress $X(t)$ as $\widetilde{X(s)}$. We also utilize the fact on the strain that

$$L[x_1] = (\partial U/\partial x) = A(s/v)e^{(s^{x/v})} - B(s/v)e^{-(s^{x/v})}. \tag{5.58}$$

If we rewrite Eq. (5.57) as $s_{11}^E\widetilde{X(s)} = L[x_1] - d_{31}\tilde{E}$, and using Eq. (5.58), we obtain

$$s_{11}^E\widetilde{X(s)} = L[x_1] + \frac{k_{31}^2}{(1 - k_{31}^2)}\frac{2}{L}U(s, L). \tag{5.59}$$

Different from the short-circuit condition, the external stress under the open-circuit condition generates the total displacement term (second term of Eq. (5.59)) originated from the induced electric field E_z.

Using Eqs. (5.55) and (5.58), and $U(s, L) = -U(s, 0)$ yields the following two equations for the condition at $x = 0$ and L to solve the parameters A and B:

$$A(s/v) - B(s/v) - \frac{k_{31}^2}{(1 - k_{31}^2)}\frac{2}{L}[A + B] = s_{11}^E\tilde{X}, \tag{5.60a}$$

$$A(s/v)e^{(sL/v)} - B(s/v)e^{-(sL/v)} + \frac{k_{31}^2}{(1 - k_{31}^2)}\frac{2}{L}[Ae^{(sL/v)} + Be^{-(sL/v)}] = s_{11}^E\tilde{X}. \tag{5.60b}$$

Thus, we obtain

$$A = \frac{s_{11}^E\tilde{X}\left[\left(\frac{s}{v}\right)\left(1 - e^{-\frac{sL}{v}}\right) + \left(\frac{2K_{31}^2}{L}\right)\left(1 + e^{-\frac{sL}{v}}\right)\right]}{e^{sL/v}\left[\left(\frac{s}{v}\right) + (2K_{31}^2/L)\right]^2 - e^{-sL/v}\left[\left(\frac{s}{v}\right) - (2K_{31}^2/L)\right]^2}, \tag{5.61a}$$

$$B = \frac{s_{11}^E\tilde{X}\left[\left(\frac{s}{v}\right)\left(1 - e^{\frac{sL}{v}}\right) - \left(\frac{2K_{31}^2}{L}\right)\left(1 + e^{\frac{sL}{v}}\right)\right]}{e^{sL/v}\left[\left(\frac{s}{v}\right) + (2K_{31}^2/L)\right]^2 - e^{-sL/v}\left[\left(\frac{s}{v}\right) - (2K_{31}^2/L)\right]^2}, \tag{5.61b}$$

where we use a new notation

$$K_{31}^2 = \frac{k_{31}^2}{(1 - k_{31}^2)} \tag{5.62}$$

and, consequently, $U(s, L)$ is expressed as

$$U(s, x) = Ae^{\frac{sL}{v}} + Be^{-\left(\frac{sL}{v}\right)}$$
$$= \frac{s_{11}^E\tilde{X}\left\{-2\left(\frac{s}{v}\right) + e^{\frac{sL}{v}}\left[\left(\frac{s}{v}\right) + \left(\frac{2K_{31}^2}{L}\right)\right] + e^{-\frac{sL}{v}}\left[\left(\frac{s}{v}\right) - \left(\frac{2K_{31}^2}{L}\right)\right]\right\}}{e^{sL/v}\left[\left(\frac{s}{v}\right) + (2K_{31}^2/L)\right]^2 - e^{-sL/v}\left[\left(\frac{s}{v}\right) - (2K_{31}^2/L)\right]^2}. \tag{5.63}$$

It is important to note that when we consider the $e^{-\frac{sL}{v}}$ term, which is essential to sum up the time shift functions for future intervals with the unit of $T/2 = L/v$, the term is composed of $[s - (2K_{31}^2v/L)]$, not merely of s. When we take into account s on the (angular) frequency domain replaced by ω, the new ω at the antiresonance condition $[\omega_A - (2K_{31}^2v/L)]$ seems to be equal to the resonance frequency ω_R, which was discussed in the previous section; that is, the antiresonance frequency is higher than the resonance frequency, and the difference is proportional to K_{31}^2 (modified electromechanical coupling factor). Since further calculations are complicated, only the expected results are described here. The precise calculation for sinusoidal operation is described in the Section 5.3.2. Figure 5.11 illustrates the transient displacement ΔL produced by a rect-

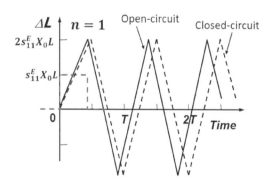

Fig. 5.11: Transient displacement ΔL produced by a rectangular pulse stress under open- and short-circuit conditions with the pulse width close to $(T/2)$.

angular pulse stress under open-circuit conditions excited by the pulse width slightly shorter than $(T/2) = L/v^E$, that is, a half of the fundamental antiresonance time period $T_A/2 = \frac{L}{v^E}/\left(1 + \frac{4}{\pi^2}k_{31}^2\right)$. We expect similar triangular (linear) displacement change to the case in the short-circuit condition. The voltage/electric field change with time should be a similar triangular shape expected from Eq. (5.52).

We discussed in Sections 5.3.1.2.1 and 5.3.1.2.2 under short-circuit and open-circuit conditions, where we cannot expect any electric energy harvesting. In order to cultivate the energy, a suitable electrical impedance should be connected in the external circuit.

5.3.1.2.3 Impedance matching load condition

An external electrical impedance Z is connected to a piezoelectric k_{31} plate (Fig. 5.12). When we assume impulse input stress $X = X_0(t)$, the output electric charge Q (i.e., no loss, no time lag) under the vibration ringing-down process at almost the fundamental resonance frequency can be described as

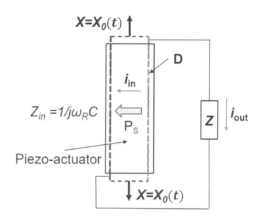

Fig. 5.12: Electric energy harvesting model under the external electrical impedance Z on a piezoelectric actuator.

$$Q = w \int_0^L D_3 dx = w \int_0^L [d_{31} X_1 + \varepsilon_0 \varepsilon_{33}^X E_z] dx. \quad (5.64)$$

On the contrary, from $X_1 = x_1/s_{11}^E - (d_{31}/s_{11}^E)E_z$, we obtain

$$\int_0^L X_1 dx = \frac{1}{s_{11}^E} \int_0^L (x_1 - d_{31} E_z) dx \quad (5.65)$$

Knowing that $\int_0^L x_1 dx = 2u(t, L)$ and $E_z(t) = $ constant in terms of the coordinate x, and inserting Eq. (5.65) into Eq. (5.64), we obtain

$$Q = w \left\{ \left(\frac{d_{31}}{s_{11}^E} \right) 2u(t, L) + \varepsilon_0 \varepsilon_{33}^X (1 - k_{31}^2) E_z(t) L \right\} \quad (5.66)$$

Note here that the motional capacitance enhancement can be neglected in this scenario due to no vibration enhancement via the mechanical quality factor Q_m during the ringing-down process after an impulse stress. Taking Laplace transform formulation for the force $\tilde{X}_0$ and total polarization $\tilde{Q}$,

$$\tilde{Q} = d_{31} \tilde{X}_0$$

$$I = \frac{\partial Q}{\partial t} = i_{in} + i_{out};$$

$$Z_{in} i_{in} = Z i_{out};$$

$$\tilde{I} = s\tilde{Q} = Cs\tilde{V} + \frac{\tilde{V}}{Z} = d_{31} \tilde{X}_0$$

Thus, the effective impedance and the voltage/force relation is obtained as

$$\frac{\tilde{V}}{\tilde{I}} = \frac{1}{sC + \frac{1}{Z}} \quad (5.67)$$

$$\frac{\tilde{V}}{\tilde{X}_0} = \frac{d_{31}}{sC + \frac{1}{Z}} \quad (5.68)$$

- When we adopt the step force, that is, $\tilde{X}_0 = X_0/s$, we can obtain the voltage

$$\tilde{V} = \frac{d_{31} X_0}{s \left(sC + \frac{1}{Z} \right)} = \frac{d_{31} X_0}{C} \frac{1}{\left(s + \frac{1}{2ZC} \right)^2 - \left(\frac{1}{2ZC} \right)^2}$$

Taking an inverse Laplace transform, we obtain the voltage change with time for the first impact as

$$V = d_{31} X_0 Z (1 - e^{-t/ZC}) \quad (5.69)$$

However, this analysis misleads the reader to a wrong direction: we do not consider the vibration and voltage ringing with the resonance frequency, but consider just one-time polarization generation.

- When we approximate the vibration ringing after the impulse force as cyclic (sinusoidal) natural (resonance) vibration (though the actual displacement behavior is linear, not sinusoidal), we adopt Fourier transform by replacing the above Laplace form with $s = j\omega$. Here, ω is considered to be the natural resonance frequency under Z-shunt condition. The dynamic impedance of the piezo component (off-resonance) becomes $1/j\omega C$.

The output electric energy is described as

$$|P|_{out} = \frac{1}{2}Zi_{out}^2 = \frac{1}{2}Z\frac{(\omega dX_0)^2}{(1+(\omega CZ)^2)}. \tag{5.70}$$

The maximum power energy $|P| = \frac{1}{4}\frac{\omega d^2 X_0^2}{C}$ can be obtained when the external impedance is adjusted to

$$Z = 1/\omega C \tag{5.71}$$

In other words, the "stored" electric energy can be spent maximum by a half when the external resistive load impedance matches exactly to the internal impedance. This is called "*electrical impedance matching*".

We calculate now the "input mechanical energy" under Z-shunt condition from the second constitutive equation:

$$x = d_{31}E + s_{11}^E X = -d_{31}\left(\frac{V}{b}\right) + s_{11}^E X = -\left(\frac{d_{31}}{b}\right)\left[\frac{j\omega d_{31}X_0}{\frac{1}{Z}+j\omega C}\right] + s_{11}^E X \tag{5.72}$$

The last transformation used Eq. (5.68). We obtained effective elastic compliance as

$$s_{\text{eff}}^E = \frac{x}{X} = s^E\left[1 - k_{31}^2\frac{j\omega CZ}{(1+j\omega CZ)}\right] \tag{5.73}$$

You can verify the above "effective elastic compliance" is equal to s^E or $s^D = s^E[1 - k_{31}^2]$, when $Z = 0$ or ∞, respectively. On the contrary, under $Z = 1/\omega C$,

$$s_{\text{eff}}^E = s^E\left(1 - \frac{1}{2}k_{31}^2 + \frac{j}{2}k_{31}^2\right) \tag{5.74}$$

The resonance period T may be estimated as

$$T = 2L\sqrt{\rho s_{\text{eff}}^E} = T_0\left(1 - \frac{1}{4}k_{31}^2\right) \tag{5.75}$$

which is shorter than the short-circuit condition T_0, but longer than the open-circuit condition $T_0\left(1 - \frac{4}{\pi^2}k_{31}^2\right)$.

Let us calculate the energy spent in the external impedance Z, which should be the matched external impedance of the piezoelectric energy harvesting system. The "impulse" vibration energy U_M (time averaged by $\int_{half\ cycle}\frac{1}{2}\frac{x_0^2}{s^E}dt$) is transformed into electric energy U_E by the factor of k^2 ($U_E = U_M \times k^2$), then a half of this energy can be spent for accumulating into a rechargeable battery ($(1/2)k^2 \times U_M$) when we take the exact matching impedance condition for the piezo component. For the next half cycle (i.e., opposite voltage and current during shrinkage), we start from the mechanical energy $(1 - k^2/2)U_M$, without taking into account elastic loss in the system. As the square of the amplitude is equivalent to the amount of energy, the amplitude decreases at a rate of $(1 - k^2/2)^{1/2}$ times with every vibration repeated. If the resonance period is taken to be T_0, the number of vibrations for t sec is $2t/T_0$. Consequently, the amplitude in t sec is $(1 - k^2/2)^{t/T_0}$. If the residual vibration period is taken to be T_0, the damping in the amplitude of vibration

Table 5.4:   Electromechanical properties of the commercial PZT ceramics.

Sequence n	Mechanical energy	Electrical energy	Spent/accumulated energy
0	U_M	$k^2 U_M$	$\frac{1}{2}k^2 U_M$
1	$(1-\frac{1}{2}k^2)\,e^{-\pi/2Q_M}U_M$	$k^2\left(1-\frac{1}{2}k^2\right)e^{-\pi/2Q_M}U_M$	$\frac{1}{2}k^2\left(1-\frac{1}{2}k^2\right)e^{-\pi/2Q_M}U_M$
3	$\left[(1-\frac{1}{2}k^2)\,e^{-\frac{\pi}{2Q_M}}\right]^2 U_M$	$k^2\left[(1-\frac{1}{2}k^2)\,e^{-\frac{\pi}{2Q_M}}\right]^2 U_M$	$\frac{1}{2}k^2\left[(1-\frac{1}{2}k^2)\,e^{-\frac{\pi}{2Q_M}}\right]^2 U_M$
$\vdots$	$\vdots$	$\vdots$	$\vdots$
n	$\left[(1-\frac{1}{2}k^2)\,e^{-\frac{\pi}{2Q_M}}\right]^n U_M$	$k^2\left[(1-\frac{1}{2}k^2)\,e^{-\frac{\pi}{2Q_M}}\right]^n U_M$	$\frac{1}{2}k^2\left[(1-\frac{1}{2}k^2)\,e^{-\frac{\pi}{2Q_M}}\right]^n U_M$

is t sec which can be expressed as follows:

$$(1 - k^2/2)^{t/T_0} = e^{-t/\tau}. \tag{5.76}$$

Thus, the following relationship for the time constant of the vibration damping is obtained.

$$\tau = -\frac{T_0}{\ln(1 - k^2/2)} \tag{5.77}$$

If we consider the elastic loss (which corresponds to the damping constant τ_S), the energy decay will be $(1-k^2/2)e^{-T_0/2\tau_S}U_M$ every half cycle. Note that elastic loss or its inverse "mechanical quality factor" Q_M is related with τ_S as

$$\frac{T_0}{\tau_S} = \frac{\pi}{Q_M} \tag{5.76}$$

A similar process will be repeated every half cycle during the ringing-down process after the step force is applied. The sequential energy ring-down process can be summarized in Table 5.4. The total loss energy (after n time sequences) can be calculated as

$$\frac{1}{2}k^2 U_M \sum_{n=0}^{n}\left[\left(1-\frac{1}{2}k^2\right)e^{-\frac{\pi}{2Q_M}}\right]^n = \frac{1}{2}k^2 U_M\frac{1-\left[\left(1-\frac{1}{2}k^2\right)e^{-\frac{\pi}{2Q_M}}\right]^n}{1-\left(1-\frac{1}{2}k^2\right)e^{-\frac{\pi}{2Q_M}}} \tag{5.77}$$

This indicates the following results according to the mechanical quality factor:

- High $Q_M(\sim 10{,}000) \rightarrow$ Since the energy loss is small, the total energy accumulated will reach the input mechanical energy U_M after more than 10,000 ring-down processes.
- Low $Q_M(\sim 0.5) \rightarrow$ Since the original damping is large, the total energy loss will be $\frac{1}{2}k^2\frac{1}{0.96+0.02k^2}$, just slightly larger than $\frac{1}{2}k^2$ (only the first pulse electric energy).

Figure 5.13 compares the transient displacement ΔL produced by a rectangular pulse stress under Z-shunt and short-circuit conditions with the pulse width close to $(T/2)$. When we use a matching impedance, the vibration amplitude decay rate is determined by k^2. The total energy accumulation reaches the input mechanical energy U_M after multiple ring-down processes. Figure 3.3 is the vibration damping result after the impulse force for a bimorph with $Q_M \sim 20$. For this level of Q_M, the total loss energy is around $0.92k^2U_M$ (summation roughly from the first 20 pulses) which is between $\frac{1}{2}k^2U_M$ and U_M.

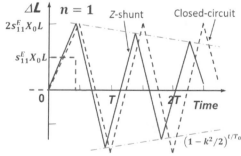

Fig. 5.13:   Transient displacement ΔL produced by a rectangular pulse stress under Z-shunt and short-circuit conditions with the pulse width close to $(T/2)$.

5.3.2 *Sinusoidal Operation*

5.3.2.1 *Piezoelectric dynamic equation for the k_{31} mode plate*

Let us consider the same piezo ceramic k_{31} plate, as shown in Fig. 5.7. Sinusoidal force/pressure F and $-F$ (angular frequency ω) are applied on the plate ends at $x = 0$ and L along the length direction x. Thus,

$$F(t) = -bwX_1(t). \tag{5.78}$$

If the polarization is in the z-direction and x-y planes are the planes of the electrodes, the extensional vibration in the x (length) direction is represented by the following dynamic equations (when the length L is more than 4–6 times the width w or the thickness b; we can neglect the coupling modes with width or thickness vibrations):

$$\rho(\partial^2 u/\partial t^2) = (\partial X_{11}/\partial x), \tag{5.79}$$

where ρ is the density of the piezo ceramic, u is the displacement of a small volume element in the ceramic plate in the x-direction. Only the following two equations are essential to solve the dynamic equation around the resonance/antiresonance frequencies:

$$x_1 = s_{11}^E X_1 + d_{31} E_z, \tag{5.80a}$$

$$D_3 = d_{31} X_1 + \varepsilon_0 \varepsilon_{33}^X E_z. \tag{5.80b}$$

Since Eq. (5.80a) is transformed into $X_1 = x_1/s_{11}^E - (d_{31}/s_{11}^E)E_z$, we obtain

$$\frac{\partial X_1}{\partial x} = \frac{1}{s_{11}^E}\frac{\partial x_1}{\partial x} - \frac{d_{31}}{s_{11}^E}\frac{\partial E_z}{\partial x}. \tag{5.81}$$

Because of the equal potential on each electrode, $\partial E_z/\partial x = 0$, and knowing the strain definition $x_1 = \partial u/\partial x$ along the $1(x)$ direction (non-suffix x corresponds to the Cartesian coordinate). Eq. (5.79) is transformed into

$$\rho\frac{\partial^2 u}{\partial t^2} = \frac{1}{s_{11}^E}\frac{\partial^2 u}{\partial x^2}. \tag{5.82}$$

If we assume a *harmonic vibration* equation, u under a sinusoidal force application, we can simplify Eq. (5.82):

$$-\omega^2 \rho s_{11}^E u = \partial^2 u/\partial x^2 \quad \text{or} \quad -\left(\frac{\omega}{v_{11}^E}\right)^2 u(x) = \frac{\partial^2 u(x)}{\partial x^2}. \tag{5.83}$$

Here, ω is the angular frequency of the sinusoidal force, u displacement, and v_{11}^E is the *sound velocity* along the length x direction in the piezo ceramic plate, expressed by

$$v_{11}^E = 1/\sqrt{\rho s_{11}^E}. \tag{5.84}$$

The reader can easily understand that the above process is almost the same as in the Laplace transform's case, except for the replacement from s to $j\omega$, which corresponds to the Fourier transform. Supposing the displacement u also vibrates with the frequency of ω, a general solution of Eq. (5.83) is expressed by

$$u(x) = A\sin\left(\frac{\omega}{v_{11}^E}x\right) + B\cos\left(\frac{\omega}{v_{11}^E}x\right). \tag{5.85}$$

From Eq. (5.85), the strain $x_1(x)$ is given by

$$x_1(x) = \frac{\partial u}{\partial x} = A\frac{\omega}{v_{11}^E}\cos\left(\frac{\omega}{v_{11}^E}x\right) - B\frac{\omega}{v_{11}^E}\sin\left(\frac{\omega}{v_{11}^E}x\right), \tag{5.86}$$

In order to determine the above two parameters, A and B, the boundary condition is imposed: $X_1 = X_0 e^{j\omega t}$ at $x = 0$ and L (both plate ends). We will consider here again three cases: (1) Short-circuit condition of the piezo-plate, where $E_z = 0$, (2) Open-circuit condition of the piezo-plate, where the current $I = \frac{\partial Q}{\partial t} = 0$, and (3) Matching impedance shunt condition.

5.3.2.2 *Solution under short-circuit condition*

Since $E_z = 0$, $x_1 = s_{11}^E X_1$ and $D_3 = d_{31} X_1$ are the necessary equations. From the first strain equation (5.86) at $x = 0$ and L, we obtain the following two equations:

$$\begin{cases} A\dfrac{\omega}{v_{11}^E} = s_{11}^E X_0 \\ A\dfrac{\omega}{v_{11}^E}\cos\left(\dfrac{\omega}{v_{11}^E}L\right) - B\dfrac{\omega}{v_{11}^E}\sin\left(\dfrac{\omega}{v_{11}^E}L\right) = s_{11}^E X_0 \end{cases}.$$

Thus,

$$\begin{cases} A = \left(\dfrac{v_{11}^E}{\omega}\right) s_{11}^E X_0 \\ B = -\left(\dfrac{v_{11}^E}{\omega}\right) s_{11}^E X_0 \dfrac{\sin\left(\dfrac{\omega}{2v_{11}^E}L\right)}{\cos\left(\dfrac{\omega}{2v_{11}^E}L\right)} \end{cases}.$$

$$\text{(Displacement)}\ u(x) = \left(\frac{v_{11}^E}{\omega}\right) s_{11}^E X_0 \frac{\sin\left[\frac{\omega(2x-L)}{2v_{11}^E}\right]}{\cos\left(\frac{\omega L}{2v_{11}^E}\right)} \tag{5.87}$$

$$\text{(Strain)}\ x_1 = \partial u/\partial x = s_{11}^E X_0 \left(\frac{\cos\left[\frac{\omega(2x-L)}{2v_{11}^E}\right]}{\cos\left(\frac{\omega L}{2v_{11}^E}\right)}\right) \tag{5.88}$$

First, the displacement and strain are proportional to the external stress X_0. Second, their distributions in terms of x in Eqs. (5.87) and (5.88) are anti-symmetrically and symmetrically sinusoidal in respect to $x = L/2$ position (the numerator becomes maximum, $\cos(0) = 1$), and the maximum strain (i.e., *nodal line*) exists on this line. It is important that the difference of the above equations from Eqs. (5.32a) and (5.33a) is sin, cos, in comparison with sinh and cosh. Note that $\omega \to 0$ (i.e., pseudo-DC) makes Eq. (5.88) to $x_1 = s_{11}^E X_0$, that is, uniform strain distribution on the whole piezo-plate. For $\omega \to \left(\frac{\omega L}{2v_{11}^E}\right) = n\frac{\pi}{2}$ [$n : 1, 3, 5, \ldots$], the denominator of both Eqs. (5.87) and (5.88) approaches to 0 and the strain becomes infinite, which is called the "resonance" frequency. The resonance frequency under closed-circuit is given by

$$f_R = \frac{v_{11}^E}{2L} \tag{5.89}$$

Now, we use another set of constitutive equations with respect to electric displacement $D_3 = d_{31} X_1$. Taking into account $E_z = 0$, we calculate the total electric current under AC stress $X_1 = X_0 e^{j\omega t}$:

$$I = \frac{\partial Q}{\partial t} = j\omega w \int_0^L D_3 dx = j\omega w d_{31} \int_0^L X_1 dx = j\omega w \frac{d_{31}}{s_{11}^E} \int_0^L x_1 dx. \tag{5.90}$$

From Eq. (5.88), we obtain

$$\int_0^L x_1 dx = \Delta L = s_{11}^E X_0 \frac{2v_{11}^E}{\omega}\tan\left(\frac{\omega L}{2v_{11}^E}\right).$$

Thus, the total current is represented by

$$I = j\omega w \frac{d_{31}}{s_{11}^E} s_{11}^E X_0 \frac{2v_{11}^E}{\omega} \tan\left(\frac{\omega L}{2v_{11}^E}\right) = j\left(\frac{2wd_{31}}{s_{11}^E}\right) s_{11}^E X_0 v_{11}^E \tan\left(\frac{\omega L}{2v_{11}^E}\right). \tag{5.91}$$

Here, $\left(\frac{2wd_{31}}{s_{11}^E}\right)$ is the "*force factor*" to convert the mechanical input to the electrical input, and "j" stands for $90°$ phase ahead of the total displacement change. When ω is small, the current increases in proportion to the frequency ω,

$$I = j\omega \left(\frac{wd_{31}}{s_{11}^E}\right) s_{11}^E X_0 L \tag{5.92}$$

and for $\omega \to \left(\frac{\omega L}{2v_{11}^E}\right) = n\frac{\pi}{2}$ $[n : 1, 3, 5, \ldots]$, that is, approaching to the resonance, I becomes infinite, similar to the total displacement behavior.

5.3.2.3 *Solution under open-circuit condition*

Since $E_z \neq 0$, we should use the original $x_1 = s_{11}^E X_1 + d_{31}E_z$, and $D_3 = d_{31}X_1 + \varepsilon_0\varepsilon_{33}^X E_z$, and calculate the voltage generated on the electrode under AC stress on both ends of the piezo-plate. Open-circuit condition means $I = \frac{\partial Q}{\partial t} = 0$ (Q: total charge). Integrating the constitutive equation of the electric displacement, $D_3 = d_{31}X_1 + \varepsilon_0\varepsilon_{33}^X E_z$, with respect to the electrode area

$$Q = w\int_0^L D_3 dx = w\int_0^L [d_{31}X_1 + \varepsilon_0\varepsilon_{33}^X E_z]dx. \tag{5.93}$$

On the contrary, from $X_1 = x_1/s_{11}^E - (d_{31}/s_{11}^E)E_z$, we obtain

$$\int_0^L X_1 dx = \frac{1}{s_{11}^E}\int_0^L (x_1 - d_{31}E_z)dx. \tag{5.94}$$

Knowing that $\int_0^L x_1 dx = \Delta L = 2u(tL)$ and $E_z(t) = $ constant in terms of the coordinate x, inserting Eq. (5.94) into Eq. (5.93), we obtain

$$Q = w\left\{\left(\frac{d_{31}}{s_{11}^E}\right)[2u(t, L) - d_{31}E_z L] + \varepsilon_0\varepsilon_{33}^X E_z L\right\}$$

$$= w\left\{\left(\frac{d_{31}}{s_{11}^E}\right)2u(t, L) + \varepsilon_0\varepsilon_{33}^X(1 - k_{31}^2)E_z(t)L\right\}. \tag{5.95}$$

Equation (5.95) indicates that the total Q consists of the charge accumulated in the "damped" (clamped) capacitance (the second term of the right-hand formula) and the additional charge induced by the mechanical vibration (the 1st term, which is called "motional" capacitance). Note the "force factor" $\left(2w\frac{d_{31}}{s_{11}^E}\right)$ which converts the mechanical parameter u to electrical parameter Q (or equivalently, vibration velocity $\dot{u}$ to current I, Eq. (5.92)). Open-circuit condition, $I = \frac{\partial Q}{\partial t} = j\omega Q = 0$, results in the relation between the electric field E_z and the total displacement $2u(t, L)$ as

$$E_z = -\frac{k_{31}^2}{(1 - k_{31}^2)}\frac{1}{Ld_{31}}2u(t, L) \tag{5.96}$$

Note also that $u(t, L) = -u(t, 0)$, symmetric for the displacement profile.

Taking into account the electric field generated on the piezo-plate, we derive a mechanical resonance frequency (which corresponds to the piezoelectric "antiresonance" frequency). The displacement u and strain $x_1(x)$ are assumed to be expressed by

$$u(x) = A \sin\left(\frac{\omega}{v_{11}^E} x\right) + B \cos\left(\frac{\omega}{v_{11}^E} x\right). \tag{5.97}$$

$$x_1(x) = \frac{\partial u}{\partial x} = A \frac{\omega}{v_{11}^E} \cos\left(\frac{\omega}{v_{11}^E} x\right) - B \frac{\omega}{v_{11}^E} \sin\left(\frac{\omega}{v_{11}^E} x\right), \tag{5.98}$$

We start from $x_1 = s_{11}^E X_1 + d_{31} E_z$, or $X_1 = x_1/s_{11}^E - (d_{31}/s_{11}^E) E_z$. This equation should be satisfied under $X_0(t)$ at $x = 0$ and L, and we obtain the following two equations:

$$\begin{cases} s_{11}^E X_0 = A\dfrac{\omega}{v_{11}^E} - \dfrac{k_{31}^2}{(1-k_{31}^2)}\dfrac{1}{L} 2u(t,0) = A\dfrac{\omega}{v_{11}^E} - B\dfrac{k_{31}^2}{(1-k_{31}^2)}\dfrac{2}{L} \\[2mm] s_{11}^E X_0 = A\dfrac{\omega}{v_{11}^E}\cos\left(\dfrac{\omega}{v_{11}^E}L\right) - B\dfrac{\omega}{v_{11}^E}\sin\left(\dfrac{\omega}{v_{11}^E}L\right) + \dfrac{k_{31}^2}{(1-k_{31}^2)}\dfrac{2}{L}\left[A\sin\left(\dfrac{\omega}{v_{11}^E}L\right) + B\cos\left(\dfrac{\omega}{v_{11}^E}L\right)\right] \end{cases}.$$

Thus,

$$\begin{cases} A = s_{11}^E X_0 \Big/ \left[\dfrac{\omega}{v_{11}^E} + \dfrac{k_{31}^2}{(1-k_{31}^2)}\dfrac{2}{L} tan\left(\dfrac{\omega L}{2v_{11}^E}\right)\right] \\[3mm] B = -s_{11}^E X_0 \tan\left(\dfrac{\omega L}{2v_{11}^E}\right) \Big/ \left[\dfrac{\omega}{v_{11}^E} + \dfrac{k_{31}^2}{(1-k_{31}^2)}\dfrac{2}{L}\tan\left(\dfrac{\omega L}{2v_{11}^E}\right)\right] \end{cases}. \tag{5.99}$$

$$\text{(Displacement)}\ u(x) = s_{11}^E X_0 \frac{\sin\left[\dfrac{\omega(2x-L)}{2v_{11}^E}\right]}{\left[\dfrac{\omega}{v_{11}^E}\cos\left(\dfrac{\omega L}{2v_{11}^E}\right) + \dfrac{k_{31}^2}{(1-k_{31}^2)}\dfrac{2}{L}\sin\left(\dfrac{\omega L}{2v_{11}^E}\right)\right]} \tag{5.100}$$

$$\text{(Strain)}\ x_1 = \frac{\partial u}{\partial x} = s_{11}^E X_0 \frac{\left(\dfrac{\omega}{v_{11}^E}\right)\cos\left[\dfrac{\omega(2x-L)}{2v_{11}^E}\right]}{\left[\dfrac{\omega}{v_{11}^E}\cos\left(\dfrac{\omega L}{2v_{11}^E}\right) + \dfrac{k_{31}^2}{(1-k_{31}^2)}\dfrac{2}{L}\sin\left(\dfrac{\omega L}{2v_{11}^E}\right)\right]} \tag{5.101}$$

The mechanical resonance, that is, the displacement or strain maximum (∞) can be obtained when the denominator of Eqs. (5.100) and (5.101) $[\frac{\omega}{v_{11}^E}\cos\left(\frac{\omega L}{2v_{11}^E}\right) + \frac{k_{31}^2}{(1-k_{31}^2)}\frac{2}{L}sin\left(\frac{\omega L}{2v_{11}^E}\right)]$ is equal to zero, which is obtained by the following equation:

$$(\omega_A L/2v_{11}^E)\cot(\omega_A L/2v_{11}^E) = -k_{31}^2/(1-k_{31}^2) = -d_{31}^2/\varepsilon_{33}^{LC} s_{11}^E, \tag{5.102}$$

where ω_A is the "antiresonance" frequency. You are reminded that under the short-circuit condition, the denominator of the displacement formula $\cos\left(\frac{\omega L}{2v_{11}^E}\right) = 0$, then the resonance condition $\left(\frac{\omega L}{2v_{11}^E}\right) = \frac{\pi}{2}$, and

$$\begin{cases} f_R = v_{11}^E/2L \\[2mm] f_A = (v_{11}^E/2L)\left(1 + \dfrac{4}{\pi^2}k_{31}^2\right) \ \text{(for small } k_{31}) \end{cases} \tag{5.103}$$

Example Problem 5.1.

In a piezo ceramic k_{31} plate as shown in Fig. 5.7, the fundamental resonance frequency is given by $f_R = v_{11}^E/2L$, while the fundamental antiresonance frequency is approximated as $f_A = f_R(1 + \frac{4}{\pi^2}k_{31}^2)$. Derive this formula from the exact solution equation: $(\omega_A L/2v_{11}^E)\cot(\omega_A L/2v_{11}^E) = -k_{31}^2/(1-k_{31}^2)$.

Solution:

Introducing $\Delta\omega = (\omega_A - \omega_R)$, and knowing $\omega_R = \pi v_{11}^E/L$, the equation $(\omega_A L/2v_{11}^E)\cot(\omega_A L/2v_{11}^E) = -k_{31}^2/(1-k_{31}^2)$ is transformed into

$$-\frac{k_{31}^2}{(1-k_{31}^2)} = \left(\frac{\pi}{2} + \frac{\Delta\omega L}{2v_{11}^E}\right)\frac{\cos\left(\frac{\pi}{2} + \frac{\Delta\omega L}{2v_{11}^E}\right)}{\sin\left(\frac{\pi}{2} + \frac{\Delta\omega L}{2v_{11}^E}\right)} = -\left(\frac{\pi}{2} + \frac{\Delta\omega L}{2v_{11}^E}\right)\tan\left(\frac{\Delta\omega L}{2v_{11}^E}\right). \quad (P5.1.1)$$

Supposing that $\left(\frac{\Delta\omega L}{2v_{11}^E}\right) = \Delta\omega/4f_R \ll 1$, we use the Taylor expansion to $\tan(x) = x + \frac{x^3}{3} + \frac{2x^5}{15} + \cdots$ into Eq. (P5.1.1):

$$\left(\frac{\pi}{2} + \frac{\Delta\omega L}{2v_{11}^E}\right)\tan\left(\frac{\Delta\omega L}{2v_{11}^E}\right) = \left(\frac{\pi}{2} + \Delta\omega/4f_R\right)\left[(\Delta\omega/4f_R) + \frac{1}{3}(\Delta\omega/4f_R)^3 + \cdots\right] = \frac{k_{31}^2}{(1-k_{31}^2)}. \quad (P5.1.2)$$

Then,

$$\frac{\pi}{2}(\Delta\omega/4f_R) = \frac{k_{31}^2}{(1-k_{31}^2)}. \quad (P5.1.3)$$

Finally, we obtain the following relation for the case $k_{31}^2 \ll 1$:

$$f_A = f_R\left(1 + \frac{4}{\pi^2}k_{31}^2\right). \quad (P5.1.4)$$

5.3.2.4 *Resonance and antiresonance vibration modes*[12]

The resonance and antiresonance states are both mechanical resonance states with amplified strain/ displacement states under external stress excitation, but they are different from the electric constraint condition, that is, short-circuit or open-circuit. The mode difference is described by the following intuitive model. In a high electromechanical coupling material with k almost equal to 1, the resonance or antiresonance states appear for $\tan(\omega L/2v) = \infty$ or 0 (i.e., $\omega L/2v = (m-1/2)\pi$ or $m\pi$ (m: integer)), respectively. The strain amplitude x_1 distribution for each state calculated from Eqs. (5.88) and (5.101)) is illustrated in Fig. 5.14 (the figure is slightly off-resonance to escape from the infinite amplitude). In the resonance state (top-left in Fig. 5.14), the strain distribution is basically sinusoidal with the maximum at the center of plate ($x = L/2$) (see the numerator) with small exciting strain $s_{11}^E X_0$ at two edges. When ω is close to ω_R, $(\omega_R L/2v) = \pi/2$, leading to the denominator $\cos(\omega_R L/2v) \to 0$, and significant strain amplification is obtained. To the contrary, at the antiresonance (top-right in Fig. 5.14), the strain induced in the device compensates completely (because extension and compression are compensated in one wave on the specimen length, and the plate ends become the node), with both edges as nodes in addition to the plate center. Thus, for a high k_{31} material, the first antiresonance frequency f_A should be almost twice as large as the first resonance frequency f_R.

However, in a typical PZT case, where $k_{31} = 0.3$, the antiresonance state varies from the previously mentioned mode and becomes closer to the resonance mode (top-center in Fig. 5.14). The low-coupling material exhibits an antiresonance mode where the capacitance change due to the size change (i.e., *motional capacitance*) is compensated completely by the current required to charge up the static capacitance (called *damped capacitance*). Note that above the resonance frequency, the motional capacitance phase changes by $180°$ owing to $\tan(\omega L/2v) < 0$. Thus, the antiresonance frequency f_A will approach the resonance frequency f_R [Recall the relation, $f_A = f_R\left(1 + \frac{4}{\pi^2}k_{31}^2\right)$]. According to the external electrical impedance connected, the mechanical resonance is changed between the piezoelectric resonance and antiresonance frequencies in a practical energy harvesting system, which is discussed in the next section.

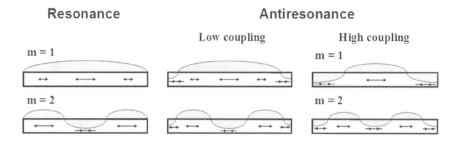

Fig. 5.14: Strain distribution in the resonance and antiresonance states for a k_{31}-type piezoelectric plate.

5.3.2.5 *Solution under Z-shunt condition*

An external electrical impedance Z is connected to a piezoelectric k_{31} plate (Fig. 5.12). We start from the constitutive equations: $x_1 = s_{11}^E X_1 + d_{31}E_z$, and $D_3 = d_{31}X_1 + \varepsilon_0\varepsilon_{33}^X E_z$. When we assume sinusoidal input force $F(t) = -bwX_1(t)$, and stress $X_1(t) = X_0 e^{j\omega t}$ at both ends of the piezo-plate symmetrically, we can also assume the displacement $u(x)e^{j\omega t}$ and strain $x_1(x)e^{j\omega t}$ as

$$u(x) = A\sin\left(\frac{\omega}{v_{11}^E}x\right) + B\cos\left(\frac{\omega}{v_{11}^E}x\right). \tag{5.104}$$

$$x_1(x) = \frac{\partial u}{\partial x} = A\frac{\omega}{v_{11}^E}\cos\left(\frac{\omega}{v_{11}^E}x\right) - B\frac{\omega}{v_{11}^E}sin\left(\frac{\omega}{v_{11}^E}x\right), \tag{5.105}$$

Knowing $E_z(t) = $ constant in terms of the coordinate x, owing to the surface electrode, at $x = 0$ and L, $X_1 = x_1/s_{11}^E - (d_{31}/s_{11}^E)E_z = X_0$:

$$\begin{cases} s_{11}^E X_0 = A\dfrac{\omega}{v_{11}^E} - d_{31}E_z \\ s_{11}^E X_0 = A\dfrac{\omega}{v_{11}^E}cos\left(\dfrac{\omega}{v_{11}^E}L\right) - B\dfrac{\omega}{v_{11}^E}sin\left(\dfrac{\omega}{v_{11}^E}L\right) - d_{31}E_z \end{cases}. \tag{5.106}$$

The output electric charge Q (i.e., no loss, no time lag) can be described as

$$Q = w\int_0^L D_3 dx = w\int_0^L [d_{31}X_1 + \varepsilon_0\varepsilon_{33}^X E_z]dx. \tag{5.107}$$

On the contrary, from $X_1 = x_1/s_{11}^E - (d_{31}/s_{11}^E)E_z$, we obtain

$$\int_0^L X_1 dx = \frac{1}{s_{11}^E}\int_0^L (x_1 - d_{31}E_z)dx.$$

Knowing that $\int_0^L x_1 dx = 2u(L)$ and $E_z = $ constant, we obtain

$$I = \dot{Q} = \left\{ \left(\frac{2wd_{31}}{s_{11}^E}\right)\dot{u}(L) + \varepsilon_0\varepsilon_{33}^X(1 - k_{31}^2)\dot{E}_z wL \right\}$$

$$= j\omega\left(\frac{2wd_{31}}{s_{11}^E}\right)u(L) + j\omega\varepsilon_0\varepsilon_{33}^X(1 - k_{31}^2)E_z wL.$$

Using $V = bE_z = Z \cdot I$ and $C_d = \varepsilon_0\varepsilon_{33}^X(1 - k_{31}^2)wL/b$, Eq. (5.107) leads to the following relations:

$$j\omega\left(\frac{2wd_{31}}{s_{11}^E}\right)\left[A\sin\left(\frac{\omega}{v_{11}^E}L\right) + B\cos\left(\frac{\omega}{v_{11}^E}L\right)\right] + j\omega C_d E_z b - \frac{b}{Z}E_z = 0. \tag{5.108}$$

From Eqs. (5.106) and (5.108), we can derive A, B and E_z as follows:

$$
\begin{pmatrix}
\dfrac{\omega}{v_{11}^E} & 0 & -d_{31} \\
\dfrac{\omega}{v_{11}^E}cs & -\dfrac{\omega}{v_{11}^E}sn & -d_{31} \\
j\omega\left(\dfrac{2wd_{31}}{s_{11}^E}\right)sn & j\omega\left(\dfrac{2wd_{31}}{s_{11}^E}\right)cs & b\left(j\omega C_d - \dfrac{1}{Z}\right)
\end{pmatrix}
\begin{pmatrix} A \\ B \\ E_z \end{pmatrix}
=
\begin{pmatrix} s_{11}^E X_0 \\ s_{11}^E X_0 \\ 0 \end{pmatrix},
\qquad (5.109)
$$

where $sn = \sin\left(\dfrac{\omega}{v_{11}^E}L\right)$ and $cn = \cos\left(\dfrac{\omega}{v_{11}^E}L\right)$. Thus,

$$
\begin{cases}
A = \dfrac{s_{11}^E X_0 b\left(j\omega C_d - \dfrac{1}{Z}\right)\cos\left(\dfrac{\omega L}{2v_{11}^E}\right)}{\Delta} \\[3mm]
B = -\dfrac{s_{11}^E X_0 b\left(j\omega C_d - \dfrac{1}{Z}\right)\sin\left(\dfrac{\omega L}{2v_{11}^E}\right)}{\Delta} \\[3mm]
E_z = -\dfrac{s_{11}^E X_0 \left(j\omega \dfrac{2wd_{31}}{s_{11}^E}\right)\sin\left(\dfrac{\omega L}{2v_{11}^E}\right)}{\Delta} \\[3mm]
\Delta = \left(\dfrac{\omega}{v_{11}^E}\right)b\left(j\omega C_d - \dfrac{1}{Z}\right)\cos\left(\dfrac{\omega L}{2v_{11}^E}\right) + d_{31}j\omega\left(\dfrac{2wd_{31}}{s_{11}^E}\right)\sin\left(\dfrac{\omega L}{2v_{11}^E}\right)
\end{cases}
\qquad (5.110)
$$

The last equation can be transformed to

$$
\Delta = j\omega 2w\varepsilon_0\varepsilon_{33}^X\left[(1-k_{31}^2)\left(\dfrac{\omega L}{2v_{11}^E}\right)\cos\left(\dfrac{\omega L}{2v_{11}^E}\right) + k_{31}^2\sin\left(\dfrac{\omega L}{2v_{11}^E}\right)\right] + \left(\dfrac{\omega}{v_{11}^E}\right)\dfrac{b}{Z}\cos\left(\dfrac{\omega L}{2v_{11}^E}\right). \qquad (5.111)
$$

- When $Z \to 0$ (i.e., short-circuited), the last term $(\dfrac{\omega}{v_{11}^E})\dfrac{b}{Z}\cos(\dfrac{\omega L}{2v_{11}^E})$ is the major contribution, which leads to the resonance condition of $\cos(\dfrac{\omega L}{2v_{11}^E}) = 0$, or $\dfrac{\omega L}{2v_{11}^E} = \dfrac{\pi}{2}$. This is the resonance mode condition in the k_{31} plate.
- When $Z \to \infty$ (i.e., open-circuited), since the last term is neglected, and the resonance condition should be

$$
(1-k_{31}^2)\left(\dfrac{\omega L}{2v_{11}^E}\right)\cos\left(\dfrac{\omega L}{2v_{11}^E}\right) + k_{31}^2\sin\left(\dfrac{\omega L}{2v_{11}^E}\right) = 0, \ \text{or}\ \left(\dfrac{\omega L}{2v_{11}^E}\right)\cot\left(\dfrac{\omega L}{2v_{11}^E}\right) = -\dfrac{k_{31}^2}{(1-k_{31}^2)},
$$

which is the familiar formula for calculating the antiresonance frequency.

Now, by connecting $Z = 1/\omega C_d$, we consider the minimization of the magnitude of for obtaining the resonance condition.

$$
\Delta = j\omega 2w\varepsilon_0\varepsilon_{33}^X\left[(1-k_{31}^2)\left(\dfrac{\omega L}{2v_{11}^E}\right)\cos\left(\dfrac{\omega L}{2v_{11}^E}\right) + k_{31}^2\sin\left(\dfrac{\omega L}{2v_{11}^E}\right) + j(1-k_{31}^2)\left(\dfrac{\omega L}{2v_{11}^E}\right)\cos\left(\dfrac{\omega L}{2v_{11}^E}\right)\right]. \qquad (5.112)
$$

- For a small ω (much lower than the resonance frequency),

$$
\Delta = j\omega 2w\varepsilon_0\varepsilon_{33}^X\left[(1-k_{31}^2)\left(\dfrac{\omega L}{2v_{11}^E}\right) + k_{31}^2\left(\dfrac{\omega L}{2v_{11}^E}\right) + j(1-k_{31}^2)\left(\dfrac{\omega L}{2v_{11}^E}\right)\right]
$$

$$
= j\omega 2w\varepsilon_0\varepsilon_{33}^X\left(\dfrac{\omega L}{2v_{11}^E}\right)[1 + j(1-k_{31}^2)]. \qquad (5.113)
$$

The apparent dissipation factor $\tan\varphi = (1-k_{31}^2)$ is quite high under $Z = 1/\omega C_d$ resistive shunt case.

- For a frequency around the resonance frequency ($\frac{\omega_R L}{2v_{11}^E} = \frac{\pi}{2}, \Delta\omega = \omega - \omega_R$), taking $\frac{\omega L}{2v_{11}^E} = (\frac{\pi}{2} + \frac{\Delta\omega L}{2v_{11}^E})$, $\cos(\frac{\omega L}{2v_{11}^E}) = -\sin(\frac{\Delta\omega L}{2v_{11}^E})$, and $\sin(\frac{\omega L}{2v_{11}^E}) = \cos(\frac{\Delta\omega L}{2v_{11}^E})$ into account,

$$\Delta = j(\omega_R + \Delta\omega)2w\varepsilon_0\varepsilon_{33}^X \left[-\left(1 - k_{31}^2\right)\left(\frac{\pi}{2} + \frac{\Delta\omega L}{2v_{11}^E}\right)\sin\left(\frac{\Delta\omega L}{2v_{11}^E}\right) + k_{31}^2\cos\left(\frac{\Delta\omega L}{2v_{11}^E}\right) \right.$$
$$\left. -j\left(1 - k_{31}^2\right)\left(\frac{\pi}{2} + \frac{\Delta\omega L}{2v_{11}^E}\right)\sin\left(\frac{\Delta\omega L}{2v_{11}^E}\right) \right]$$
$$\approx j\omega_R 2w\varepsilon_0\varepsilon_{33}^X \left[-\left(1 - k_{31}^2\right)\left(\frac{\pi}{2}\right)\left(\frac{\Delta\omega L}{2v_{11}^E}\right) + k_{31}^2 - j\left(1 - k_{31}^2\right)\left(\frac{\pi}{2}\right)\left(\frac{\Delta\omega L}{2v_{11}^E}\right) \right]. \tag{5.114}$$

If you recall the discussion in Section 2.7.3.1, the external impedance Z connection is equivalent to the loss tangent increase, and

$$\tan\varphi = \left(1 - k_{31}^2\right)\left(\frac{\omega L}{2v_{11}^E}\right)\cos\left(\frac{\omega L}{2v_{11}^E}\right) / \left[\left(1 - k_{31}^2\right)\left(\frac{\omega L}{2v_{11}^E}\right)\cos\left(\frac{\omega L}{2v_{11}^E}\right) + k_{31}^2\sin\left(\frac{\omega L}{2v_{11}^E}\right)\right]$$
$$\approx -\left(1 - k_{31}^2\right)\left(\frac{\pi}{2}\right)\left(\frac{\Delta\omega L}{2v_{11}^E}\right) / \left[-\left(1 - k_{31}^2\right)\left(\frac{\pi}{2}\right)\left(\frac{\Delta\omega L}{2v_{11}^E}\right) + k_{31}^2\right]$$
$$\approx -\frac{(1 - k_{31}^2)}{k_{31}^2}\left(\frac{\pi}{2}\right)\left(\frac{\Delta\omega L}{2v_{11}^E}\right). \tag{5.115}$$

It is very interesting that the resistive shunt contributes largely to the dissipation, but its contribution is small around the mechanical resonance frequency range.

The displacement $u(x)$ and strain $x_1(x)$ are summarized here:

$$\begin{cases} u(x) = \dfrac{s_{11}^E X_0 bw C_d(j-1)\cos\left(\frac{\omega L}{2v_{11}^E}\right)}{\Delta}\sin\left(\frac{\omega}{v_{11}^E}x\right) + \dfrac{s_{11}^E X_0 bw C_d(j-1)\sin\left(\frac{\omega L}{2v_{11}^E}\right)}{\Delta}\cos\left(\frac{\omega}{v_{11}^E}x\right) \\[3mm] x_1(x) = \dfrac{s_{11}^E X_0 bw C_d(j-1)\cos\left(\frac{\omega L}{2v_{11}^E}\right)}{\Delta}\dfrac{\omega}{v_{11}^E}\cos\left(\frac{\omega}{v_{11}^E}x\right) - \dfrac{s_{11}^E X_0 bw C_d(j-1)\sin\left(\frac{\omega L}{2v_{11}^E}\right)}{\Delta}\dfrac{\omega}{v_{11}^E}\sin\left(\frac{\omega}{v_{11}^E}x\right) \end{cases}. \tag{5.116}$$

The energy spent in the resistive shunt $Z = 1/\omega C_d$ can be calculated from Eq. (5.110) as

$$|P_{out}| = \text{Re}\left[\frac{1}{2}\frac{VV^*}{Z}\right] = \text{Re}\left[\frac{b^2}{2Z}\left[\frac{s_{11}^E X_0\left(j\omega\frac{2wd_{31}}{s_{11}^E}\right)\sin\left(\frac{\omega L}{2v_{11}^E}\right)}{\left(\frac{\omega}{v_{11}^E}\right)b\left(j\omega C_d - \frac{1}{Z}\right)\cos\left(\frac{\omega L}{2v_{11}^E}\right) + d_{31}j\omega\left(\frac{2wd_{31}}{s_{11}^E}\right)\sin\left(\frac{\omega L}{2v_{11}^E}\right)}\right]^2\right]. \tag{5.117}$$

If we consider a small ω (much lower than the resonance frequency),

$$|P_{out}| = \frac{b^2}{2}\omega\frac{Lw}{b}\varepsilon_0\varepsilon_{33}^X(1 - k_{31}^2)\text{Re}\left[\frac{s_{11}^E X_0(j\omega\frac{2wd_{31}}{s_{11}^E})}{\left(\frac{2}{L}\right)bw\frac{Lw}{b}\varepsilon_0\varepsilon_{33}^X\left(1 - k_{31}^2\right)(j-1) + d_{31}j\omega\left(\frac{2wd_{31}}{s_{11}^E}\right)}\right]^2$$
$$= \frac{1}{2}\omega(Lwb)\varepsilon_0\varepsilon_{33}^X\left(1 - k_{31}^2\right)\text{Re}\left[\frac{d_{31}X_0(j\omega 2w)}{\omega 2w\varepsilon_0\varepsilon_{33}^X\left(1 - k_{31}^2\right)(j-1) + j\omega 2w\varepsilon_0\varepsilon_{33}^X\left(\frac{d_{31}^2}{\varepsilon_0\varepsilon_{33}^X s_{11}^E}\right)}\right]^2$$
$$= \frac{1}{2}\omega(Lwb)\text{Re}\left[\frac{(d_{31}X_0)^2\left(1 - k_{31}^2\right)}{\varepsilon_0\varepsilon_{33}^X[-(1 - k_{31}^2) + j]}\right] = \frac{1}{2}\omega(Lwb)\frac{(d_{31}X_0)^2}{\varepsilon_0\varepsilon_{33}^X}\frac{1}{\left(1 - k_{31}^2\right) + \frac{1}{(1 - k_{31}^2)}}. \tag{5.118}$$

In Equation (5.118), because $d_{31}X_0 = P_3$, and $\frac{1}{2}\frac{(d_{31}X_0)^2}{\varepsilon_0\varepsilon_{33}^X}$ corresponds to the electric energy per unit volume converted via the piezoelectric effect with some calibration factor by the electromechanical coupling factor, $\frac{1}{(1-k_{31}^2)+\frac{1}{(1-k_{31}^2)}} = \frac{1}{2+\frac{k_{31}^4}{(1-k_{31}^2)}}$. The power obtained from this piezoelectric k_{31} specimen is expressed by the product of frequency ω and volume (Lwb). When k_{31}^2 is not large ($k_{31} < 30\%$), $|P_{out}|$ becomes roughly $1/2$ of the converted energy via the resistance Z which matches the piezoelectric damped capacitance.

In the previous Section 5.3.1.2.3, with a simple assumption, we derived the output electric energy described as

$$|P|_{out} = \text{Re}\left[\frac{1}{2}Zi_{out}(i_{out})^*\right] = \frac{1}{2}Z\frac{(\omega dX_0)^2}{(1+(\omega CZ)^2)}. \tag{5.119}$$

The maximum power energy $|P| = \frac{1}{4}\frac{\omega d^2 X_0^2}{C}$ can be obtained when the external impedance is adjusted to

$$Z = 1/\omega C \tag{5.120}$$

The reader should recognize two differences of Eq. (5.120), in which we did not consider the difference depending on the mechanical constraint (i.e., ε_{33}^X or $\varepsilon_{33}^{x_1}$) from the exact solution in Eq. (5.118):

(1) Matching impedance is replaced by the "damped" capacitance C_d, rather free capacitance C_0.
(2) The calibration factor $\frac{1}{(1-k_{31}^2)+\frac{1}{(1-k_{31}^2)}}$ is more precise, rather than simple $1/2$.

Refresh your mind that the mechanical excitation of a piezoelectric plate under short-circuit condition exhibits so-called "piezoelectric resonance mode" with the elastic compliance s_{11}^E, while the mechanical excitation of a piezoelectric plate under open-circuit condition exhibits so-called "piezoelectric antiresonance mode" with the lower "effective" elastic compliance $s_{11,\text{eff}}^E$ (i.e., higher peak frequency, though the device elastic compliance should be s_{11}^E theoretically). We calculate now the "input mechanical energy" under Z-shunt condition from the second constitutive equation:

$$x = d_{31}E + s_{11}^E X = -d_{31}\left(\frac{V}{b}\right) + s_{11}^E X = -\left(\frac{d_{31}}{b}\right)\left[\frac{j\omega d_{31}X_0}{\frac{1}{Z}+j\omega C}\right] + s_{11}^E X$$

The last transformation used Eq. (5.68). We obtained effective elastic compliance as

$$s_{\text{eff}}^E = \frac{x}{X} = s^E\left[1 - k_{31}^2\frac{j\omega CZ}{(1+j\omega CZ)}\right] \tag{5.121}$$

You can verify that the above "effective elastic compliance" is equal to s_{11}^E or $s_{11}^D = s_{11}^E(1-k_{31}^2)$ [no electrode on the surface ideally], when $Z = 0$ or ∞, respectively. On the contrary, under $Z = 1/\omega C$,

$$s_{\text{eff}}^E = s^E\left(1 - \frac{1}{2}k_{31}^2 + \frac{j}{2}k_{31}^2\right) \tag{5.122}$$

The resonance frequency f may be estimated as

$$f = 1/2L\sqrt{\rho s_{\text{eff}}^E} = f_R\left(1 + \frac{1}{4}k_{31}^2\right) \tag{5.123}$$

which is higher than the short-circuit condition f_R, but lower than the open-circuit condition $f_A(1+\frac{4}{\pi^2}k_{31}^2)$.

5.4 Design Optimization

5.4.1 *Piezoelectric Device Design and Electromechanical Coupling Factor*

Five different definitions were introduced in Section 2.6.2 on the *electromechanical coupling factor k*, which corresponds to the rate of electromechanical transduction: the input electric energy to the output mechanical energy, and vice versa.[11] We review them here for the sake of further discussion. Notice the difference between the static k_v and the dynamics k_{vn} which are used for "off-resonance" and "resonance" applications, respectively.

5.4.1.1 *Mason's definition*[13]

When we apply the electric field on a piezoelectric material or when we apply the mechanical force on the sample pseudo-statically or off-resonance ($\omega \to 0$), the electromechanical coupling factors defined, respectively, by Mason as follows

$$k^2 = (\text{Stored mechanical energy/Input electrical energy}), \tag{5.124a}$$

$$k^2 = (\text{Stored electrical energy/Input mechanical energy}), \tag{5.124b}$$

can be calculated by

$$k^2 = d^2/\varepsilon_0\varepsilon^X \cdot s^E, \tag{5.125}$$

where d is the piezoelectric constant, $\varepsilon_0\varepsilon^X$ permittivity under stress constant, and s^E is elastic compliance under electric field constant condition. Recall Fig. 2.22 for the derivation process.

5.4.1.2 *Definition in materials*

The internal energy (per unit volume) U of a piezoelectric is given by summation of the mechanical energy $U_M (= \int x \, dX)$ and the electrical energy $U_E (= \int D \, dE)$. When linear piezoelectric constitutive equations are applicable, U is calculated as follows:

$$
\begin{aligned}
U &= U_M + U_E \\[6pt]
&= \left[(1/2) \sum_{i,j} s_{ij}^E X_j X_i + (1/2) \sum_{m,i} d_{mi} E_m X_i \right] \\[6pt]
&\quad + \left[(1/2) \sum_{m,i} d_{mi} X_i E_m + (1/2) \sum_{k,m} \varepsilon_0 \varepsilon_{mk}^X E_k E_m \right] \\[6pt]
&= U_{MM} + 2U_{ME} + U_{EE} \\[6pt]
&= (1/2) \sum_{i,j} s_{ij}^E X_j X_i + 2 \cdot (1/2) \sum_{m,i} d_{mi} E_m X_i + (1/2) \sum_{k,m} \varepsilon_0 \varepsilon_{mk}^X E_k E_m .
\end{aligned}
\tag{5.126}
$$

The s and ε terms represent purely mechanical and electrical energies (U_{MM} and U_{EE}), respectively, and the d term denotes the energy transduced from electrical to mechanical energy or vice versa through the piezoelectric effect (U_{ME}). The electromechanical coupling factor k is defined by

$$k^2 = U_{ME}^2/U_{MM}U_{EE}. \tag{5.127}$$

Table 5.5: The characteristics of various piezoelectric resonators with different shapes and sizes.

	Factor	Boundary Conditions	Resonator Shape	Definition
a	k_{31}	$X_1 \neq 0,\ X_2 = X_3 = 0$ $x_1 \neq 0,\ x_2 \neq 0,\ x_3 \neq 0$		$\dfrac{d_{31}}{\sqrt{s_{11}^E \varepsilon_0 \varepsilon_{33}^X}}$
b	k_{33}	$X_1 = X_2 = 0,\ X_3 \neq 0$ $x_1 = x_2 \neq 0,\ x_3 \neq 0$	 **Fundamental Mode**	$\dfrac{d_{33}}{\sqrt{s_{33}^E \varepsilon_0 \varepsilon_{33}^X}}$
c	k_p	$X_1 = X_2 \neq 0,\ X_3 = 0$ $x_1 = x_2 \neq 0,\ x_3 \neq 0$	 **Planar/Radial Mode**	$k_{31}\sqrt{\dfrac{2}{1-\sigma}}$
d	k_t	$X_1 = X_2 \neq 0,\ X_3 \neq 0$ $x_1 = x_2 = 0,\ x_3 \neq 0$	 **Thickness Mode**	$k_{33}\sqrt{\dfrac{\varepsilon_0 \varepsilon_{33}^x}{c_{33}^D}}$
e	$k_{24} = k_{15}$	$X_1 = X_2 = X_3 = 0,\ X_4 \neq 0$ $x_1 = x_2 = x_3 = 0,\ x_5 \neq 0$	 **Shear Mode**	$\dfrac{d_{15}}{\sqrt{s_{55}^E \varepsilon_0 \varepsilon_{11}^X}}$
f	k_{eff}	d_{31} Type Bimorph Bending		$\sqrt{\dfrac{3}{4}}k_{31}$

Using the above energy terms (Eq. (5.126)) with s^E, $\varepsilon_0\varepsilon^X$ and d notations,

$$k^2 = \frac{\left(\frac{1}{2}dEX\right)^2}{\left(\frac{1}{2}s^E X^2\right)\left(\frac{1}{2}\varepsilon_0\varepsilon^X E^2\right)} = \frac{d^2}{s^E \varepsilon^X \varepsilon_0} \tag{5.128}$$

5.4.1.3 *Definition in devices*

Though the constitutive equations can be derived from the internal energy in Eq. (5.126) in general, since the key equations are limited depending on the specimen geometry, there are several definitions according to the mode or specimen geometry in consideration:

$$\begin{bmatrix} x \\ D \end{bmatrix} = \begin{bmatrix} s^E & d \\ d & \varepsilon_0\varepsilon^X \end{bmatrix} \begin{bmatrix} X \\ E \end{bmatrix}, \tag{5.129}$$

the electromechanical coupling factor is defined by

$$k^2 = \frac{(\text{Coupling factor})^2}{(\text{Product of the diagonal parameters})} = \frac{(d)^2}{(s^E \varepsilon_0\varepsilon^X)}. \tag{5.130}$$

Table 5.5 summarizes the electromechanical coupling k_{ij} for typical piezoelectric resonators with different shapes and sizes.

(1) When the vibration mode can be expressed directly by the same constitutive piezoelectric equations, the electromechanical coupling factor k of the resonator is equal to the material's electromechanical coupling factor, which is represented merely by the material's constants such as piezoelectric constant, dielectric permittivity and elastic compliance (and Poisson's ratio).

Examples can be found:

- k_{31} mode — $\begin{bmatrix} x_1 \\ D_3 \end{bmatrix} = \begin{bmatrix} s_{11}^E & d_{31} \\ d_{31} & \varepsilon_0\varepsilon_{33}{}^X \end{bmatrix}\begin{bmatrix} X_1 \\ E_3 \end{bmatrix}$ provides $k_{31} = \dfrac{d_{31}}{\sqrt{s_{11}^E \varepsilon_0\varepsilon_{33}{}^X}}$,

- k_{33} mode $- \begin{bmatrix} x_3 \\ D_3 \end{bmatrix} = \begin{bmatrix} s_{33}^E & d_{33} \\ d_{33} & \varepsilon_0 \varepsilon_{33}^X \end{bmatrix} \begin{bmatrix} X_3 \\ E_3 \end{bmatrix}$ provides $k_{33} = \frac{d_{33}}{\sqrt{s_{33}^E \varepsilon_0 \varepsilon_{33}^X}}$,

- k_p mode $- \begin{bmatrix} x_1 + x_2 \\ D_3 \end{bmatrix} = \begin{bmatrix} 2(s_{11}^E + s_{12}^E) & 2d_{31} \\ 2d_{31} & \varepsilon_0 \varepsilon_{33}^X \end{bmatrix} \begin{bmatrix} X_p \\ E_3 \end{bmatrix}$ provides $k_p = k_{31} \cdot \sqrt{2/(1-\sigma)}$.

Because the piezoelectric Poisson's ratio $|d_{31}/d_{33}| \approx 1/3$ in PZTs, supposing the difference in elastic compliances s_{11}^E and s_{33}^E is just around 10%, the reader can easily understand that $k_{33} \gg k_{31}$ by the factor of 3. You can conclude similarly that $k_p > k_{31}$ by the factor of $\sqrt{3}(\sigma \approx 1/3)$.

Example Problem 5.2.

PZT 5AH exhibits the following physical parameters: elastic, dielectric and piezoelectric constants. Calculate the electromechanical coupling factors k_{33}, k_{31} and k_p on this PZT.

Parameter	s_{11}^E	s_{12}^E	s_{33}^E	ε_{33}^X	d_{31}	d_{33}
Unit	$10^{-12}\,\mathrm{m^2/N}$	$10^{-12}\,\mathrm{m^2/N}$	$10^{-12}\,\mathrm{m^2/N}$	$8.854 \times 10^{-12}\,\mathrm{F/m}$	$10^{-12}\,\mathrm{C/N}$	$10^{-12}\,\mathrm{C/N}$
Value	16.4	-4.7	20.8	3630	-274	593

Solution:

Necessary numerical values are

$$\text{Elastic Poisson's ratio: } \sigma = \left| \frac{s_{12}^E}{s_{11}^E} \right| = \frac{4.7}{16.4} = 0.29, \tag{P5.2.1}$$

$$\text{Piezoelectric Poisson's ratio: } \left| \frac{d_{31}}{d_{33}} \right| = \frac{274}{593} = 0.46, \tag{P5.2.2}$$

$$k_{31} = \frac{d_{31}}{\sqrt{s_{11}^E \varepsilon_0 \varepsilon_{33}^X}} = \frac{274}{\sqrt{16.4 \times 8.85 \times 3630}} = 0.37, \tag{P5.2.3}$$

$$k_{33} = \frac{d_{33}}{\sqrt{s_{33}^E \varepsilon_0 \varepsilon_{33}^X}} = \frac{593}{\sqrt{20.8 \times 8.85 \times 3630}} = 0.73, \tag{P5.2.4}$$

$$k_p = \frac{d_{31}}{\sqrt{s_{11}^E \varepsilon_0 \varepsilon_{33}^X}} \cdot \sqrt{\frac{2}{1-\sigma}} = 0.37 \times \sqrt{2/(1-0.29)} = 0.62. \tag{P5.2.5}$$

(2) When the structure of a vibrator is complicated, the electromechanical coupling factor k of the vibrator is dependent on the specimen geometry/size, in addition to the material's constants such as piezoelectric constant, dielectric permittivity and elastic compliance (and Poisson's ratio).

An example can be found:

- Bimorph: k is dependent on the elastic ship material and thickness. The derivation is introduced in Section 5.4.3. Only when the piezo-plate bimorph without shim, $k_{\mathrm{eff}} = \sqrt{3/4} \cdot k_{31}$.

5.4.1.4 *Constraint condition method*

From the relations between the E-constant, D-constant elastic compliances, s^E, s^D, stiffness c^E, c^D; and stress-free, strain-free permittivity $\varepsilon_0 \varepsilon^X$, $\varepsilon_0 \varepsilon^x$, inverse permittivity $\kappa_0 \kappa^X$, $\kappa_0 \kappa^x$ (refer to Eqs. (2.20)–(2.25))

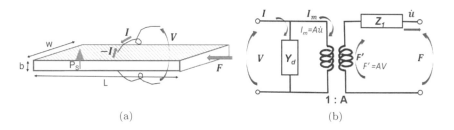

Fig. 5.15: (a) k_{31}-type piezoelectric plate geometry; (b) 4 terminal equivalent circuit for k_{31} piezo-plate.

$$1 - k^2 = \frac{s^D}{s^E} = \frac{c^E}{c^D} = \frac{\varepsilon^x}{\varepsilon^X} = \frac{\kappa^X}{\kappa^x} \tag{5.131}$$

5.4.1.5 *Dynamic definition*

In the 4-terminal EC (see Fig. 5.15(b)) of the k_{31} piezoelectric plate (Fig. 5.15(a)), the electric terminal parameters voltage V and current I, and the mechanical terminal parameters force F and vibration velocity $\dot{u}$ related to each other as

$$\begin{bmatrix} F \\ I \end{bmatrix} = \begin{bmatrix} Z_1 & -A \\ A & Y_1 \end{bmatrix} \begin{bmatrix} \dot{u} \\ V \end{bmatrix} \tag{5.132}$$

The dynamic electromechanical coupling factor k_v^2 is defined by [(Complex power in the mechanical branch)/(complex power in the electrical branch)] under short-circuit condition of mechanical terminal, or [(Complex power in the electrical branch)/(complex power in the mechanical branch)] under short-circuit condition of electrical terminal, which leads to

$$k_v^2 = \left| \frac{\left(\frac{A^2}{Z_1 Y_1} \right)}{1 + \left(\frac{A^2}{Z_1 Y_1} \right)} \right|. \tag{5.133}$$

Since $Z_1 = j Z_0 \tan \left(\frac{\omega L}{2v} \right)$, $Y_1 = j\omega C_d$, $A = \frac{2 d_{31} w}{s_{11}^E}$ and $Z_0 = wb\rho v = \frac{wb}{v s_{11}^E}$ in the k_{31} mode, k_v^2 is ω dependent. By taking $\omega \to 0$, $k_v^2 \to k_{31}^2 = \frac{d_{31}^2}{s^E \varepsilon_0 \varepsilon^X}$. This dynamic k definition is particularly useful to consider/calculate a complex structured piezo transducer, which is detailed in the next subsection.

5.4.2 *Dynamic Electromechanical Coupling Factor*

We reconsider the electromechanical coupling factor with an EC expressed by Eq. (5.133) in this subsection.

5.4.2.1 *Constitutive equations vs. equivalent circuit — reconsideration*[13]

Referring to Fig. 5.15(a), the piezoelectric constitutive equations are given by

$$\begin{bmatrix} x_1 \\ D_3 \end{bmatrix} = \begin{bmatrix} s_{11}^E & d_{31} \\ d_{31} & \varepsilon_0 \varepsilon_{33}{}^X \end{bmatrix} \begin{bmatrix} X_1 \\ E_3 \end{bmatrix}, \quad \text{or} \quad \begin{bmatrix} X_1 \\ E_3 \end{bmatrix} = \begin{bmatrix} s_{11}^E & d_{31} \\ d_{31} & \varepsilon_0 \varepsilon_{33}^X \end{bmatrix}^{-1} \begin{bmatrix} x_1 \\ D_3 \end{bmatrix} \tag{5.134}$$

which originates from the material's parameters directly (we call this "Stress-Strain-Field" based description as "derivative" type). You are reminded that the electromechanical coupling factor k_{31} is given from the above piezoelectric constitutive equations by

$$k_{31} = \frac{d_{31}}{\sqrt{s_{11}^E \varepsilon_0 \varepsilon_{33}{}^X}}. \tag{5.135}$$

When we integrate in terms of length or area, such as

$$
\begin{cases}
\iint x_1 dx = u\text{(displacement)}, \quad \int D_3 dS = Q \text{ (charge)}, \\
\int X_1 dS = -F\text{(pressure)}, \text{ and } \int E_3 dz = -V \text{ (voltage)},
\end{cases}
\tag{5.136}
$$

we can obtain the following formula:

$$
\begin{bmatrix} F \\ Q \end{bmatrix} = \begin{bmatrix} c_1 & -A_0 \\ A_0 & C_d \end{bmatrix} \begin{bmatrix} u \\ V \end{bmatrix},
\tag{5.137}
$$

which with a "force–displacement–voltage"-based description is called "integrated (in terms of space)" type. Here, c_1 is macroscopic elastic stiffness constant (or spring constant) between force F and displacement u. Knowing that the electric current I measurement is more practical than the charge Q, and taking into account $I = \frac{\partial Q}{\partial t} = j\omega Q$ and vibration velocity $\dot{u} = \frac{\partial u}{\partial t} = j\omega u$ in parallel (instead of displacement u), we can introduce an alternative "integrated" formula with "force–velocity–voltage"-based description:

$$
\begin{bmatrix} F \\ I \end{bmatrix} = \begin{bmatrix} Z_1 & -A \\ A & Y_d \end{bmatrix} \begin{bmatrix} \dot{u} \\ V \end{bmatrix}.
\tag{5.138}
$$

Here, A is called *force factor*, which converts "mechanical to electrical" or "electrical to mechanical" parameters. Y_d and Z_1 are *electrical admittance* and *mechanical impedance*, respectively. The mechanical impedance here is a macroscopic ("integrated") relation between velocity (dynamic flow) and force (potential), while the mechanical impedance discussed in Chapter 4 is a microscopic ("derivative") formula in a material.

It is important to note here that we can obtain a special displacement/strain dynamic mode such as linear or quadratic time dependence according to the transient applied voltage wave in the "derivative" formula treatment, while in the "integrated" formula, the output displacement/strain shows only the sinusoidal shape because only discrete electronic components (L, C and R) are handled. Comparing Eqs. (5.137) and (5.138), we can derive the relations (based on harmonic oscillation)

$$
A = A_0, \quad Y_d = j\omega C_d, \quad \text{and} \quad Z_1 = c_1/j\omega
\tag{5.139}
$$

From Eq. (5.138), it is not difficult to structure an EC: (1) We can consider 2 ports (or 4 terminals calculating the ground lines): electrical port with voltage V and current I, and mechanical port with force F (compressive pressure on the both ends symmetrically) and velocity $\dot{u}$ (vibration velocity at the both ends) (refer to Fig. 5.15(a)). (2) We can use a "transformer" from "voltage to force" or "velocity to current" exchange with a sort of step-up ratio (called *force factor*) A. Figure 5.15(b) illustrates an EC model for the k_{31} piezo-plate. Remember that the equivalency of EC to the actual piezoelectric device is valid merely under the harmonic oscillation.

We consider next the electromechanical coupling calculation from the "integrated" formula Eq. (5.138) (or EC) under the voltage application process. Because the electromechanical coupling factor k is defined by

$$
k^2 = \frac{\text{Induced mechanical power}}{\text{input electrical power}}
\tag{5.140}
$$

under the condition $F_1 = 0$ (zero mechanical potential), and $\dot{u} = \frac{A}{Z_1}V$, the induced mechanical power (kinetic energy) can be calculated by

$$
W_M = \frac{1}{2}Z_1\dot{u}^2 = \frac{1}{2}\frac{A^2}{Z_1}V^2,
\tag{5.141}
$$

while the input electrical power is

$$W_E = \frac{1}{2}VI = \frac{1}{2}V\left(A\dot{u} + Y_d V\right) = \frac{1}{2}Y_d\left(1 + \frac{A^2}{Z_1 Y_d}\right)V^2 \tag{5.142}$$

Thus, the electromechanical coupling factor k_v in the EC can be obtained as

$$k_v^2 = \frac{W_M}{W_E} = \frac{\frac{A^2}{Z_1 Y_d}}{\left(1 + \frac{A^2}{Z_1 Y_d}\right)} \tag{5.143}$$

The reader can easily translate Eq. (5.142) such that the input electrical energy is composed of pure electric energy $\frac{1}{2}Y_d V^2$ and converted mechanical energy $\frac{1}{2}\frac{A^2}{Z_1}V^2$. It is worth noting that since the electrical admittance Y_d and mechanical impedance Z_1 are frequency dependent, as seen from Eq. (139), electromechanical coupling factor k_v may also be frequency dependent, in general. We can also derive the fact that $k_v \rightarrow$ material's k with $\omega \rightarrow 0$, as verified below.

Let us derive the k_v in terms of the material's constants in the case of k_{31} piezo-plate in order to verify the above conclusion. We will put $X_1 = F/wb$, $E_1 = V/b$, $x_1 = \frac{\dot{u}}{j\omega L}$, and $D_3 = \frac{I}{j\omega wL}$ into

$$X_1 = \frac{1}{s_{11}^E}x_1 - \frac{d_{31}}{s_{11}^E}E_3 \tag{5.144}$$

$$D_3 = d_{31}X_1 + \varepsilon_0\varepsilon_{33}^X E_3 = \frac{d_{31}}{s_{11}^E}x_1 + \varepsilon_0\varepsilon_{33}^X(1 - \frac{d_{31}^2}{\varepsilon_0\varepsilon_{33}^X s_{11}^E})E_3, \tag{5.145}$$

we obtain

$$F = \frac{wb}{j\omega L s_{11}^E}\dot{u} - \frac{wd_{31}}{s_{11}^E}V \tag{5.146}$$

$$I = \frac{wd_{31}}{s_{11}^E}\dot{u} + j\omega\left(\frac{Lw}{b}\right)\varepsilon_0\varepsilon_{33}^X\left(1 - \frac{d_{31}^2}{\varepsilon_0\varepsilon_{33}^X s_{11}^E}\right)V \tag{5.147}$$

From Eqs. (5.138), (5.146) and (5.147), under a static ($\omega \rightarrow 0$) condition,

$$A = \frac{wd_{31}}{s_{11}^E}, Z_1 = \frac{wb}{j\omega L s_{11}^E}, \quad \text{and} \quad Y_d = j\omega\left(\frac{Lw}{b}\right)\varepsilon_0\varepsilon_{33}^X\left(1 - \frac{d_{31}^2}{\varepsilon_0\varepsilon_{33}^X s_{11}^E}\right). \tag{5.148}$$

Thus, we obtain the vibrator electromechanical coupling factor k_v from Eq. (5.143):

$$k_v^2 = \frac{\frac{A^2}{Z_1 Y_d}}{\left(1 + \frac{A^2}{Z_1 Y_d}\right)} = \frac{d_{31}^2}{\varepsilon_0\varepsilon_{33}^X s_{11}^E}, \tag{5.149}$$

which is exactly the same as the material's k_{31}.

5.4.2.2 *Equivalent circuit construction*

Under the stress-free condition $F_1 = 0$ (zero mechanical potential), and $\dot{u} = \frac{A}{Z_1}V$, let us now consider the electrical admittance Y. Since Eq. (5.147) is transformed into

$$I = A \cdot \dot{u} + Y_d \cdot V = \left(Y_d + \frac{A^2}{Z_1}\right)V, $$

the admittance is obtained as

$$Y = \frac{I}{V} = \left(Y_d + \frac{A^2}{Z_1}\right). \tag{5.150}$$

Recall the content in Section 2.8.2, where the admittance of the k_{31} mode piezo-plate was described as

$$Y = j\omega C_d \left[1 + \frac{k_{31}^2}{1 - k_{31}^2} \frac{\tan(\omega L/2v_{11}^E)}{(\omega L/2v_{11}^E)} \right]. \tag{5.151}$$

Correlating the above with Eq. (5.150), we obtain the damped admittance

$$Y_d = j\omega C_d, \tag{5.152}$$

and motional admittance,

$$\frac{A^2}{Z_1} = j\omega C_d \frac{k_{31}^2}{1 - k_{31}^2} \frac{\tan(\omega L/2v_{11}^E)}{(\omega L/2v_{11}^E)}, \tag{5.153}$$

and with *force constant* $A = \frac{wd_{31}}{s_{11}^E}$, *dynamic mechanical impedance* as

$$Z_1 = jZ_0 \cdot \tan(\omega L/2v_{11}^E), \quad Z_0 = wb\rho v_{11}^E = wb\sqrt{\frac{\rho}{s_{11}^E}}. \tag{5.154}$$

Note that Z_0 is the integrated value of material's mechanical impedance $\rho v_{11}^E = \sqrt{\frac{\rho}{s_{11}^E}}$.

Taking *Mittag-Leffler's theorem* of $\frac{\tan(\omega L/2v_{11}^E)}{(\omega L/2v_{11}^E)}$ around $\omega_{R,n}$(nth resonance mode), we get

$$\frac{\tan(\omega L/2v_{11}^E)}{(\omega L/2v_{11}^E)} = \sum_{n:odd}^{\infty} \left(\frac{8}{n^2\pi^2} \right) / [1 - ((\omega L/2v_{11}^E)/(\omega_{R,n}L/2v_{11}^E))^2] \tag{5.155}$$

When we use an LC series connection EC model in Fig. 5.15(b) on the motional branch just around the resonance frequency peak region, we convert the mass contribution to L and elastic compliance to C, and create L_n and C_n series connections as shown in Fig. 5.16(a). Note that the two-terminal EC is enough for the force-free condition. Each pair of (L_1, C_1), (L_3, C_3), ... (L_n, C_n) contributes to the fundamental, second and the n-th resonance vibration mode, respectively. Remember that n is only for the odd-number, or even-number n does not show up in the piezoelectric resonance, which corresponds basically to the antiresonance mode. Though each branch is activated only at its own n-th resonance frequency, the capacitance's contribution remains even at an inactive frequency range, in particular, at a low-frequency range (Note the impedance of capacitance, $1/j\omega C$, and inductance, $j\omega L$. Under a low-frequency region, the inductance contribution will disappear). Now, we can express the total admittance as

$$Y = j\omega C_d + \sum_{n:odd}^{\infty} 1/j \left(\omega L_n - \frac{1}{\omega C_n} \right)$$

$$= j\omega C_d \left[1 + \sum_{n:odd}^{\infty} \left(\frac{\frac{C_n}{C_d}}{1 - \frac{\omega^2}{\omega_{R,n}^2}} \right) \right] [\omega_{R,n}^2 = 1/L_n C_n] \tag{5.156}$$

5.4.2.3 *Capacitance factor and capacitance ratio*[14]

Knowing the relation $C_d = (1 - k_v^2)C_0$, where C_d and C_0 are the "damped" and "free" capacitances, respectively, we define the "*capacitance factor*" P_n by

$$P_n = \left(\frac{1}{k_v^2} \right) \frac{C_n}{C_0} \tag{5.157}$$

Since $\sum_{n:odd}^{\infty} C_n = k_v^2 C_0$, $\sum_{n:odd}^{\infty} P_n = 1$, and P_n is also expressed by

$$P_n = \left(\frac{1 - k_v^2}{k_v^2} \right) \frac{C_n}{C_d}. \tag{5.158}$$

Then, we can express the admittance with capacitance factors as

$$Y = j\omega C_d \left[1 + \frac{k_{31}^2}{1 - k_{31}^2} \sum_{n:odd}^{\infty} \left(\frac{P_n}{1 - \frac{\omega^2}{\omega_{R,n}^2}} \right) \right] \tag{5.159}$$

The admittance Y can be expressed by the basic EC parameters, C_d, k_v, P_n and $\omega_{R,n}$. The *capacitance factor* P_n has been already determined for various piezoelectric vibrators,[14] and $P_n = \frac{8}{n^2\pi^2}$, $P_1 = 8/\pi^2$ for longitudinal vibrators (such as k_{31} and k_{33} modes) from Eq. (5.155).

We now introduce another parameter *capacitance ratio* γ_n defined by

$$\gamma_n = \frac{C_d}{C_n}. \tag{5.160}$$

This "capacitance ratio" is the key parameter to provide the maximum of the bandwidth of the piezo transducer in filters. Both electromechanical coupling factor k_v and capacitance ratio γ_n are essential to the high-performance transducer, and these parameters are mutually correlated as

$$k_v^2 = \frac{1}{1 + P_n \gamma_n} \tag{5.161}$$

With increasing k_v, γ_n decreases; also, the following relations are important:

$$k_v^2 = 1 - \frac{C_d}{C_0}, 1 + P_n \gamma_n = \frac{C_0}{C_0 - C_d}. \tag{5.162}$$

Let us introduce how to derive the EC parameters such as C_d, C_1, L_1, k_v from the experimental data with an impedance analyzer measure on a piezoelectric k_{31} plate. We need first the admittance value at an off-resonance frequency (such as 10 kHz), which should correspond to $j\omega C_0$, and C_0 is obtained (typically μF unit). Second, we obtain the fundamental resonance (f_R) and antiresonance frequency (f_A) ($n = 1$) from the admittance maximum and minimum frequencies (typically 40 kHz range). In this case, the EC should be re-written as Fig. 5.16(b), where the initial damped capacitance C_d is replaced by ($C_0 - C_1$), because higher-order C_n's contribute just to capacitance, in addition to the original capacitance C_d [Note ($C_0 - C_1$) > C_d]. As the reader is familiar, the resonance (motional branch circuit) and antiresonance (closed-loop circuit) frequencies of the EC in Fig. 5.16(b) are given by

$$\begin{cases} 2\pi f_{R1} = 1/\sqrt{L_1 C_1} \\ 2\pi f_{A1} = 1/\sqrt{L_1 / \left(\frac{1}{C_1} + \frac{1}{C_0 - C_1} \right)} \end{cases} \tag{5.163}$$

Since the following equation is obtained by eliminating L_1,

$$f_{A1} = f_{R1} \sqrt{\left(1 + \frac{C_1}{C_0 - C_1} \right)} \tag{5.164}$$

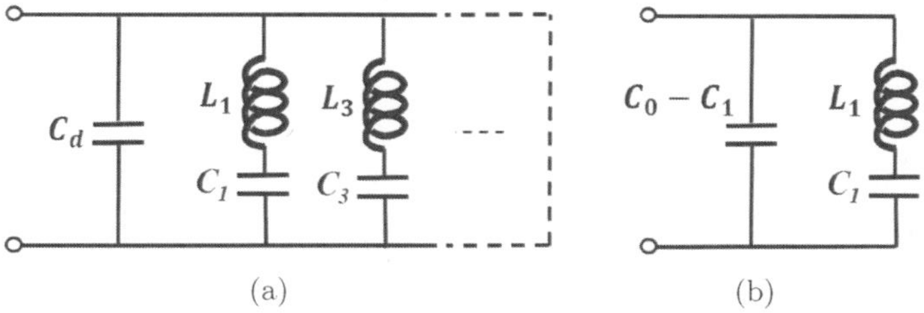

Fig. **5.16:** (a) 2-terminal equivalent circuit for the k_{31}-type piezoelectric plate geometry; (b) 2-terminal equivalent circuit for the fundamental mode in the k_{31} piezo-plate.

we can determine C_1 value. Then, from the resonance frequency expressed by Eq. (5.163), L_1 can be calculated. Finally, using the capacitance factor P_1, the electromechanical coupling factor k_v is given by

$$k_v^2 = \frac{C_n}{P_n C_0}\Big|_{n=1} = \frac{1}{P_1}\left(\frac{f_{A1}^2 - f_{R1}^2}{f_{A1}^2}\right) \tag{5.165}$$

Because $P_1 = \frac{8}{\pi^2}$ for the longitudinal vibration like k_{31}, the k_v is obtained from the resonance and antiresonance frequencies as

$$k_v^2 = \frac{\pi^2}{8}\left(\frac{f_{A1}^2 - f_{R1}^2}{f_{A1}^2}\right) \tag{5.166}$$

You can understand that all necessary EC parameters can be obtained without knowing any material's physical parameters such as permittivity, elastic compliance or piezoelectric constants.

5.4.2.4 *Dynamic electromechanical coupling factor k_{vn}*

When we consider the piezoelectric transducer used at a higher-order harmonic mode, we need to consider a different electromechanical coupling factor (usually smaller than the material's k or static k_v discussed so far).

Let us express the higher-order harmonic modes separately in a 4-terminal EC for the k_{31}-type piezoelectric plate in Fig. 5.17. Accordingly, the constitutive equations are described as

$$\begin{cases} F_n = Z_n \dot{u}_n - A_n V \\ I = \sum_{n=1}^{\infty} A_n \dot{u}_n + Y_d V \end{cases} \tag{5.167}$$

Under no load ($F_n = 0$) condition, $\dot{u}_n = \frac{A_n}{Z_n}V$, then

$$Y = Y_d + \sum_{n=1}^{\infty} \frac{A_n^2}{Z_n}. \tag{5.168}$$

Thus, the input electric power can be obtained by

$$P_i = \frac{1}{2}YV^2 = \frac{1}{2}\left(Y_d + \sum_{n=1}^{\infty} \frac{A_n^2}{Z_n}\right)V^2. \tag{5.169}$$

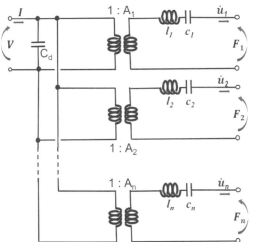

Fig. 5.17: 4-terminal equivalent circuit for the k_{31}-type piezoelectric plate in terms of higher-order harmonic modes.

On the other hand, the output mechanical power of the nth mode in the mechanical branch is estimated by

$$P_{o,n} = \frac{1}{2}Z_n \dot{u}_n^2 = \frac{1}{2}\frac{A_n^2}{Z_n}V^2. \tag{5.170}$$

Now, we define the *dynamic electromechanical coupling factor k_{vn}* by $\left|\frac{P_{o,n}}{P_i}\right|$:

$$k_{vn}^2 = \left|\frac{P_{o,n}}{P_i}\right| = \frac{\frac{A_n^2}{Z_n}}{Y_d + \sum_{n=1}^{\infty} \frac{A_n^2}{Z_n}} = \frac{C_n k_v^2}{C_0} = P_n k_v^2. \tag{5.171}$$

In conclusion, n-*th* mode *dynamic* electromechanical coupling factor k_{vn}^2 can be obtain by multiplying the capacitance factor P_n ($P_n = \frac{8}{n^2\pi^2}$ in the case of longitudinal vibration) on the static electromechanical coupling factor k_v^2. Note again that since $\sum_{n=1}^{\infty} P_n = \frac{8}{\pi^2} \sum_{n=1}^{\infty} \frac{1}{n^2} = 1$, the total mechanical energy by summing up all harmonic vibration becomes equal to k_v^2 of the input electrical energy.

Example Problem 5.3.

In a piezo ceramic k_{33} rod as shown in Fig. 5.18, construct the EC, started from the "integrated" constitutive equations described as

$$\begin{bmatrix} F \\ I \end{bmatrix} = \begin{bmatrix} Z_1 & -A \\ A & Y_d \end{bmatrix} \begin{bmatrix} \dot{u} \\ V \end{bmatrix},$$ (P5.3.1)

and calculate the electromechanical coupling factor from the obtained EC.

Hint:

The k_{33} piezo rod (via d_{33}) does not sustain the electric field E or voltage V constant, but keeps the electric displacement D or current I constant in the specimen. Therefore, variable parameters should be $\begin{bmatrix} \dot{u} \\ I \end{bmatrix}$, rather than $\begin{bmatrix} \dot{u} \\ V \end{bmatrix}$ in the k_{31} piezo-plate where V is constant along the vibration direction.

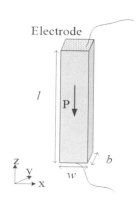

Electrode

Fig. 5.18: k_{33}-type piezo-rod.

Solution:

Because the current I is constant in the longitudinal k_{33} specimen,

$$\begin{cases} F = \left(Z_1 + \frac{A^2}{Y_d} \right) \dot{u} - \frac{A}{Y_d} I = Z_1' \dot{u} + A' I \\ V = -\frac{A}{Y_d} \dot{u} + \frac{1}{Y_d} I = A' \dot{u} + Z_d I \\ Z_1' = \left(Z_1 + \frac{A^2}{Y_d} \right), A' = -\frac{A}{Y_d}, Z_d = \frac{1}{Y_d} \end{cases}$$ (P5.3.2)

we can use the following equation

$$\begin{bmatrix} F \\ V \end{bmatrix} = \begin{bmatrix} Z_1' & A' \\ A' & Z_d \end{bmatrix} \begin{bmatrix} \dot{u} \\ I \end{bmatrix}.$$ (P5.3.3)

Under the no-load ($F = 0$) condition, $\dot{u} = -\frac{A'}{Z_1'} I$, and $V = (Z_d - \frac{A'^2}{Z_1'}) I$. Thus, admittance Y and impedance Z are obtained, respectively, as

$$Y = \frac{1}{(Z_d - \frac{A'^2}{Z_1'})} \quad \text{and} \quad Z = Z_d - \frac{A'^2}{Z_1'}.$$ (P5.3.4)

Taking an analogy to Eq. (5.150) in the case of k_{31} mode, since the second motional term of the impedance, $-\frac{A'^2}{Z_1'}$, is represented by tangent expression of frequency, we will consider *Mittag-Leffler's theorem* around the antiresonance frequency (refer to Eq. (5.155)) which satisfies infinite motional impedance. Considering the static electromechanical coupling factor expressed by $k_v^2 = \frac{\frac{A^2}{Z_1 Y_d}}{(1 + \frac{A^2}{Z_1 Y_d})}\Big|_{\omega \to 0}$, we can derive the following impedance expression:

$$Z = \frac{1}{Y_d} - \frac{A'^2}{Z_1'} = \frac{1}{Y_d} \left[1 - \frac{\frac{A^2}{Z_1 Y_d}}{\left(1 + \frac{A^2}{Z_1 Y_d} \right)} \right] = \frac{1}{j\omega C_d} \left[1 - k_v^2 \sum_{n:odd}^{\infty} \left(\frac{P_n}{1 - \frac{\omega^2}{\omega_{A,n}^2}} \right) \right]$$ (P5.3.5)

where P_n is the *capacitance factor* given by $P_n = \frac{C_n}{C_0}$. Note here that the impedance maximum (i.e., antiresonance mode) corresponds to the primary mechanical resonance mode, and the antiresonance frequency is given by $f_A = v_{33}^D/2l$. Further, taking into account their residues, Eq. (P5.3.5) can be modified into

$$
\begin{cases}
Z = \dfrac{1}{j\omega C_d} + \dfrac{R_0}{\omega} + \displaystyle\sum_{n:odd}^{\infty}\left(\dfrac{R_{n+}}{\omega-\omega_{A,n}} + \dfrac{R_{n-}}{\omega+\omega_{A,n}}\right) \\[2ex]
R_0 = \omega\left[-\dfrac{k_v^2}{j\omega C_d}\displaystyle\sum_{n:odd}^{\infty}\left(\dfrac{P_n}{1-\frac{\omega^2}{\omega_{A,n}^2}}\right)\right] = -\dfrac{k_v^2}{jC_d} \\[3ex]
R_{n+} = R_{n-} = (\omega\pm\omega_{A,n})\left[-\dfrac{k_v^2}{j\omega C_d}\displaystyle\sum_{n:odd}^{\infty}\left(\dfrac{P_n}{1-\frac{\omega^2}{\omega_{A,n}^2}}\right)\right] = \dfrac{1}{2}\dfrac{k_v^2 P_n}{jC_d}
\end{cases}
\tag{P5.3.6}
$$

$$
Z = \dfrac{(1-k_v^2)}{j\omega C_d} + \sum_{n:odd}^{\infty}\left(\dfrac{j\omega}{\omega_{A,n}^2-\omega^2}\right)\left(\dfrac{k_v^2 P_n}{C_d}\right) = \dfrac{1}{j\omega C_0} + \sum_{n:odd}^{\infty}\left(\dfrac{1}{j\omega C_n'+\frac{1}{j\omega L_n'}}\right),
\tag{P5.3.7}
$$

Equation (P5.3.7) leads to a possible EC with a series connection of the impedance components: free-capacitance C_0 and multiple units, each of which is composed of a parallel connection of capacitance C_n' and inductance L_n'. Here,

$$
\begin{cases}
C_n' = \dfrac{C_d}{k_v^2 P_n} \\[2ex]
L_n' = \dfrac{1}{\omega_{A,n}^2 C_n'}
\end{cases}
.
\tag{P5.3.8}
$$

Figure 5.19(a) visualizes the above impedance series connection type EC.

When we translate the k_v^2 in terms of the material's constants,

$$
k_v^2 = \left.\dfrac{\frac{A^2}{Z_1 Y_d}}{\left(1+\frac{A^2}{Z_1 Y_d}\right)}\right|_{\omega\to 0} = k_{33}^2 = \dfrac{d_{33}^2}{s_{33}^E \varepsilon_0 \varepsilon_{33}^X}.
\tag{P5.3.9}
$$

In order to determine the EC components, we initially measure the free-capacitance C_0, resonance and antiresonance frequencies. Since we obtain the following relations from Fig. 5.19(a),

$$
\begin{cases}
\omega_{A1}^2 = \dfrac{1}{L_1' C_1'} \\[2ex]
\omega_{R1}^2 = \dfrac{1}{L_1'(C_0+C_1')}
\end{cases}
,
\tag{P5.3.10}
$$

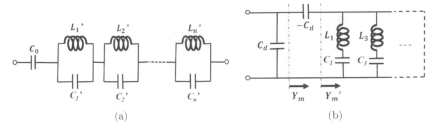

(a)   (b)

Fig. 5.19: Two equivalent circuit models of the k_{33}-type piezo-rod: (a) Impedance series connection and (b) admittance parallel connection types.

we can determine C_1', then L_1' from the antiresonance frequency. Finally, the electromechanical coupling factor k_v^2 can be calculated from

$$\frac{k_v^2}{1-k_v^2} = \frac{1}{P_1}\left(\frac{\omega_{A1}^2 - \omega_{R1}^2}{\omega_{R1}^2}\right)\left[P_1 = \frac{8}{\pi^2}\right]. \qquad \text{(P5.3.11)}$$

Compare the above expression for the k_{33} mode with Eq. (5.165) in the k_{31} type.

We can consider an alternative EC with the admittance parallel connection type. For this purpose, taking an inverse of Eq. (P5.3.7), we obtain the following expression:

$$Y = \frac{j\omega C_d}{\left[1 - k_v^2 \sum_{n:odd}^{\infty}\left(\frac{P_n}{1-\frac{\omega^2}{\omega_{A,n}^2}}\right)\right]} = j\omega C_d + \frac{j\omega C_d}{\left[-1 + 1/k_v^2 \sum_{n:odd}^{\infty}\left(\frac{P_n}{1-\frac{\omega^2}{\omega_{A,n}^2}}\right)\right]}$$

$$= j\omega C_d + \frac{1}{\left[-\frac{1}{j\omega C_d} + \frac{1}{j\omega C_d k_v^2}\sum_{n:odd}^{\infty}\left(\frac{P_n}{1-\frac{\omega^2}{\omega_{A,n}^2}}\right)\right]}. \qquad \text{(P5.3.12)}$$

Equation (P5.3.12) can be translated as a parallel connection of two admittance components, "damped capacitance" and the "second term component". Further, the second term is composed of a series connection of the impedance (which is equal to the "negative" damped capacitance) and multiple units of the L_n and C_n combination (pure motional admittance). Figure 5.19(b) illustrates the admittance parallel connection type EC model of the k_{33}-type piezo rod, where Y_m (motional admittance in a wide meaning) corresponds to the second term of Eq. (P5.3.12),

$$\frac{1}{\left[-\frac{1}{j\omega C_d} + \frac{1}{j\omega C_d k_v^2}\sum_{n:odd}^{\infty}\left(\frac{P_n}{1-\frac{\omega^2}{\omega_{A,n}^2}}\right)\right]},$$

while Y_m' represents the "pure motional admittance" expressed by $j\omega C_d k_v^2 \sum_{n:odd}^{\infty}\left(\frac{P_n}{1-\frac{\omega^2}{\omega_{A,n}^2}}\right)$. Note that the EC of the impedance series-connection type does not include the "negative" capacitance, but that of the admittance parallel connection type only includes the "negative" capacitance. The physical meaning of the "negative capacitance" comes from the "depolarization field" originated from the D-constant condition in the k_{33} mode.

Though we intentionally used different notations L_n, C_n and L_n', C_n' in the admittance or impedance connection ECs, "prime" and "non-prime" can be exactly the same in practice.

5.4.3 *Electromechanical Coupling Factor in Composite Piezo Components*

The EC approach is easier for analyzing composite piezo components; we describe the process for calculating the electromechanical coupling factor in structures such as bimorph, cymbal and multilayer & hinge lever in this subsection.

5.4.3.1 *Bimorph*

The bimorph/unimorph is one of the most popular piezoelectric designs in energy harvesting systems. Though the author indicated that this is the lowest energy harvesting piezo design, as long as the reader needs to use it because of the simple and inexpensive design and the low resonance frequency, we had better analyze it in detail in order to improve the energy harvesting rate as much as possible.

5.4.3.1.1 Derivation of dynamic equation[14]

Figure 5.20 shows a piezoelectric bimorph, which consists of two piezo material plates (thickness t_1) and an elastic shim (usually metal) plate (thickness $2t_2$) sandwiched. When a voltage is applied on the top and bottom electrodes, since the electric field direction is downward and upward, the top and bottom piezo-plates extend and shrink along x-direction (length direction) via the piezoelectric d_{31} effect, respectively, taking into account the polarization directions. Because these

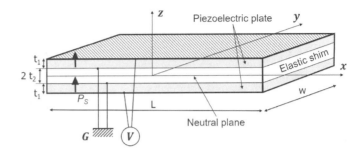

Fig. 5.20:   Piezoelectric bimorph structure.

plates are bonded together with the shim, this bimorph will bend with the tip displacement downward along z-direction. In order to analyze the bending motion, we adopt a simple classical beam theory. The bonding layer is not taken into account in the following analysis. Note that the neutral plane (i.e., stress/strain zero plane) exists in the center of the elastic shim, because of the laminate structure symmetry under bending vibration.

The necessary piezoelectric constitutive equations are the same as the k_{31} type already discussed:

$$\begin{bmatrix} x_1 \\ D_3 \end{bmatrix} = \begin{bmatrix} s_{11}^E & d_{31} \\ d_{31} & \varepsilon_0\varepsilon_{33}{}^X \end{bmatrix} \begin{bmatrix} X_1 \\ E_3 \end{bmatrix}. \tag{5.172}$$

The strain x_1 can be expressed by

$$x_1 = -z\left(\frac{\partial^2\xi}{\partial x^2}\right), \tag{5.173}$$

where ξ is the plate transversal displacement along z-direction. Refer to Example Problem 5.4 for the derivation of Eq. (5.173).

Equation (5.172) leads to the following transformation:

$$\begin{aligned} D_3 &= d_{31}X_1 + \varepsilon_0\varepsilon_{33}^X E_3 = \frac{d_{31}}{s_{11}^E}x_1 + \left(\varepsilon_0\varepsilon_{33}^X - \frac{d_{31}^2}{s_{11}^E}\right)E_3 \\ &= \frac{d_{31}}{s_{11}^E}\left(-z\frac{\partial^2\xi}{\partial x^2}\right) + \varepsilon_0\varepsilon_{33}{}^X(1 - k_{31}^2)E_3 \left[k_{31}^2 = \frac{d_{31}^2}{s_{11}^E\varepsilon_0\varepsilon_{33}{}^X}\right]. \end{aligned} \tag{5.174}$$

Supposing that $\mathrm{div}(D) = \frac{\partial D_3}{\partial z} = 0$, $E_3 = -\frac{\partial V}{\partial z}$ and $\frac{\partial^2\xi}{\partial x^2}$ is not dependent on z, we obtain

$$-\frac{d_{31}}{s_{11}^E}\left(\frac{\partial^2\xi}{\partial x^2}\right) - \varepsilon_0\varepsilon_{33}{}^X\left(1 - k_{31}^2\right)\frac{\partial^2 V}{\partial z^2} = 0, \quad \text{or} \quad \frac{\partial^2 V}{\partial z^2} = -\frac{1}{d_{31}}\frac{k_{31}^2}{\left(1 - k_{31}^2\right)}\left(\frac{\partial^2\xi}{\partial x^2}\right). \tag{5.175}$$

Thus, we can obtain a general solution of V as

$$V = -\frac{1}{d_{31}}\frac{k_{31}^2}{\left(1 - k_{31}^2\right)}\left(\frac{\partial^2\xi}{\partial x^2}\right)\left(\frac{1}{2}z^2\right) + A_1 z + A_2. \tag{5.176}$$

Taking into account the boundary conditions, $V = 0$ at $z = \pm t_2$, and $V = V$ at $z = \pm(t_1 + t_2)$, A_1 and A_2 can be determined as

$$\begin{cases} A_1 = \dfrac{1}{d_{31}}\dfrac{k_{31}^2}{\left(1 - k_{31}^2\right)}\left(\dfrac{\partial^2\xi}{\partial x^2}\right)\dfrac{(t_1 + 2t_2)}{2} + \dfrac{1}{t_1}V \\[4mm] A_2 = -\dfrac{1}{d_{31}}\dfrac{k_{31}^2}{\left(1 - k_{31}^2\right)}\left(\dfrac{\partial^2\xi}{\partial x^2}\right)\dfrac{t_2\left(t_1 + t_2\right)}{2} - \dfrac{t_2}{t_1}V \end{cases}. \tag{5.177}$$

Now, we can calculate the electric displacement D_3 by putting $E_3 = -\frac{\partial V}{\partial z} = \frac{1}{d_{31}} \frac{k_{31}^2}{(1-k_{31}^2)} \left(\frac{\partial^2 \xi}{\partial x^2}\right)(z) - A_1$ into Eq. (5.174):

$$D_3 = -\frac{(t_1 + 2t_2)}{2} \frac{d_{31}}{s_{11}^E} \left(\frac{\partial^2 \xi}{\partial x^2}\right) - \frac{1}{t_1} \varepsilon_0 \varepsilon_{33}^X (1 - k_{31}^2)V. \tag{5.178}$$

Note here that the direct "z" dependence in the term $\left(-z\frac{\partial^2 \xi}{\partial x^2}\right)$ of Eq. (5.174) disappears in the above equation. On the other hand, regarding the vibration equation, we obtain

$$X_1 = \frac{1}{s_{11}^E} x_1 - \frac{d_{31}}{s_{11}^E} E_3 = \frac{1}{s_{11}^E}\left(-z\frac{\partial^2 \xi}{\partial x^2}\right) - \frac{d_{31}}{s_{11}^E}\left[\frac{1}{d_{31}} \frac{k_{31}^2}{(1-k_{31}^2)}\left(\frac{\partial^2 \xi}{\partial x^2}\right)(z) - A_1\right]$$

$$= \frac{1}{s_{11}^E(1-k_{31}^2)}\left(-\frac{\partial^2 \xi}{\partial x^2}\right)\left[z - k_{31}^2 \frac{(t_1 + 2t_2)}{2}\right] - \frac{d_{31}}{s_{11}^E} \frac{1}{t_1} V \tag{5.179}$$

We understand that Eqs. (5.179) and (5.178) lead to the bending vibration and the electric admittance equations. Note here that the electromechanical coupling factor of the bimorph device (via the d_{31} transverse piezoelectric effect) is based on the k_{31}, which originates from the k_v of the base piezoelectric plate.

The analysis of bending vibration requires further mechanical knowledge of the *classical beam theory*, such as *bending moment*, *second moment of area* and *shear force*. Try Example Problem 5.4 for learning these issues.

Example Problem 5.4.

(a) We consider a thin beam structure with a plate normal direction z, as shown in Fig. 5.20. When this beam is bent slightly, verify that the strain x_1 along the beam length x-axis can be expressed by

$$x_1 = -z\left(\frac{\partial^2 \xi}{\partial x^2}\right), \tag{P5.4.1}$$

where ξ is the beam perpendicular displacement along z-axis.

(b) Verify that the shear force F on the beam cross-section is given by the moment M derivative with respect to the coordinate x along the beam length.

$$F = \frac{\partial M}{\partial x}. \tag{P5.4.2}$$

Solution:

(a) We can calculate the strain in a bent beam as follows. Referring to Fig. 5.21(a), let us first obtain the radius of curvature R in a bent beam. On the *neutral plane*, N-N' distance ds is given by $Rd\theta$, while $\tan\theta\, dx = d\xi$, where ξ is the beam bent displacement along z-direction:

$$\begin{cases} \dfrac{d\theta}{ds} = \dfrac{1}{R} \\ \dfrac{d\xi}{dx} = \tan\theta \end{cases} \tag{P5.4.3}$$

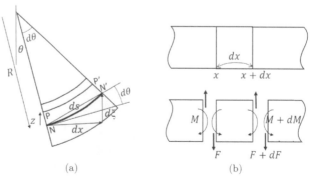

(a)   (b)

Fig. 5.21: (a) Bent beam and radius of curvature. (b) Shear force and moment relation.

Thus, $\sec^2\theta \cdot \frac{d\theta}{ds} = \frac{d^2\xi}{dx^2} \cdot \frac{dx}{ds} = \frac{d^2\xi}{dx^2} \cdot \cos\theta$ and

$$\frac{1}{R} = \frac{d\theta}{ds} = \frac{d^2\xi}{dx^2}\frac{1}{\sec^3\theta} = \frac{d^2\xi}{dx^2}\frac{1}{(1+\tan^2\theta)^{3/2}} = \frac{d^2\xi}{dx^2} \Big/ \left(1 + \left(\frac{d\xi}{dx}\right)^2\right)^{3/2} \approx \frac{d^2\xi}{dx^2}. \qquad (P5.4.4)$$

The strain along x-axis is zero on the neutral plane N–N' ($z = 0$). On the plane P–P' at $z = z$, the shrinkage occurs. Since N–$N' = Rd\theta$ and P–$P' = (R$–$z)d\theta$, the strain x_1 along x-axis is given by

$$x_1 = \frac{PP' - NN'}{NN'} = -\frac{z}{R} \approx -z\frac{d^2\xi}{dx^2}. \qquad (P5.4.5)$$

(b) Let us consider the shear stress F and moment M for a small division between x and $(x + dx)$ on a beam. Referring to Fig. 5.21(b), we can describe the relation as

$$(M + dM) = M + Fdx \qquad (P5.4.6)$$

Thus, we can derive

$$F = \frac{dM}{dx}. \qquad (P5.4.7)$$

In the case of a uniform metal beam, using its *Young's modulus* Y_M, let us calculate the bending moment.

Since the stress along x-axis at $z = z$ is given by

$$X_1 = Y_M x_1 = -Y_M z\frac{d^2\xi}{dx^2}. \qquad (P5.4.8)$$

the moment is calculated as

$$M = \iint_A X_1 z dA = -Y_M \frac{d^2\xi}{dx^2}\iint_A z^2 dA = -K\frac{d^2\xi}{dx^2}. \qquad (P5.4.9)$$

Here, *bending stiffness* $K = Y_M \cdot I$, and *second moment of area(area moment of inertia)* $I = \iint_A z^2 dA$. The *shear force* is expressed now by

$$F = \frac{dM}{dx} = -Y_M I\frac{d^3\xi}{dx^3}. \qquad (P5.4.10)$$

The *bending moment* of the beam at a cross-section can be calculated by $\int$ (normal stress) $\cdot zdz$ because the structure is symmetrical with respect to $z = 0$. Though the beam length strain x_1 is given by $\left(-z\frac{\partial^2\xi}{\partial x^2}\right)$, the normal force is different among the elastic shim (Young's modulus Y_M) and piezo-plate. Taking into account the symmetry with respect to $z = 0$, the bending moment M can be calculated as

$$\begin{aligned}
M &= 2\int_0^w dy\left[\int_0^{t_2}\left(-Y_M z\frac{\partial^2\xi}{\partial x^2}\right)zdz + \int_{t_2}^{t_2+t_1} X_1 zdz\right]\\[4pt]
&= 2w\left[-\frac{1}{3}Y_M\left(\frac{\partial^2\xi}{\partial x^2}\right)t_2^3 + \int_{t_2}^{t_2+t_1}\left\{\frac{1}{s_{11}^E\left(1-k_{31}^2\right)}\left(-\frac{\partial^2\xi}{\partial x^2}\right)[z - k_{31}^2\frac{(t_1+2t_2)}{2}] - \frac{d_{31}}{s_{11}^E}\frac{1}{t_1}V\right\}zdz\right]\\[4pt]
&= M_V - K_C\left(\frac{\partial^2\xi}{\partial x^2}\right). \qquad (5.180)
\end{aligned}$$

Here,

$$M_V = \frac{d_{31}}{s_{11}^E} w(t_1 + 2t_2)V, \tag{5.181}$$

$$K_C = \frac{2}{3} w \left[\frac{1}{s_{11}^E (1 - k_{31}^2)} t_1 \left(t_1^2 + 3t_1 t_2 + 3t_2^2 \right) + Y_M t_2^3 \right] - w k_{31}^2 \frac{1}{s_{11}^E (1 - k_{31}^2)} t_1 (t_1 + 2t_2)^2. \tag{5.182}$$

Supposing that the shear force F along z-axis on the cross-section is given by the first derivative of the moment M in terms of x (see Example Problem 5.4),

$$F = \frac{\partial M}{\partial x}, \tag{5.183}$$

The acceleration $\left(\frac{\partial^2 \xi}{\partial t^2} \right)$ of a narrow volume mass $\rho \bar{S} dx$ can provide the force difference:

$$(F + dF) - F = \rho \bar{S} dx \left(\frac{\partial^2 \xi}{\partial t^2} \right). \tag{5.184}$$

Here, we introduce the average surface density $\rho \bar{S} = 2w(\rho_1 t_1 + \rho_2 t_2)$. Thus,

$$\rho \bar{S} \left(\frac{\partial^2 \xi}{\partial t^2} \right) = \frac{\partial F}{\partial x} = \frac{\partial^2 F}{\partial x^2} = -K_C \left(\frac{\partial^4 \xi}{\partial x^4} \right). \tag{5.185}$$

Under an assumption of sinusoidal vibration $\xi \propto e^{-i\omega t}$, we introduce $\left(\frac{\partial^2 \xi}{\partial t^2} \right) = -\omega^2 \xi$, then

$$\begin{cases} \left(\frac{\partial^4 \xi}{\partial x^4} \right) - n^4 \xi = 0 \\ n^4 = \frac{\rho \bar{S} \omega^2}{K_C} \end{cases}. \tag{5.186}$$

We introduce a *frequency constant* α so as to express the resonance frequency ω follows:

$$\begin{cases} \omega = \frac{\alpha^2}{L^2} \sqrt{\frac{K_C}{\rho \bar{S}}} \\ \alpha^4 = (nL)^4 = \frac{\rho \bar{S} L^4 \omega^2}{K_C} \end{cases}. \tag{5.187}$$

The frequency constant $\alpha = nL$, which is equal to $\left(\frac{\omega L}{v_{11}^E} \right)$ for the simple k_{31} mode. Thus, $n = 2\pi/\lambda$, which corresponds to the wave vector in the k_{31} mode. Note, however, that in this bending mode, α and n do not have such a simple meaning.

5.4.3.1.2 Resonance mode under free-free condition

If we assume a solution in a form of e^{ax}, when $a = \pm n$, or $\pm jn$, Eq. (5.187) is satisfied. Thus, knowing $e^{\pm nx} = \cosh(nx) \pm \sinh(nx)$, and $e^{\pm jnx} = \cos(nx) \pm \sin(nx)$, a general solution for Eq. (5.186) can be expressed by this summation:

$$\xi = A \cosh(nx) + B \sinh(nx) + C \cos(nx) + D \sin(nx) \tag{5.188}$$

We introduce the boundary condition for both-end free of the bimorph:

$$M = 0 \quad \text{and} \quad F = 0 \text{ at } x = \pm L/2.$$

That results in

$$\begin{cases} M = 0 \rightarrow \left(\dfrac{\partial^2 \xi}{\partial t^2}\right) = \dfrac{M_V}{K_C} \\ F = 0 \rightarrow \left(\dfrac{\partial^3 \xi}{\partial t^3}\right) = 0 \end{cases}. \tag{5.189}$$

Using a *frequency constant* α introduced as $\alpha = nL$, we can generate the necessary four equations for determining A, B, C and D:

$$\begin{pmatrix} \cosh\left(\frac{\alpha}{2}\right) & \sinh\left(\frac{\alpha}{2}\right) & -\cos\left(\frac{\alpha}{2}\right) & -\sin\left(\frac{\alpha}{2}\right) \\ \sinh\left(\frac{\alpha}{2}\right) & \cosh\left(\frac{\alpha}{2}\right) & \sin\left(\frac{\alpha}{2}\right) & -\cos\left(\frac{\alpha}{2}\right) \\ \cosh\left(\frac{\alpha}{2}\right) & -sinh\left(\frac{\alpha}{2}\right) & -\cos\left(\frac{\alpha}{2}\right) & \sin\left(\frac{\alpha}{2}\right) \\ \sinh\left(\frac{\alpha}{2}\right) & -\cosh\left(\frac{\alpha}{2}\right) & \sin\left(\frac{\alpha}{2}\right) & \cos\left(\frac{\alpha}{2}\right) \end{pmatrix} \begin{pmatrix} A \\ B \\ C \\ D \end{pmatrix} = \begin{pmatrix} \frac{M_V}{n^2 K_C} \\ 0 \\ \frac{M_V}{n^2 K_C} \\ 0 \end{pmatrix}. \tag{5.190}$$

Denoting the determinant

$$\begin{vmatrix} \cosh\left(\frac{\alpha}{2}\right) & \sinh\left(\frac{\alpha}{2}\right) & -\cos\left(\frac{\alpha}{2}\right) & -\sin\left(\frac{\alpha}{2}\right) \\ \sinh\left(\frac{\alpha}{2}\right) & \cosh\left(\frac{\alpha}{2}\right) & \sin\left(\frac{\alpha}{2}\right) & -\cos\left(\frac{\alpha}{2}\right) \\ \cosh\left(\frac{\alpha}{2}\right) & -\sinh\left(\frac{\alpha}{2}\right) & -\cos\left(\frac{\alpha}{2}\right) & \sin\left(\frac{\alpha}{2}\right) \\ \sinh\left(\frac{\alpha}{2}\right) & -\cosh\left(\frac{\alpha}{2}\right) & \sin\left(\frac{\alpha}{2}\right) & \cos\left(\frac{\alpha}{2}\right) \end{vmatrix} = \Delta. \tag{5.191}$$

we can obtain A, B, C and D; for example,

$$B = \frac{1}{\Delta} \begin{vmatrix} \cosh\left(\frac{\alpha}{2}\right) & \frac{M_V}{n^2 K_C} & -\cos\left(\frac{\alpha}{2}\right) & -\sin\left(\frac{\alpha}{2}\right) \\ \sinh\left(\frac{\alpha}{2}\right) & 0 & \sin\left(\frac{\alpha}{2}\right) & -\cos\left(\frac{\alpha}{2}\right) \\ \cosh\left(\frac{\alpha}{2}\right) & \frac{M_V}{n^2 K_C} & -\cos\left(\frac{\alpha}{2}\right) & \sin\left(\frac{\alpha}{2}\right) \\ \sinh\left(\frac{\alpha}{2}\right) & 0 & \sin\left(\frac{\alpha}{2}\right) & \cos\left(\frac{\alpha}{2}\right) \end{vmatrix} = 0. \tag{5.192}$$

Similarly, $D = 0$. This leads to a conclusion that only the odd modes can be excited in the both-end-free bimorph as in Fig. 5.20, without exhibiting even modes, as the symmetry can suggest.

We now obtain

$$\xi = \frac{M_V}{K_C} \frac{L^2}{\alpha^2} \frac{\sin\left(\frac{\alpha}{2}\right)\cosh\left(\frac{\alpha}{L}x\right) - \sinh\left(\frac{\alpha}{2}\right)\cos\left(\frac{\alpha}{L}x\right)}{\sin\left(\frac{\alpha}{2}\right)\cosh\left(\frac{\alpha}{2}\right) + \sinh\left(\frac{\alpha}{2}\right)\cos\left(\frac{\alpha}{2}\right)}. \tag{5.193}$$

The resonance frequencies are obtained from the displacement infinite condition:

$$\sin\left(\frac{\alpha}{2}\right)\cosh\left(\frac{\alpha}{2}\right) + \sinh\left(\frac{\alpha}{2}\right)\cos\left(\frac{\alpha}{2}\right) = 0. \tag{5.194}$$

Note that different from simple k_{31} or k_{33} modes, the bimorph does not show the resonance/antiresonance frequencies provided by simple integer multiplication of the fundamental frequency, but these can be calculated as inserted in Fig. 5.22(a), where the bending displacement profiles are also illustrated. Though the resonance frequency constant α is not related with the piezoelectric constant in the case of k_{31} mode, it should be pointed out that the even-order α_2 mode does not appear in the piezo bimorph. Though the analysis on the cantilever (one-end clamp) mode is skipped, Fig. 5.22(b) illustrates the bending displacement

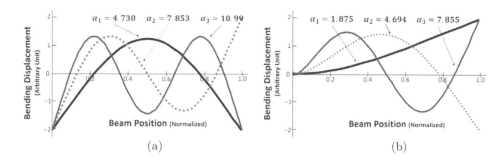

Fig. 5.22: Bimorph bending vibration resonance modes: (a) Both-end-free condition. (b) Cantilever.

profiles and the resonance frequency constants [Try Chapter 5 Problem 5.1]. The even-order α_2 mode DOES appear in the piezo bimorph under the cantilever (one-end clamp) support condition. The bimorph tip displacement exhibits the maximum for all odd- and even-order modes.

The resonance frequency of a beam (one-phase beam), regardless of free-free or one-end clamp cantilever types, is described by a general formula:

$$f = \frac{\alpha_m^2}{2\sqrt{3}(2\pi)}\left(\frac{t}{L^2}\right)\sqrt{\frac{Y}{\rho}}, \tag{5.195}$$

where L is the length of beam, t, its thickness, Y, Young's modulus, ρ, the mass density, and α_m is *frequency constant* determined by the vibration mode. In a two-phase (e.g., with metal shim) bimorph, Young's modulus and mass density should be replaced by the average values.

We now calculate the admittance of the bimorph (free-free condition). From Eq. (5.178) and knowing two piezo-plates in this bimorph, we can calculate the input current as

$$I = \frac{\partial Q}{\partial t} = j\omega 2w\left[-\int_{-\frac{L}{2}}^{\frac{L}{2}}\left\{\frac{1}{t_1}\varepsilon_0\varepsilon_{33}{}^X\left(1-k_{31}^2\right)V\right\}dx - \int_{-\frac{L}{2}}^{\frac{L}{2}}\frac{(t_1+2t_2)}{2}\frac{d_{31}}{s_{11}^E}\left(\frac{\partial^2\xi}{\partial x^2}\right)dx\right]$$

$$= -j\omega\varepsilon_0\varepsilon_{33}{}^X\left(1-k_{31}^2\right)\frac{2wL}{t_1}V - j\omega w\left(t_1+2t_2\right)\frac{d_{31}}{s_{11}^E}\left(\frac{\partial\xi}{\partial x}\right)\Big|_{-L/2}^{L/2}. \tag{5.196}$$

Using Eq. (5.193), admittance $Y=\left(-\frac{I}{V}\right)$ can be calculated as

$$Y = Y_d + Y_m$$

$$= j\omega\varepsilon_0\varepsilon_{33}{}^X(1-k_{31}^2)\frac{2wL}{t_1} + j\omega\left(\frac{d_{31}}{s_{11}^E}\right)^2\left(\frac{1}{K_c}\right)(t_1+2t_2)^2 w^2 L\left[\left(\frac{1}{\alpha}\right)\frac{4\tanh\left(\frac{\alpha}{2}\right)\tan\left(\frac{\alpha}{2}\right)}{\tanh\left(\frac{\alpha}{2}\right)+\tan\left(\frac{\alpha}{2}\right)}\right] \tag{5.197}$$

The maximum admittance (i.e., the resonance frequency) is obtained when $\tanh\left(\frac{\alpha}{2}\right)+\tan\left(\frac{\alpha}{2}\right)=0$, which is the same as the condition of Eq. (5.194).

5.4.3.1.3 Electromechanical coupling factor k_{eff} of the bimorph
When we define the electromechanical coupling factor k_{eff} by the damped and motional admittance ratio as $(1-k_{\text{eff}}^2):k_{\text{eff}}^2 = j\omega\varepsilon_0\varepsilon_{33}{}^X(1-k_{31}^2)\frac{2wL}{t_1}:j\omega(\frac{d_{31}}{s_{11}^E})^2(\frac{1}{K_c})(t_1+2t_2)^2 w^2 L[\frac{1}{\alpha}\frac{4\tanh\left(\frac{\alpha}{2}\right)\tan\left(\frac{\alpha}{2}\right)}{\tanh\left(\frac{\alpha}{2}\right)+\tan\left(\frac{\alpha}{2}\right)}]$, and knowing

$\left[\frac{1}{\alpha} \frac{4 \tanh\left(\frac{\alpha}{2}\right) \tan\left(\frac{\alpha}{2}\right)}{\tanh\left(\frac{\alpha}{2}\right) + \tan\left(\frac{\alpha}{2}\right)} \right] \to 1$ for the static condition $\omega \to 0$, we obtain the following equation:

$$
\begin{cases}
\dfrac{k_{\text{eff}}^2}{\left(1 - k_{\text{eff}}^2\right)} = \left(\dfrac{d_{31}}{s_{11}^E}\right)^2 \left(\dfrac{1}{K_c}\right) \left(t_1 + 2t_2\right)^2 w^2 L / \varepsilon_0 \varepsilon_{33}^X \left(1 - k_{31}^2\right) \dfrac{2wL}{t_1} = \dfrac{k_{31}^2}{\left(1 - k_{31}^2\right)} \dfrac{\left(t_1 + 2t_2\right)^2 t_1 w}{2 s_{11}^E} \left(\dfrac{1}{K_c}\right) \\[4mm]
K_C = \dfrac{2}{3} w \left[\dfrac{1}{s_{11}^E \left(1 - k_{31}^2\right)} t_1 \left(t_1^2 + 3t_1 t_2 + 3t_2^2\right) + Y_M t_2^3 \right] - w k_{31}^2 \dfrac{1}{s_{11}^E \left(1 - k_{31}^2\right)} t_1 \left(t_1 + 2t_2\right)^2
\end{cases}
$$

$$(5.198)$$

Equation (5.198) provides the k_{eff}, which is directly dependent on the size.

Let us calculate k_{eff} for three extreme cases:

- No metal shim ($t_2 = 0$): Since $\dfrac{k_{\text{eff}}^2}{\left(1 - k_{\text{eff}}^2\right)} = \dfrac{k_{31}^2}{2\left(\frac{2}{3} - k_{31}^2\right)}$ is obtained.

$$
k_{\text{eff}} = \sqrt{\frac{3}{4}} \frac{k_{31}}{\sqrt{1 - \frac{3}{4} k_{31}^2}} \approx \sqrt{\frac{3}{4}} k_{31}
\tag{5.199}
$$

Without shim, the piezo bimorph has an electromechanical coupling factor determined by the pure physical parameters, $\sqrt{\frac{3}{4}} k_{31}$, which is, however, smaller than k_{31}.

- $t_1 = t_2$, and we also assume $s_{11}^E \approx 1/Y_M$ just from the simplicity in this case.

$$
\frac{k_{\text{eff}}^2}{\left(1 - k_{\text{eff}}^2\right)} = \frac{k_{31}^2}{\left(1 - k_{31}^2\right)} \frac{9 t_1^3 w}{2 s_{11}^E} \left(\frac{1}{K_c}\right)
$$

$$
K_c = \frac{w t_1^3 \left(\frac{16}{3} - \frac{29}{3} k_{31}^2\right)}{s_{11}^E \left(1 - k_{31}^2\right)}.
$$

Then,

$$
\frac{k_{\text{eff}}^2}{\left(1 - k_{\text{eff}}^2\right)} = \frac{27}{32} \frac{k_{31}^2}{\left(1 - \frac{29}{16} k_{31}^2\right)} \to k_{\text{eff}} \approx \sqrt{\frac{27}{32}} k_{31}.
\tag{5.200}
$$

The k_{eff} is slightly higher than the case of no shim, though it is still smaller than the pure k_{31} mode.

- Thin piezo-plate ($t_1/t_2 \ll 1$) — Most popular in the published researches: Since $K_C \approx \frac{2}{3} w Y_M t_2^2$,

$$
\frac{k_{\text{eff}}^2}{\left(1 - k_{\text{eff}}^2\right)} = \frac{k_{31}^2}{\left(1 - k_{31}^2\right)} \left(\frac{3}{Y_M s_{11}^E}\right) \left(\frac{t_1}{t_2}\right).
\tag{5.201}
$$

The electromechanical coupling factor of the bimorph depends on both the elastic property difference among the piezo s_{11}^E and Y_M and the thickness ratio $\left(\frac{t_1}{t_2}\right)$. If we suppose that mechanical energy ($\propto k_{31}^2$) converted from the input electric energy is transferred to both piezo-plates and metal plates, the k_{eff}^2 decrease with the volume ratio (which is the same as thickness ratio) seems to be reasonable. When we take an example PZT4 with $s_{11}^E = 12.3 \times 10^{-12} \text{m}^2/\text{N}$, depending on the shim materials, steel ($Y_M = 210 \times 10^9 \text{N/m}^2$), brass ($Y_M = 92 \times 10^9 \text{N/m}^2$) or aluminum ($Y_M = 75 \times 10^9 \text{N/m}^2$), the product $Y_M s_{11}^E$ becomes 2.58, 1.13 or 0.92, respectively. Thus, a rough estimation is made by

$$
k_{\text{eff}} \approx 1.2 \sim 1.8 \sqrt{\left(\frac{t_1}{t_2}\right)} k_{31}.
\tag{5.202}
$$

A simple conclusion: the elastic shim thickness should be adjusted in the same range with the piezoelectric plate, so that the electromechanical coupling factor is maintained in a similar order (though lower) to the

k_{31} value. When the shim thickness is 10 times thicker than the piezo-plate, the bimorph k_{eff} becomes around only 10%, in comparison with a typical $k_{31} = 34\%$. Try Example Problems 5.5 and 5.6 for further understanding.

Example Problem 5.5.

A unimorph bending actuator can be fabricated by bonding one piezo ceramic plate to a metallic shim.[1,15] The tip deflection, δ, of the unimorph supported in a cantilever configuration is given by

$$\delta = \frac{d_{31} E L^2 Y_c t_c}{(Y_m[t_o^2 - (t_o - t_m)^2] + Y_c[(t_o + t_c)^2 - t_o^2])} \tag{P5.5.1}$$

Here, E is the electric field applied to the piezoelectric ceramic, d_{31}, the piezoelectric strain coefficient, L, the length of the unimorph, Y, Young's modulus for the ceramic or the metal, and t is the thickness of each material. The subscripts c and m denote the ceramic and the metal, respectively. The quantity, t_0, is the distance between the *strain-free neutral plane* and the bonding surface, and is defined according to the following:

$$t_0 = \frac{t_c t_m^2 (3t_c + 4t_m) Y_m + t_c^4 Y_c}{6 t_c t_m (t_c + t_m) Y_m}. \tag{P5.5.2}$$

Assuming $Y_c = Y_m$ (e.g., PZT and Brass usage popular in speaker/microphone application), calculate the optimum (t_m/t_c) ratio that will maximize the deflection, δ, under the following conditions:

(a) A fixed ceramic thickness (because you are a purchaser of a PZT plate, you need to choose the manufacturer's standard samples), t_c, and

(b) a fixed total thickness (you need to keep the resonance frequency, because you are a speaker and buzzer manufacturer for a particular resonance frequency), $t_c + t_m$.

Solution:

Setting $Y_c = Y_m$, Equations (P5.5.1) and (P5.5.2) become

$$\delta = \frac{d_{31} E L^2 t_c}{([t_o^2 - (t_o - t_m)^2] + [(t_o + t_c)^2 - t_o^2])}, \tag{P5.5.3}$$

$$t_0 = \frac{t_c t_m^2 (3t_c + 4t_m) + t_c^4}{6 t_c t_m (t_c + t_m)}. \tag{P5.5.4}$$

Substituting t_0 as it is expressed in Eq. (P5.5.4) into Eq. (P5.5.3) yields

$$\delta = \frac{d_{31} E L^2 3 t_m t_c}{(t_m + t_c)^3}. \tag{P5.5.5}$$

(a) The function $f(t_m) = (t_m t_c)/(t_m + t_c)^3$ must be maximized for a fixed ceramic thickness, t_c.

$$\frac{df(t_m)}{dt_m} = \frac{(t_c - 2t_m)t_c}{(t_m + t_c)^4} = 0. \tag{P5.5.6a}$$

The metal plate thickness should be $t_m = t_c/2$ and $t_o = t_c/2$.

(b) Equation (P5.5.5) is maximized under a fixed total thickness, $t_{tot} = t_c + t_m$:

$$\frac{df(t_m)}{dt_m} = \frac{(t_{tot} - 2t_m)}{t_{tot}^3} = 0 \tag{P5.5.6b}$$

Thus, it is determined that both the metal and ceramic plate thickness should be $t_m = t_c = t_{tot}/2$ and $t_o = t_{tot}/3$.

5.4.3.1.4 Bimorph equivalent circuit

To create an EC for a bimorph in Fig. 5.20, we start from the following fundamental equations of electromechanical transduction:

$$\begin{bmatrix} F \\ I \end{bmatrix} = \begin{bmatrix} Z_1 & -A \\ A & Y_d \end{bmatrix} \begin{bmatrix} \dot{u} \\ V \end{bmatrix}. \tag{5.203}$$

To create a two-terminal EC, a parallel connection of two admittance branches, Y_d and $\frac{A^2}{Z_1}$, is the solution, while a four-terminal (two-port) EC is created by Y_d in the electric admittance branch (connected in parallel) and Z_1 in the mechanical impedance branch (connected in series) with a transformer with a step-up ration A (i.e., force factor). Let us obtain the force factor A in a bimorph structure. The A is obtained by the ratio of the force F required to compensate the displacement u and velocity $\dot{u}$ at the plate ends over the voltage applied on the bimorph.

$$\begin{cases} M = M_V - K_C \left(\dfrac{\partial^2 \xi}{\partial x^2} \right) = 0 \rightarrow \left(\dfrac{\partial^2 \xi}{\partial t^2} \right) = \dfrac{M_V}{K_C} \\ F' = \dfrac{\partial M}{\partial x} = -K_C \left(\dfrac{\partial^3 \xi}{\partial x^3} \right) = \dfrac{F}{2}. \end{cases} \tag{5.204}$$

Under this force condition, we can obtain the displacement profile as

$$\xi_F = -\frac{F}{2K_C} \frac{1}{n^3} \frac{\cos\left(\frac{\alpha}{2}\right)\cosh\left(\frac{\alpha}{L}x\right) + \cosh\left(\frac{\alpha}{2}\right)\cos\left(\frac{\alpha}{L}x\right)}{\sin\left(\frac{\alpha}{2}\right)\cosh\left(\frac{\alpha}{2}\right) + \sinh\left(\frac{\alpha}{2}\right)\cos\left(\frac{\alpha}{2}\right)}. \tag{5.205}$$

If we assume the above ξ_F equivalent to the expression ξ_V in Eq. (5.204), the force $F(F' = F/2)$ required at $x = \pm L/2$ is represented by

$$F = -M_V \frac{\alpha}{L}\left(\tan\left(\frac{\alpha}{2}\right) - \tanh\left(\frac{\alpha}{2}\right)\right), \quad \text{here} \quad M_V = \frac{d_{31}}{s_{11}^E} w\left(t_1 + 2t_2\right) V. \tag{5.206}$$

Thus, the force factor A is given by

$$A = \frac{F}{V} = -\frac{d_{31}}{s_{11}^E} w\left(t_1 + 2t_2\right) \frac{\alpha}{L}\left(\tan\left(\frac{\alpha}{2}\right) - \tanh\left(\frac{\alpha}{2}\right)\right). \tag{5.207}$$

Using the following *Mittag-Leffler theorem* around the resonance frequencies

$$\frac{\tanh\left(\frac{\alpha}{2}\right)\tan\left(\frac{\alpha}{2}\right)}{\tanh\left(\frac{\alpha}{2}\right) + \tan\left(\frac{\alpha}{2}\right)} = \sum_{n=1}^{\infty} \left\{ \frac{4\alpha^3}{\alpha^4 - \alpha_n^4} \tanh\left(\frac{\alpha_n}{2}\right)\tan\left(\frac{\alpha_n}{2}\right)\right\}. \tag{5.208}$$

the motional admittance term in Eq. (5.197) is expanded as

$$\begin{aligned} Y_m &= \left(\frac{d_{31}\left(t_1 + 2t_2\right)w\alpha}{s_{11}^E L}\right)^2 \sum_{n=1}^{\infty} j\omega \left(\frac{4L^3}{K_c}\right)\left\{\frac{4}{\alpha^4 - \alpha_n^4}\tanh\left(\frac{\alpha_n}{2}\right)\tan\left(\frac{\alpha_n}{2}\right)\right\} \\ &= \left(\frac{d_{31}\left(t_1 + 2t_2\right)w\alpha}{s_{11}^E L}\right)^2 \sum_{n=1}^{\infty} \left(\frac{4L^3}{K_c}\right)\left\{\frac{4\tanh\left(\frac{\alpha_n}{2}\right)\tan\left(\frac{\alpha_n}{2}\right)}{L^4\left(\frac{\rho \bar{A}}{K_c}\right)}\right\}\frac{j\omega}{\omega^2 - \omega_n^2} \\ &= \sum_{n=1}^{\infty} \frac{-j\omega/L_n}{\omega^2 - \omega_n^2}. \end{aligned} \tag{5.209}$$

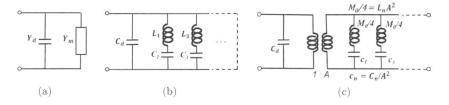

Fig. 5.23: Equivalent circuits for a bimorph bender with the d_{31} effect: (a)Basic idea. (b) 2-terminal circuit. (c) 4-terminal circuit.

Here,

$$
\begin{cases}
L_n = \dfrac{\frac{\rho \bar{A} L}{4}}{\left[\frac{d_{31}(t_1+2t_2)w\alpha n}{s_{11}^E L}\left\{\tanh\left(\frac{\alpha_n}{2}\right)-\tan\left(\frac{\alpha_n}{2}\right)\right\}\right]^2} = \dfrac{M_0}{4}/A^2 \\[4mm]
C_n = \frac{1}{\omega_n^2 L_n} = c_n A^2. \\[2mm]
M_0 = \rho \bar{A} L
\end{cases}
\tag{5.210}
$$

The corresponding ECs are shown in Figs. 5.23(a)–5.23(c).

5.4.3.1.5 Bimorph with k_{33}-Type Piezo-Plates[15]

In the previous subsection, we learned that the bimorph based on the k_{31} piezo-plates exhibits the effective electromechanical coupling factor k_{eff} in proportion to the k_{31}. Though this type of bimorph is popularly used for piezo energy harvesting because of the simple assembly and reasonably low resonance frequency which can be adjusted to the ambient mechanical vibration such as industrial electromagnetic motors, the significant problem is its low electromechanical coupling which reduces the harvesting efficiency. It is, thus, reasonable to utilize k_{33} (roughly 3 times higher than k_{31}) piezo-plates for improving the effective k_{eff}.

We consider here a no-shim bimorph illustrated in Fig. 5.24. The problem in this design is very small C_d, which is practically overcome by the interdigitated electrode in a narrow electrode pitch on the side, like a multilayer structure. The necessary piezoelectric constitutive equations are the same as the k_{31} type already discussed:

$$
\begin{bmatrix} x_1 \\ D_1 \end{bmatrix} = \begin{bmatrix} s_{33}^E & d_{33} \\ d_{33} & \varepsilon_0 \varepsilon_{33}{}^X \end{bmatrix} \begin{bmatrix} X_1 \\ E_1 \end{bmatrix}.
\tag{5.211}
$$

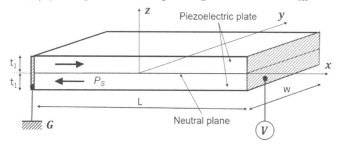

Fig. 5.24: Bimorph bender with the d_{33} effect.

The strain x_1 can be expressed exactly in the same formula as (Eq. 5.173) by

$$
x_1 = -z\left(\frac{\partial^2 \xi}{\partial x^2}\right),
\tag{5.212}
$$

where ξ is the plate-bending transversal displacement along z-axis. Similar to the analysis in the k_{31}-type bimorph, the displacement ξ can be expressed by the even function (because of the symmetry with respect to the x-axis) as

$$
\xi = A\cos\left(\frac{\alpha}{L}x\right) + B\cosh\left(\frac{\alpha}{L}x\right).
\tag{5.213}
$$

The moment M on a certain cross-section is given by

$$M = 2w \int_0^{t_1} X_1 z \, dz = 2w \left\{ \left(-\frac{1}{s_{33}^E \left(1-k_{33}^2\right)} \frac{1}{3} t_1^3 \frac{\partial^2 \xi}{\partial x^2} \right) - \frac{d_{33}}{s_{33}^E \varepsilon_0 \varepsilon_{33}{}^X \left(1-k_{33}^2\right)} \int_0^{t_1} D_1 z \, dz \right\}. \quad (5.214)$$

Regarding the electric potential, since $div\,D = \frac{\partial D_1}{\partial x} = 0$ and $V = 0$ and V at $x = -L/2$ and $L/2$, respectively,

$$\begin{cases} V = -\frac{d_{33}}{s_{33}^E \varepsilon_0 \varepsilon_{33}{}^X \left(1-k_{33}^2\right)} z \frac{\partial \xi}{\partial x} + L_1 x + L_2 \\ L_1 = \frac{V}{L} + \frac{d_{33}}{L s_{33}^E \varepsilon_0 \varepsilon_{33}{}^X \left(1-k_{33}^2\right)} \frac{2\alpha}{L} \left[-A \sin\left(\frac{\alpha}{2}\right) + B \sinh\left(\frac{\alpha}{2}\right) \right] z. \end{cases} \quad (5.215)$$

Now, we can calculate D_1,

$$D_1 = -\frac{d_{33}}{s_{33}^E} z \left(\frac{\partial^2 \xi}{\partial x^2} \right) + \varepsilon_0 \varepsilon_{33}{}^X \left(1-k_{33}^2\right) \left(-\frac{\partial V}{\partial x} \right)$$

$$= \frac{V}{L} \varepsilon_0 \varepsilon_{33}{}^X \left(1-k_{33}^2\right) - \frac{d_{33}}{L s_{33}^E} \frac{2\alpha}{L} \left[-A \sin\left(\frac{\alpha}{2}\right) + B \sinh\left(\frac{\alpha}{2}\right) \right] z,$$

$$\int_0^{t_1} D_1 z \, dz = -\frac{V}{L} \varepsilon_0 \varepsilon_{33}{}^X \left(1-k_{33}^2\right) \frac{t_1^2}{2} - \frac{d_{33}}{L s_{33}^E} \frac{2\alpha}{L} \left[-A \sin\left(\frac{\alpha}{2}\right) + B \sinh\left(\frac{\alpha}{2}\right) \right] \frac{t_1^3}{3}. \quad (5.216)$$

Since

$$\begin{cases} M = 2w \left\{ -\frac{t_1^3}{3} \frac{1}{s_{33}^E \left(1-k_{33}^2\right)} \frac{\partial^2 \xi}{\partial x^2} + \frac{V}{L} \frac{t_1^2}{2} \frac{d_{33}}{s_{33}^E} + \frac{d_{33}^2}{L s_{33}^{E2} \varepsilon_0 \varepsilon_{33}{}^X \left(1-k_{33}^2\right)} \frac{2\alpha}{L} \left[-A \sin\left(\frac{\alpha}{2}\right) + B \sinh\left(\frac{\alpha}{2}\right) \right] \frac{t_1^3}{3} \right\} \\ F = \frac{\partial M}{\partial x} = -2w \frac{t_1^3}{3} \frac{1}{s_{33}^E \left(1-k_{33}^2\right)} \frac{\partial^3 \xi}{\partial x^3} \\ \frac{\partial^2 \xi}{\partial x^2} \big|_{M=0} = \frac{V}{L} \frac{t_1^2}{2} \frac{d_{33}}{s_{33}^E K_0} + \frac{d_{33}^2}{s_{33}^{E2} \varepsilon_0 \varepsilon_{33}{}^X \left(1-k_{33}^2\right) K_0} \frac{1}{L} \frac{2\alpha}{L} \frac{t_1^3}{3} \left[-A \sin\left(\frac{\alpha}{2}\right) + B \sinh\left(\frac{\alpha}{2}\right) \right] \\ K_0 = \frac{t_1^3}{3} \frac{1}{s_{33}^E \left(1-k_{33}^2\right)}. \end{cases} \quad (5.217)$$

We can describe the relationship for obtaining the parameters A and B:

$$\begin{cases} \begin{pmatrix} R \sin\left(\frac{\alpha}{2}\right) - \cos\left(\frac{\alpha}{2}\right) & \cosh\left(\frac{\alpha}{2}\right) - R \sinh\left(\frac{\alpha}{2}\right) \\ \sin\left(\frac{\alpha}{2}\right) & \sinh\left(\frac{\alpha}{2}\right) \end{pmatrix} \begin{pmatrix} A \\ B \end{pmatrix} = \begin{pmatrix} \frac{d_{33}}{2 s_{33}^E} t_1^2 \left(\frac{L}{\alpha}\right)^2 \frac{V}{L K_0} \\ 0 \end{pmatrix} \\ R = \frac{d_{33}^2}{s_{33}^{E2} \varepsilon_0 \varepsilon_{33}{}^X \left(1-k_{33}^2\right)} \frac{t_1^3}{3 K_0} \frac{2}{\alpha} \\ \alpha^2 = \omega L^2 \sqrt{\rho s_{33}^E \left(1-k_{33}^2\right)} \cdot \sqrt{\frac{2 w t_1}{\frac{w}{12} (2 t_1)^3}} \end{cases} \quad (5.218)$$

The resonance status is obtained from the determinant zero condition under $V = 0$:

$$\Delta = \begin{vmatrix} R \sin\left(\frac{\alpha}{2}\right) - \cos\left(\frac{\alpha}{2}\right) & \cosh\left(\frac{\alpha}{2}\right) - R \sinh\left(\frac{\alpha}{2}\right) \\ \sin\left(\frac{\alpha}{2}\right) & \sinh\left(\frac{\alpha}{2}\right) \end{vmatrix} = 0. \quad (5.219)$$

Because of R, which includes piezoelectric constant d_{33}, the resonance frequency is not as simple as the k_{31} type in Eq. (5.159), which is irrelevant to the piezoelectricity. Now, the parameters A and B are

$$\begin{cases} A = \frac{1}{\Delta} \left(\frac{d_{33}}{2 s_{33}^E} t_1^2 \left(\frac{L}{\alpha}\right)^2 \frac{V}{L K_0} \right) \sinh\left(\frac{\alpha}{2}\right) \\ B = \frac{1}{\Delta} \left(\frac{d_{33}}{2 s_{33}^E} t_1^2 \left(\frac{L}{\alpha}\right)^2 \frac{V}{L K_0} \right) \left(-\sin\left(\frac{\alpha}{2}\right) \right) \end{cases}. \quad (5.220)$$

Then, the admittance is expressed by

$$Y = j\omega \frac{2wt_1}{L} \varepsilon_0 \varepsilon_{33}{}^X \left(1 - k_{33}^2\right) + j\omega \frac{2wt_1^4 d_{33}^2}{2\alpha s_{33}^{E2}} \left(\frac{1}{K_0 L}\right) \left[\frac{-2\sinh\left(\frac{\alpha}{2}\right)\sin\left(\frac{\alpha}{2}\right)}{\Delta}\right]. \tag{5.221}$$

Taking into account that $\left[\frac{-2\sinh\left(\frac{\alpha}{2}\right)\sin\left(\frac{\alpha}{2}\right)}{\frac{\alpha}{2}\Delta}\right] \to 1$ for $\left(\frac{\alpha}{2}\right) \to 0$, Eq. (5.221) approaches to

$$Y|_{\alpha \to 0} = j\omega \frac{2wt_1}{L} \varepsilon_0 \varepsilon_{33}{}^X \left(1 - k_{33}^2\right) + j\omega \frac{2wt_1^4 d_{33}^2}{4 s_{33}^{E2}} \left(\frac{1}{K_0 L}\right). \tag{5.222}$$

We consider the effective electromechanical coupling factor k_{eff} by the damped and motional admittance ratio as $\left(1 - k_{\text{eff}}^2\right) : k_{\text{eff}}^2 = j\omega \frac{2wt_1}{L} \varepsilon_0 \varepsilon_{33}{}^X \left(1 - k_{33}^2\right) : j\omega \frac{2wt_1^4 d_{33}^2}{4 s_{33}^{E2}} \left(\frac{1}{K_0 L}\right)$, we obtain the following equation:

$$\begin{cases} \frac{k_{\text{eff}}^2}{\left(1 - k_{\text{eff}}^2\right)} = \frac{t_1^3 d_{33}^2}{4 s_{33}^{E2}} \frac{1}{\varepsilon_0 \varepsilon_{33}{}^X \left(1 - k_{33}^2\right)} \left(\frac{1}{K_0}\right) \\ K_0 = \frac{t_1^3}{3} \frac{1}{s_{33}^E \left(1 - k_{33}^2\right)}. \end{cases} \tag{5.223}$$

Finally,

$$k_{\text{eff}} \approx \sqrt{\frac{3}{4}} k_{33}. \tag{5.224}$$

Note here that the formula is the same as Eq. (5.199) in the k_{31} type, which indicates that the factor, $\sqrt{3/4}$, is related with the original extending mode to the bending mode.

We derive now the EC. In a similar process to the pure k_{33} mode case (see Example Problem 5.3), we start from the admittance formula:

$$Y = Y_d \frac{1}{\left(1 - \frac{A^2}{Z_1 Y_d}\right)} = Y_d + \frac{1}{\left(-\frac{1}{Y_d} + \frac{Z_1}{A^2}\right)}. \tag{5.225}$$

Here, $Y_d = j\omega C_d = j\omega \frac{2wt_1}{L} \varepsilon_0 \varepsilon_{33}{}^X \left(1 - k_{33}^2\right)$. A is the force factor. We can obtain the impedance $\frac{Z_1}{A^2}$ as

$$\frac{Z_1}{A^2} = \frac{L}{j\omega 2wt_1} \frac{1}{\varepsilon_0 \varepsilon_{33}{}^X \left(1 - k_{33}^2\right)} - \frac{\Delta K_0 \alpha s_{33}^{E2}}{t_1^3 d_{33}^2 \sinh\left(\frac{\alpha}{2}\right)\sin\left(\frac{\alpha}{2}\right)}$$

$$= \frac{L}{j\omega 2wt_1} \frac{1}{\varepsilon_0 \varepsilon_{33}{}^X \left(1 - k_{33}^2\right)} \left(-\frac{1}{3}\right) + \frac{L}{j\omega 2wt_1} \frac{K_0 \alpha s_{33}^{E2}}{t_1^3 d_{33}^2} \frac{\sin\left(\frac{\alpha}{2}\right)\cosh\left(\frac{\alpha}{2}\right) + \sinh\left(\frac{\alpha}{2}\right)\cos\left(\frac{\alpha}{2}\right)}{\sinh\left(\frac{\alpha}{2}\right)\sin\left(\frac{\alpha}{2}\right)}. \tag{5.226}$$

We can understand that the impedance $\frac{Z_1}{A^2}$ is composed of the negative capacitance $-3C_d$ and the pure motional admittance $\left[\frac{L}{j\omega 2wt_1} \frac{K_0 \alpha s_{33}^{E2}}{t_1^3 d_{33}^2} \frac{\sin\left(\frac{\alpha}{2}\right)\cosh\left(\frac{\alpha}{2}\right) + \sinh\left(\frac{\alpha}{2}\right)\cos\left(\frac{\alpha}{2}\right)}{\sinh\left(\frac{\alpha}{2}\right)\sin\left(\frac{\alpha}{2}\right)}\right]^{-1}$ in series. Knowing that this motional admittance is equivalent to the k_{31}-type bimorph formula, the EC is obtained as a pair of $M_0/4$ and c_n connected in series. Figure 5.25 shows the ECs for a bimorph via the d_{33} effect: (a) basic idea; (b) 4-terminal circuit. The major difference from the pure k_{33}-type vibrator, the bimorph EC includes the negative damped admittance $-\left(\frac{3}{4}\right)C_d$, rather than $-C_d$. We summarize all components in the EC in Fig. 5.25(b):

$$\begin{cases} L_n = \frac{M_0}{4}/A^2 \\ C_n = \frac{1}{\omega_n^2 L_n} = c_n A^2 \\ A = -\alpha \left[\tan\left(\frac{\alpha}{2}\right) - \tanh\left(\frac{\alpha}{2}\right)\right] \left(\frac{wt_1}{L}\right)^2 \left(\frac{d_{33}}{s_{33}^E}\right) \end{cases} \qquad M_0 = 2t_1 w\rho L. \tag{5.227}$$

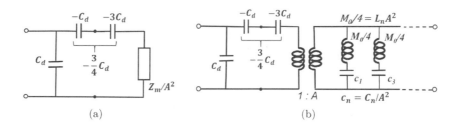

Fig. 5.25: Equivalent circuits for a bimorph with the d_{33} effect: (a) Basic idea and (b) 4-terminal circuit.

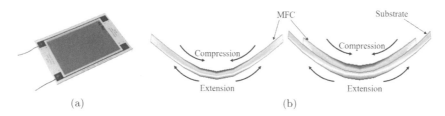

Fig. 5.26: (a) d_{31} mode macro fiber composite [MFC] [Courtesy from Smart Material Corp.]; Stress distribution of the MFC in (b-left) single plate, and (b-right) with an elastic substrate.

5.4.3.1.6 Energy harvesting system — case studies — flexible transducer

The bending mode is usually used for flexible transducers. However, bending vibration induces extension *and* compression stresses at the same time in the material, and induced electric energy cannot be obtained efficiently because of the partial cancellation. In order to avoid this problem, a unimorph design was adopted by bonding an elastic substrate to the piezoelectric plate.[12]

The d_{31}-mode type Macro Fiber Composite [MFC (M8528 P2)] from Smart Material Corporation is shown in Fig. 5.26(a). The piezo ceramic fibers in the MFC are cut by 350-μm width and 170-μm thickness from a piezoelectric wafer by a computer-controlled dicing saw. The total dimensions of the MFC are 85-mm length, 28-mm width and 0.3-mm thickness. Figure 5.26 also shows the stress distribution of MFC when the mechanical force is applied. As shown in Fig. 5.26(b-left), without any substrate, the extensive stress and compressive stress occur on the top and the bottom of the MFC. In this case, the neutral line of the stress distribution is in the middle of the MFC. Because of this, the electrical output is almost canceled, and very small. However, in Fig. 5.26(b-right), where an additional substrate is bonded on the bottom of the MFC, the neutral line is shifted down in the substrate. Therefore, the electrical output is increased because the MFC has only a compressive or tensile stress in the whole volume. Though the uniformity of the compressive stress is increased with the elastic substrate thickness, the mechanical energy distribution to the MFC is decreased, leading to the optimum thickness for harvesting electrical energy. In addition, the thickness and material of the substrate should be carefully considered to keep the flexibility of the MFC.

The minimum thickness of the substrate was calculated by the FEM software code ATILA (distributed from Micromechatronics Inc., PA) depending on the material to shift the neutral line away from the MFC. Table 5.6 shows the neutral line position and displacement calculated.[16] Typical materials are considered to calculate the neutral line and flexural bending magnitude. For this calculation, the dimensions of MFC are 85 mm × 28 mm × 0.17 mm, and the thickness of the substrate is 0.17 mm. Note that the signs of "+" and "−" indicate the distance from the adjacent line between the MFC and substrate. In the case of steel, brass and copper, the neutral line is located in the substrate below the adjacent line. For the polymer, the neutral line is not changed from the center of the MFC, same as the MFC without any substrate. The displacement in the table is a measure of the flexibility of the MFC with a substrate. This displacement is calculated under a fixed mechanical force condition. The aluminum substrate shows the best flexibility

Table 5.6: Neutral line position and the bending displacement calculated by FEM for various substrate materials ($t = 170\ \mu$m) under a fixed mechanical force.[16]

	None	Steel	Brass	Copper	Aluminum	Polymer
Neutral Line	0.085	−0.03	−0.0025	−0.0125	0.005	0.0825
Displacement	901 μm	79 μm	114 μm	101 μm	129 μm	766 μm

as compared to any other metal substrate in Table 5.6. Therefore, aluminum (aluminum cooking foil, in practice) was selected to be used for the substrate in the experiment, even though the neutral line is located slightly above the interface line.

The method to apply the stress on the sample was based on a real situation. The mechanical shaker to make vibration at high frequency is not a close approach for the piezoelectric energy harvesting with a small mechanical source such as human motion. The MFC was excited to generate a large bending displacement by small force at a frequency around 1–5 Hz. Figure 5.27 shows an experimental setup for measuring electrical output from a flexible piezo component under a fixed bending displacement. Note that the small mechanical force used in this experiment means the minimum force which can generate maximum strain in the flexible element without crack.

Fig. 5.27: Measuring setup for electrical output from a flexible piezocomponent.

The voltage signal of the MFC is shown in Fig. 5.28(a).[16] The voltage of the MFC is considerably increased by bonding an aluminum film (i.e., cooking foil) substrate as shown in Fig. 5.28(a). This signal was generated by a small mechanical force with frequency of 5 Hz, and monitored by an oscilloscope (Tektronix, TDS 420A) under high input impedance condition. The addition of the aluminum substrate gave a lower elastic flexibility to the MFC, but the output voltage from the MFC with small bending showed a much higher voltage signal. The output voltage signal from the MFC was passed through the rectifier and charged a capacitor, and successively discharged through a resistive load. The rectified voltage and output power are shown in Fig. 5.28(b). The generated electric power from a small mechanical force at 5 Hz was around 1.5 mW at 200 kΩ. This 200 kΩ corresponds roughly to the matched impedance value of $1/\omega C$ at 5 Hz. Though the harvested energy from this small rectangular (50 mm long) specimen is small, a 20 times higher level (30 mW) can be anticipated in the final wearable energy harvesting jacket product.

Example Problem 5.6

Piezoelectric MEMS (micro electro-mechanical system) is researched popularly with the PZT thin film thickness around 1 μm for the actuators and piezo energy harvesting applications. What are the major problems to be solved on a popular design as shown in Fig. 5.29? Ti thin film is used as an adhesion layer of Platinum (Pt) electrode on silica.

Solution:

We find two major problems in this design: (1) generative mechanical power level, and (2) PZT and the substrate thickness ratio.

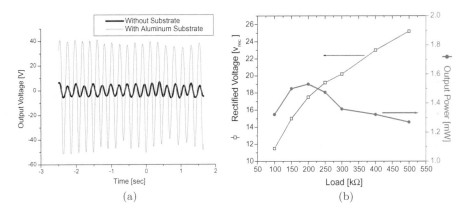

Fig. 5.28: Output of the MFC at 5 Hz. (a) Output voltage signal of the MFC, and (b) rectified voltage and output power plotted as a function of resistive load.

- *Generative Mechanical Power Level*

With accelerating the commercialization of piezo-electric actuators and transducers for portable equipment applications, we identified the bottle-neck of the piezoelectric devices; that is, signifi-cant heat generation limits the *maximum power density*. Though the problem is much smaller than the electromagnetic motors and transform-ers (i.e., Joule heat from thin conducting wires), the piezo ceramic devices become a "ceramic

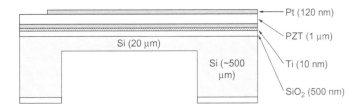

Fig. 5.29: Piezoelectric MEMS structure with the PZT thin film thickness around 1 μm for the actuators.

heater" with increasing the input/output power significantly. When used at its resonance, a piezoelec-tric device started to generate heat with increasing the vibration level primarily based on the "intensive elastic loss". The "vibration velocity", above which 20°C temperature rise (from room temperature) is observed, is called the *maximum vibration velocity* in our community, taking into account the safety to humans (e.g., a human finger is burned on a 50°C electronic component!). When used at its off-resonance, to the contrary, due to a large electric field application (though the stress is not very high), heat is gener-ated now via the dielectric loss. The current maximum handling power of a well-known hard $Pb(Zr,Ti)O_3$ (PZT) is only around $10\,W/cm^3$. Taking into account the practically required minimum power levels, 30–100 mW for charging electricity into a battery (DC-DC converter spends 2–3 mW), 10–20 mW for soak-ing blood from a human vessel, or 1–3 mW even just for sending electronic signal, minimum 1 mW han-dling is necessary, leading to the minimum PZT volume of $0.1\,mm^3$. Similar heat generation happens even under mechanical stress operation. Only for "nearly zero" load applications, such as optical beam reflectors/mirrors, the power is not very essential unless the agility is required. Accordingly, the MEMS devices with less than 1-μm thin PZT (current commercial base) films are useless from the actuator application view point. 1–10 μW level obtained from 1-μm films is usually used as a "sensor", not as an actuator. At least, 30-μm thick films should be used with minimum $3 \times 3\,mm^2$ device area, as long as we use the current existing materials (max power density ~$10\,W/cm^3$). Note that because of the ther-mal conduction from the silicon substrate, the PZT can be driven under somewhat higher input electrical energy, leading to a higher maximum vibration velocity excited. Thus, the above estimation may be a little pessimistic.

• *PZT/Substrate Thickness Ratio*

A unimorph bending actuator can be fabricated by bonding a piezo-ceramic plate to a metallic shim, as shown in Fig. 5.29. where both ends are clamped. The tip deflection, δ, of the unimorph supported in a cantilever configuration is given by (refer to Example Problem 5.5)

$$\delta = \frac{d_{31}EL^2 Y_c t_c}{(Y_m[t_o^2 - (t_o - t_m)^2] + Y_c[(t_o + t_c)^2 - t_o^2])}. \tag{P5.6.1}$$

Here, E is the electric field applied to the piezoelectric ceramic, d_{31}, the piezoelectric strain coefficient, L, the length of the unimorph, Y, Young's modulus for the ceramic or the metal, and t is the thickness of each material. The subscripts c and m denote the "ceramic" and the "metal", respectively. The quantity, t_o, is the distance between the *strain-free neutral plane* and the bonding surface, and is defined according to the following:

$$t_0 = \frac{t_c t_m^2 (3t_c + 4t_m) Y_m + t_c^4 Y_c}{6 t_c t_m (t_c + t_m) Y_m}. \tag{P5.6.2}$$

In the case of a unimorph supported at the both ends, the center displacement is 1/4 of the above displacement, and the analytical approach is similar.

Assuming $Y_c = Y_m$ just from the simplicity viewpoint, the optimum (t_m/t_c) ratio that will maximize the deflection, δ, can be calculated:

$$\delta = \frac{d_{31}EL^2 t_c}{([t_o^2 - (t_o - t_m)^2] + [(t_o + t_c)^2 - t_o^2])}, \tag{P5.6.3}$$

$$t_o = \frac{t_c t_m^2 (3t_c + 4t_m) + t_c^4}{6 t_c t_m (t_c + t_m)}. \tag{P5.6.4}$$

Substituting t_o in Eq. (P5.6.4) into Eq. (P5.6.3) yields

$$\delta = \frac{d_{31}EL^2 3 t_m t_c}{(t_m + t_c)^3} \tag{P5.6.5}$$

When the PZT film is fabricated first, we consider etching the silicon substrate thickness next. The function $f(t_m) = (t_m t_c)/(t_m + t_c)^3$ must be maximized for a fixed ceramic thickness, t_c.

$$\frac{df(t_m)}{dt_m} = \frac{(t_c - 2t_m)t_c}{(t_m + t_c)^4} = 0. \tag{P5.6.6}$$

The metal plate (in this case, substrate) thickness should be $t_m = t_c/2$ and $t_o = t_c/2$.

Anyhow, in order to obtain the effectively large displacement, the silicon substrate thickness should be in a similar dimension to the PZT thickness. Muralt calculated electromechanical coupling factor k and the resonance frequency for PZT thin films on a silicon substrate as a function of the substrate thickness on an ultrasonic stator of circular geometry (radius: 5.2 mm, mode B01. PZT with $e_{31,f}$ of $-6\,\mathrm{C/m}^2$) without considering the film stresses (see Fig. 5.30).[17] You can find that the electromechanical coupling shows the maximum when the silicon thickness is equal to the PZT

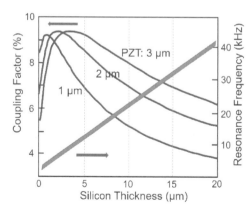

Fig. 5.30: Calculated k_{eff} and the resonance frequency for a PZT film ultrasonic circular stator as a function of silicon thickness.[17]

thickness. The maximum k_{eff} can reach close to 10%. However, as in Fig. 5.29, when the piezo film is very thin in comparison with the substrate ($t_1/t_2 \ll 1$), recall Eq. (5.166)

$$\frac{k_{\text{eff}}^2}{(1 - k_{\text{eff}}^2)} = \frac{k_{31}^2}{(1 - k_{31}^2)} \left(\frac{3}{Y_M s_{11}^E}\right) \left(\frac{t_1}{t_2}\right). \qquad \text{(P5.6.7)}$$

When the shim thickness is 20 times thicker than the piezo-plate, the bimorph k_{eff} reaches much less than 10%, in comparison with a typical $k_{31} = 34\%$ for the pure k_{31} mode resonator. PZT 1 μm on 20-μm silicon membrane design published in Fig. 5.29 may be used for sensor applications, but far from the actuator designs.

5.4.3.2 *Cymbal*

The "Cymbal" is composed of a piezo-electric disk and two metal end-caps sandwiching. Figures 5.31(a) and 5.31(b) show a photo of the cymbal, and its cross-section view, respectively. Because the base piezo component exhibits the k_p mode, much higher effective electromechanical coupling factor k_{eff} can be anticipated than the k_{31} type or its bimorph design. We discuss this cymbal design in this subsection.

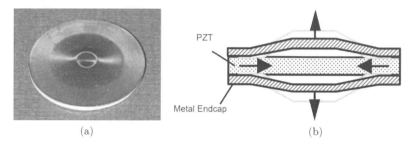

Fig. 5.31: Piezo "Cymbal" transducer: (a) photo and (b) cross-section view for showing the operation principle.

5.4.3.2.1 Analysis of a piezoelectric disk[13]

We consider first a piezo ceramic disk component (k_p planar mode) as illustrated in Fig. 5.32. Using a cylindrical coordinate system (r, θ, z), we consider planar/radial vibration k_p mode. From the symmetry, the piezoelectric constitutive equations are composed of just six components: strain x_{rr}, $x_{\theta\theta}$, stress X_{rr}, $X_{\theta\theta}$, electric displacement D_z and electric field E_z:

$$\begin{pmatrix} x_{rr} \\ x_{\theta\theta} \\ D_z \end{pmatrix} \begin{pmatrix} s_{11}^E & s_{12}^E & d_{31} \\ s_{12}^E & s_{11}^E & d_{31} \\ d_{31} & d_{31} & \varepsilon_0 \varepsilon_{33}^X \end{pmatrix} \begin{pmatrix} X_{rr} \\ X_{\theta\theta} \\ E_z \end{pmatrix}. \qquad (5.228)$$

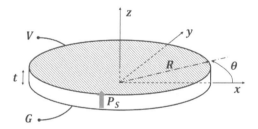

Fig. 5.32: Radial/planar k_p mode piezo-resonator.

Using Poisson's ratio $\sigma = -s_{12}^E/s_{11}^E$ and Young's modulus $Y = 1/s_{11}^E$, we transform Eq. (5.228) into

$$\begin{cases} X_{rr} = \frac{Y}{(1-\sigma^2)} x_{rr} + \frac{\sigma Y}{(1-\sigma^2)} x_{\theta\theta} - \frac{d_{31} Y}{(1-\sigma)} E_z \\ X_{\theta\theta} = \frac{Y}{(1-\sigma^2)} (x_{\theta\theta} + \sigma x_{rr}) - \frac{d_{31} Y}{(1-\sigma)} E_z \end{cases}. \qquad (5.229)$$

Because the vibration is symmetrical extension mode, only the radial displacement u_r is considered. Knowing $\frac{\partial X_{r\theta}}{\partial \theta} = \frac{\partial X_{rz}}{\partial z} = 0$, the dynamic equation is obtained as

$$\rho \ddot{u}_r = \frac{\partial X_{rr}}{\partial r} + \frac{(X_{rr} - X_{\theta\theta})}{r}. \qquad (5.230)$$

Using the following strain vs. displacement relationships in a cylindrical coordinate

$$\begin{cases} x_{rr} = \frac{\partial u_r}{\partial r} \\ x_{\theta\theta} = \frac{1}{r}\frac{\partial u_\theta}{\partial \theta} + \frac{u_r}{r} = \frac{u_r}{r} \end{cases}. \tag{5.231}$$

we can transform the right-side terms as

$$\begin{cases} \frac{\partial X_{rr}}{\partial r} = \frac{Y}{(1-\sigma^2)}\left(\frac{\partial^2 u_r}{\partial r^2} + \sigma\frac{1}{r}\frac{\partial u_r}{\partial r} - \sigma\frac{u_r}{r^2}\right) \\ X_{rr} - X_{\theta\theta} = \frac{Y(1-\sigma)}{(1-\sigma^2)}\left(\frac{\partial u_r}{\partial r} + \frac{u_r}{r}\right) \end{cases}. \tag{5.232}$$

If we assume $u_r = u(r)e^{-j\omega t}$, Eq. (5.230) becomes

$$\frac{\partial^2 u(r)}{\partial r^2} + \frac{1}{r}\frac{\partial u(r)}{\partial r} - \frac{u(r)}{r} + \frac{(1-\sigma^2)}{Y}\rho\omega^2 u(r) = 0. \tag{5.233}$$

Introducing sound velocity v and wave number n by the following relations.

$$\begin{cases} v^2 = \frac{Y}{(1-\sigma^2)\rho}, \\ n = \omega/v \end{cases} \tag{5.234}$$

we convert r by x with $x = nr$ in Eq. (5.233):

$$\frac{\partial^2 u}{\partial x^2} + \frac{1}{x}\frac{\partial u}{\partial x} + \left(1 - \frac{1}{x^2}\right)u = 0. \tag{5.235}$$

A general solution can be written by the Bessel function of the first- and second-kind with a cylindrical coordinate as

$$u(r) = A \cdot J_1\left(\frac{\omega r}{v}\right) + B \cdot N_1\left(\frac{\omega r}{v}\right). \tag{5.236}$$

A and B are parameters to be determined by the boundary conditions:

- $u(r) = 0$ at $r = 0 \rightarrow B = 0$,
- $X_{rr} = 0$ at $r = R \rightarrow \frac{\partial u(r)}{\partial r} + \sigma\frac{u(r)}{r} = d_{31}(1+\sigma)E_z$.

Here, $J_1(x)$ is the Bessel function of first-kind of order 1. When x is not large, the following expansion series match:

$$J_0(x) = 1 - \frac{x^2}{2^2} + \frac{x^4}{2^2 \cdot 4^2} - \cdots,$$

$$J_1(x) = \frac{x}{2} - \frac{x^3}{2^2 \cdot 4} + \frac{x^5}{2^2 \cdot 4^2 \cdot 6} - \cdots.$$

And, the following relation between Order 0 and 1 is useful:

$$\frac{\partial}{\partial x}J_1(x) = J_0(x) - \frac{J_1(x)}{x}.$$

From the stress-free boundary condition at $r = R$, and the following equations

$$\begin{cases} \frac{\partial u(r)}{\partial r} = A\left(\frac{\omega}{v}\right)\left\{J_0\left(\frac{\omega r}{v}\right) - \frac{J_1\left(\frac{\omega r}{v}\right)}{\frac{\omega r}{v}}\right\} \\ \sigma\frac{u(r)}{r} = A\sigma J_1\left(\frac{\omega r}{v}\right)\frac{1}{r} \end{cases}$$

Table 5.7: Frequency constant α_n vs. Poisson's ratio σ.

n \ σ	0	0.1	0.2	0.3	0.4	0.5
1	1.83	1.91	1.98	2.05	2.11	2.16
2	5.33	5.35	5.37	5.39	5.41	5.43
3	8.54	8.55	8.56	8.57	8.58	8.59
4	11.71	11.71	11.72	11.73	11.74	11.75

we can determine the parameter A as

$$A = \frac{d_{31}(1+\sigma)E_z R}{\left(\frac{\omega R}{v}\right) J_0\left(\frac{\omega R}{v}\right) - (1-\sigma)J_1\left(\frac{\omega R}{v}\right)}. \tag{5.237}$$

The resonance condition can be obtained from $u(r)$ infinite condition, or

$$\begin{cases} \left(\frac{\omega_n R}{v}\right) J_0\left(\frac{\omega_n R}{v}\right) - (1-\sigma)J_1\left(\frac{\omega_n R}{v}\right) = 0 \\ \omega_n = \frac{\alpha_n}{R}v = \frac{\alpha_n}{R}\sqrt{\frac{Y}{(1-\sigma^2)\rho}}. \end{cases} \tag{5.238}$$

Because the *frequency constant* α_n depends on Poisson's ratio σ, as summarized in Table 5.7, the resonance frequency $f_{r,n}$ also changes with Poisson's ratio.

We now calculate the admittance Y.

$$\begin{aligned} D_z &= d_{31}(X_{rr} + X_{\theta\theta}) + \varepsilon_0 \varepsilon_{33}^X E_z \\ &= d_{31}(X_{rr} + X_{\theta\theta}) + \varepsilon_0 \varepsilon_{33}^X E_z \\ &= \frac{d_{31}/s_{11}^E}{(1-\sigma)}(x_{rr} + x_{\theta\theta}) + \varepsilon_0 \varepsilon_{33}^X\left[1 - \frac{2}{(1-\sigma)}\frac{d_{31}^2/s_{11}^E}{\varepsilon_0 \varepsilon_{33}^X}\right]E_z. \end{aligned} \tag{5.239}$$

Taking into account

$$\begin{cases} x_{rr} = \frac{\partial u(r)}{\partial r} = A\left(\frac{\omega}{v}\right)\left[J_0\left(\frac{\omega r}{v}\right) - \frac{J_1\left(\frac{\omega r}{v}\right)}{\left(\frac{\omega r}{v}\right)}\right] \\ x_{\theta\theta} = \frac{A \cdot J_1\left(\frac{\omega r}{v}\right)}{r} \end{cases} \tag{5.240}$$

we obtain

$$x_{rr} + x_{\theta\theta} = A\left(\frac{\omega}{v}\right) J_0\left(\frac{\omega r}{v}\right). \tag{5.241}$$

Then, the current I can be calculated as

$$I = j\omega \iint D_z r\,dr\,d\theta = -j\omega\frac{\pi R^2}{t}\varepsilon_0 \varepsilon_{33}^X\left[1 - \frac{2}{(1-\sigma)}k_{31}^2\right]V + j\omega\frac{d_{31}/s_{11}^E}{(1-\sigma)}\iint A\left(\frac{\omega}{v}\right) J_0\left(\frac{\omega r}{v}\right) r\,dr\,d\theta.$$

From a general relationship $\int x J_0(x)dx = x J_1(x)$, and $A = \frac{d_{31}(1+\sigma)E_z R}{\left(\frac{\omega R}{v}\right)J_0\left(\frac{\omega R}{v}\right)-(1-\sigma)J_1\left(\frac{\omega R}{v}\right)}$, we obtain

$$\begin{aligned} I &= -j\omega\frac{\pi R^2}{t}\varepsilon_0 \varepsilon_{33}^X\left[1 - \frac{2}{(1-\sigma)}k_{31}^2\right]V + j\omega\frac{d_{31}/s_{11}^E}{(1-\sigma)}\iint A\left(\frac{\omega}{v}\right) J_0\left(\frac{\omega r}{v}\right) r\,dr\,d\theta \\ &= -j\omega\frac{\pi R^2}{t}\varepsilon_0 \varepsilon_{33}^X\left[1 - \frac{2}{(1-\sigma)}k_{31}^2\right]V + j\omega\frac{d_{31}/s_{11}^E}{(1-\sigma)}RJ_1\left(\frac{\omega R}{v}\right) \cdot \frac{d_{31}(1+\sigma)E_z R}{\left(\frac{\omega R}{v}\right) J_0\left(\frac{\omega R}{v}\right) - (1-\sigma)J_1\left(\frac{\omega R}{v}\right)}. \end{aligned} \tag{5.242}$$

Taking $\left(\frac{\omega R}{v}\right) \to 0$ approximation on Eq. (5.242),

$$
\begin{aligned}
I &= -j\omega \frac{\pi R^2}{t}\varepsilon_0\varepsilon_{33}^X \left[1 - \frac{2}{(1-\sigma)}k_{31}^2\right]V - j\omega\frac{d_{31}/s_{11}^E}{(1-\sigma)}R\frac{1}{2}\left(\frac{\omega R}{v}\right)\cdot\frac{d_{31}(1+\sigma)E_z R}{\left(\frac{\omega R}{v}\right) - \frac{(1-\sigma)}{2}\left(\frac{\omega R}{v}\right)} \\
&\approx -j\omega\frac{\pi R^2}{t}\varepsilon_0\varepsilon_{33}^X\left[1 - \frac{2}{(1-\sigma)}k_{31}^2\right]V - j\omega\frac{\pi R^2}{t}\left[\frac{2}{(1-\sigma)}d_{31}^2/s_{11}^E\right]V.
\end{aligned} \tag{5.243}
$$

The admittance Y can be expressed by

$$
Y = -\frac{I}{V} \approx j\omega\frac{\pi R^2}{t}\varepsilon_0\varepsilon_{33}^X\left[1 - \frac{2}{(1-\sigma)}k_{31}^2\right] + j\omega\frac{\pi R^2}{t}\left[\frac{2}{(1-\sigma)}d_{31}^2/s_{11}^E\right]. \tag{5.244}
$$

The effective electromechanical coupling factor k_{eff} can be determined by the ratio between the damped admittance and motional admittance:

$$
(1 - k_{\text{eff}}^2) : k_{\text{eff}}^2 = j\omega\frac{\pi R^2}{t}\varepsilon_0\varepsilon_{33}^X\left[1 - \frac{2}{(1-\sigma)}k_{31}^2\right] : j\omega\frac{\pi R^2}{t}\left[\frac{2}{(1-\sigma)}d_{31}^2/s_{11}^E\right].
$$

Thus,

$$
\frac{k_{\text{eff}}^2}{\left(1 - k_{\text{eff}}^2\right)} = \frac{\left[\frac{2}{(1-\sigma)}d_{31}^2/s_{11}^E\right]}{\varepsilon_0\varepsilon_{33}^X\left[1 - \frac{2}{(1-\sigma)}k_{31}^2\right]} = \frac{\frac{2}{(1-\sigma)}k_{31}^2}{\left[1 - \frac{2}{(1-\sigma)}k_{31}^2\right]}.
$$

Finally, we obtain the well-known formula

$$
k_p = \sqrt{\frac{2}{(1-\sigma)}}\,k_{31}. \tag{5.245}
$$

Note here that the k_{31}-based bimorph degrades the k_{eff}, but the planar mode k_p is enhanced significantly by the factor of $\sqrt{\frac{2}{(1-\sigma)}} \approx \sqrt{3}$, in comparison with the k_{31} mode.

Figure 5.33 shows FEM computer simulation results of the Von Mises stress distribution profile on the planar (or radial) mode $n = 1, 2$ and 3, simulated on PZT 4 disk with $24\,\text{mm}$ dia $\times 1\,\text{mm}$ thick. Because the node/nodal line concentrates the stress, the reader can identify only the center node (red-color dot), center plus one nodal ring (red-color ring) and center plus two nodal rings (blue-color rings) for $n = 1, 2$ and 3 modes, respectively. The resonance frequencies for $n = 1, 2$ and 3 are obtained as 95.45, 248.5 and $392.7\,\text{kHz}$, respectively. From the resonance frequency ratio between $n = 2$ and 1, $248.5/95.45 = 2.603$ results in Poisson's ratio $\sigma = 0.34$ from Table 5.7. This is actually the standard method for obtaining Poisson's ratio of the piezo material experimentally. Because PZT 4 has $s_{11}^E = 12.3 \times 10^{-12}\,\text{m}^2/N$, and

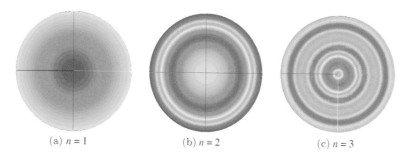

(a) $n = 1$ (b) $n = 2$ (c) $n = 3$

Fig. 5.33: Vibration profiles of higher-order modes of the radial/planar k_p piezo-resonator.

$s_{12}^E = -4.05 \times 10^{-12} \, \text{m}^2/\text{N}$, Poisson's ratio value $\sigma = -\frac{s_{12}^E}{s_{11}^E} = 0.33$ agrees very well with the above resonance frequency ratio method.

5.4.3.2.2 Analysis of a piezoelectric cymbal

The cymbal structure is considered a composite of a piezoelectric k_p-type disk (just discussed above) with a sort of "mechanical transformer", which transforms and amplifies the displacement by sacrificing the force. The amplification factor is primarily determined by the ratio of cavity part radius over cavity depth. Thus, from the unit input electrical energy, k_p^2-time energy is converted to mechanical energy, which will be distributed to the resonance modes. We introduce here the effective electromechanical coupling factor change with the endcap structure (here, the metal thickness). ATILA FEM software code (Micromechatronics Inc., PA) is adopted for simulating the model cymbals illustrated in Fig. 5.35(a), which is a simple 2D-axisymmetric design. Two 1-mm-thick "PZT 4" disks (24-mm diameter) are sandwiched by a pair of "brass" endcaps, the detailed dimensions of which are inserted in the figure. The admittance spectrum (i.e., current change under a constant voltage with sweeping the drive frequency) is important for evaluating the piezoelectric performances. The IEEE standard adopted a method for defining the dynamic electromechanical coupling factor k_{eff} from the resonance $f_{R,n}$ and antiresonance $f_{A,n}$ frequencies:[18] the dynamic electromechanical factor k_{vn} for the *n-th* harmonic vibration mode is given by

$$\begin{cases} k_{vn}^2 = 1 - \dfrac{f_{R,n}^2}{f_{A,n}^2} \ (\text{for the mode strain/stress} \perp P_S) \\[2mm] \dfrac{k_{vn}^2}{1-k_{vn}^2} = \dfrac{f_{A,n}^2}{f_{R,n}^2} - 1 \ (\text{for the mode strain/stress} \ // \ P_S) \end{cases} \tag{5.246}$$

However, as Rogacheva pointed out,[19] this method is valid only for piezo vibrators with simple geometries and free boundary conditions. It is well accepted that this method is accurate only for one-dimensional vibrators without considering mode coupling, such as tall bars or thin spherical shells. We dare to use these equations just for simple approximation's sake in this textbook, even for cymbal and hinge-lever structure complex transducers. Then, the dynamic k_{vn} is also related with the static k_v (in this case, k_p) as

$$k_{vn}^2 = P_n k_v^2, \tag{5.247}$$

where P_n is the *capacitance factor*, which is given for the k_p case by

$$P_n = \frac{2(1+\sigma)}{\alpha_n^2 - (1-\sigma^2)}. \tag{5.248}$$

The capacitance factor P_n is a sort of distribution weight of the mechanical energy to each higher-order harmonic mode. Note here that this P_n in the k_p mode is compared with the simplest formula in the case of k_{31} type given by $P_n = \frac{8}{\pi^2 n^2}$, because of the 2D vibration with Poisson's ratio. For refreshing your memory, the resonance frequencies of the n-th mode is also expressed as $\omega_n = \frac{\alpha_n}{R} v = \frac{\alpha_n}{R} \sqrt{\frac{Y}{(1-\sigma^2)\rho}}$ (here, $Y = 1/s_{11}^E$), which indicates that the resonance frequency rate among the higher-order modes is not simply "integer".

First, let us discuss on the base piezo component PZT disk itself. "PZT 4" has the following material's parameters: $d_{31} = -123 \times 10^{-12} \, \text{C/m}$, $s_{11}^E = 12.3 \times 10^{-12} \, \text{m}^2/\text{N}$, $s_{12}^E = -4.05 \times 10^{-12} \, \text{m}^2/\text{N}$, $\varepsilon_{33}^X = 1300$, $\varepsilon_0 = 8.854 \times 10^{-12} \, \text{F/m}$, and Poisson's ratio = 0.33. Thus, we can derive the static electromechanical coupling factor

$$k_p^2 = k_v^2 = \frac{2}{(1-\sigma)} k_{31}^2 = \frac{2}{(1-\sigma)} \frac{d_{31}^2}{s_{11}^E \varepsilon_0 \varepsilon_{33}^X} = 0.319. \tag{5.249}$$

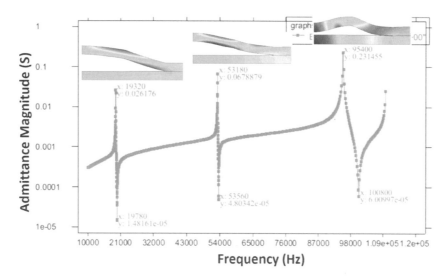

Fig. 5.34: Typical admittance spectrum on a cymbal transducer. Lower three resonance vibration modes are inserted on the top of each admittance peak.

We consider the dynamic k_{vn} for the PZT disk. We can obtain the first, second and third radial resonance frequencies at 95.3, 245.7 and 380.6 kHz, respectively, and obtain k_{vn}^2 as 0.251, 0.044 and 0.0197 from the resonance/antiresonance method. Multiplying $P_1 = \frac{2(1+\sigma)}{\alpha_n^2 - (1-\sigma^2)} = \frac{2(1+0.33)}{2.067^2 - (1-0.33^2)} = 0.786$ with $k_p^2 = 0.319$, we obtain 0.251, which is actually the same as above $k_{v1}^2 = 0.251$. By the way, the above P_1 value is not very different numerically from $P_1 = \left(\frac{8}{\pi^2}\right) = 0.810$ in the k_{31} plate (only 3% lower).

Now, we explain the cymbal by adding the metal endcaps. Figure 5.34 shows an example admittance spectrum simulated for the cymbal with 1-mm-thick endcaps. As shown in the inserted vibration mode profiles (only the top half of the cymbal motion is shown due to the symmetry), the first and second resonances are primarily from the endcap vibration with keeping the PZT disk displacement minimum and rather uniform (except for the bonding edge circumference), while the resonance at 95.4 kHz originates from the fundamental PZT disk radial vibration with large k_{vn} with large separation between $f_{R,1}$ and $f_{A,1}$ (Fig. 5.33, $n = 1$). Note that the piezo disk shows resonating strain distribution, in addition to large metal deformation, and that this frequency is very close to the fundamental resonance frequency of the PZT disk alone.

Change in the resonance frequency and dynamic electromechanical coupling factor k_{vn}^2 with the endcap thickness (0.25–1.5 mm) is summarized in Fig. 5.35(b). k_{vn}^2 is calculated from the resonance and antiresonance frequencies with $k_{vn}^2 = 1 - (f_{R,n}/f_{A,n})^2$, even for the endcap structure vibration. We plot the results only for the fundamental endcap-related and PZT-disk-related modes. The resonance frequency for the endcap-related mode increases significantly from 10 to 23 kHz with increasing the thickness (due to the elastic spring constant increase with the metal thickness), while the frequency ($\sim$95 kHz) for the PZT-disk-related mode does not change. k_{vn}^2 change is very intriguing: endcap $k_{v1-\mathrm{cap}}^2$ increases significantly from 1% to 5% with the initial thickness increase. However, endcaps with 0.75 mm t or thicker are useless because $k_{v1-\mathrm{cap}}^2$ shows the saturation tendency. On the other hand, PZT disk $k_{v1-\mathrm{PZT}}^2$ decreases also significantly from 25% to 10% with the initial thickness increase, showing a similar saturation for thicker caps. Thus, there is an optimized thickness of the metal around 0.5–0.75 mm. Roughly speaking, the input electric energy in the PZT disk is converted to the total mechanical energy by the factor of k_p^2, then this mechanical energy is transferred partially to the metal endcaps. With increasing the metal thickness, the transferring energy is increased, leading to the decrease in the disk-related 95.4 kHz vibration $k_{v1-\mathrm{PZT}}^2$.

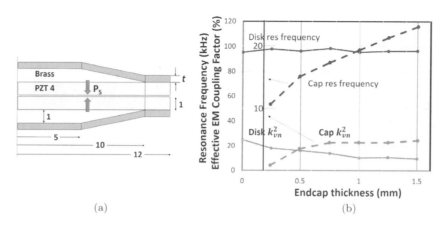

Fig. 5.35: (a) Model cymbal size and materials. (b) Change in the resonance frequency and effective electromechanical coupling factor k_{vn}^2 with the endcap thickness. [unpublished data]

In the case of underwater acoustics, the fundamental resonance frequency is used for sending and receiving the acoustic wave, and k_{v1-cap}^2 is the primary key factor. Thus, 0.5–0.75-mm thickness may be required for getting a reasonable level of k_{v1-cap}^2.

On the contrary, the energy conversion rate should be equal to the "static" k_v^2 at pseudo-static condition. As learned in Section 5.3.1, impulse force (one time) input can also realize almost k_v^2 in total, after accumulating the energy during the piezo device ring-down process. In the piezo energy harvesting system, though the vibration frequency of the piezo cymbal is under an off-resonance (i.e., much lower than the resonance frequency $\sim 100\,\text{Hz}$), the energy conversion rate is expected to be lower than the static k_v^2 owing to the mechanical energy split to the endcap parts (as long as AC, not completely static). Note that if we add all k_{vn}^2 for all resonance modes (for both PZT disk and metal endcaps), this should be equal to the static electromechanical coupling factor $k_v^2 = \frac{2}{(1-\sigma)}k_{31}^2$ for a planar coupling mode theoretically, which is the value for the PZT disk alone. How much does the metal endcap affect the 100 Hz k_v? With DC bias mechanical load (mass, force/stress), as long as the linear relationship is maintained in the piezoelectric constitutive equations, no change occurs in the k_v value at 0 Hz. However, because the endcaps are bonded on the PZT disk rigidly at the circumference, though the PZT disk vibration mode does not change (just a radial mode), the endcaps give dynamic (100 Hz) elastic spring constant change. This can be translated in an EC treatment as follows: we need to connect additional L and C to the mechanical branch of the pure PZT disk EC, corresponding to the endcaps bonding (refer to Example Problem 2.14). We may expect a degradation in the static k_v^2 with increasing the metal thickness, in principle, even at an off-resonance.

5.4.3.2.3 Analysis of a cymbal for energy harvesting[20]

We discuss in this subsection the response of the cymbal transducer under the pseudo-DC mechanical loading condition. In this case, different from the discussion in the previous section (b), the k_{vn}'s for different modes are not important, but total energy transduction $k_v^2 = \sum_n k_{vn}^2$. Since the sum of k_{v1}^2 of the endcap and PZT disk in Fig. 5.35(b) gradually decays with increasing the cap thickness, though the cap k_{v1}^2 itself increases significantly with thickness, we may find an optimized endcap thickness for the energy harvesting. Under a certain force, large deformation of the endcap is not expected, leading to small input mechanical energy.

Figure 5.36(a) shows a model, using the analogy of the "Belleville spring" or conical disk. Belleville springs resemble in shape the cap of the cymbal transducer. The stress and deflection produced in a spring of this type are not proportional to the applied load because the change in form upon the deflection significantly modifies the load-deflection and load–stress relationships. Under an applied load, P, the deflection, δ, produced along the cone height or cavity depth ($d_c = 1$ mm) is given as[21]

$$P = \frac{Y\delta}{(1-\sigma^2)Ma^2}\left[(d_c - \delta)\left(d_c - \frac{\delta}{2}\right)t_c + t_c^3\right] \qquad (5.250)$$

and the stress produced at the point x = 0 is given by the relation

$$X = \frac{-Y\delta}{(1-\sigma^2)Ma^2}\left[C_1\left(d_c - \frac{\delta}{2}\right) - C_2 t_c\right] \qquad (5.251)$$

where σ is Poisson's ratio (= 0.3), Y is Young's modulus of the cap material (= 215 GPa), t_c is the thickness of the cap (= 0.3 mm) and M, C_1 and C_2 are constants whose values are functions of a/b. In the present case $a/b = 3.4$, the magnitude of the constants are $M = 0.8$, $C_1 = 1.5$ and $C_2 = 1.88$ when load is given in pounds and length in inches in the original paper, Ref.[21] The reader is requested to convert these constants with the MKS unit by yourself. The stress X at point $x = 0$ can be resolved into x (X_1) and $y(X_2)$ components as $X \cdot \sin\theta$ and $X \cdot \cos\theta$ as shown in Fig. 5.36(b). The θ value can be computed from the dimensions, d_c, a, b, as $\theta = 80.54°$.

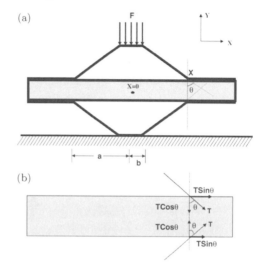

Fig. 5.36: Stress components on the cymbal transducer loaded at the top of cavity diameter. (a) Schematic representation, and (b) stress resolved into X_1 and X_2 components.

The compressive stress X_2 is mainly concentrated in the flat region where cap and ceramic are bonded together ($R_c \le x \le R$), where R and R_c are the PZT disk radius and endcap cavity part radius, respectively. Its effect is quite small on the ceramic in the cavity region ($0 \le x \le R_c$). On the other hand, the tensile stress X_1 is concentrated in the cavity region ($0 \le x \le R_c$) and its effect is quite small in the flat region ($R_c \le x \le R$). These predictions were confirmed by the finite element method (FEM) performed using the ATILA software. Once the stress fields acting on the ceramics are known, then a linear piezoelectric constitutive equation ($E = gX$, where E is the electric field and g is the piezoelectric voltage coefficient) can be used to compute the open circuit voltage generated as a result of the compressive and tensile stresses as follows:

$$V_{oc} = \left|X_1\left(\frac{R_c}{R}\right)^2 \cdot t_p \cdot g_{31}\right| + |X_2 \cdot t_p \cdot g_{33}| \qquad (5.252)$$

where the first term is averaged over the cavity surface area. Equations (5.250)–(5.252) can be simultaneously used to compute the open circuit response of the cymbal transducer under an applied load P. It should be noted here that the units of the variables in Eqs. (5.250) and (5.251) are in pounds and inches, as long as we use the M, C_1 and C_2 values in the above. In our study, a force of AC 70N was applied on the cymbal transducer corresponding to a load of 15.74 pounds. Solving Eq. (5.250) for deflection (δ) yields three solutions as $\delta = 56.9\,\mu$m. Once the magnitude of deflection is known, then the magnitude of stress, X, acting at point $x = 0$, can be computed using Eq. (5.251) and it was found to be 20.8×10^7 N/m^2 (~200 MPa, close to the critical value for the ceramic fracture).

Referring to Section 5.2.1, we tested three PZT compositions, and the material D210 (Dong Il technology, Korea) is considered for further analysis since its response is highest as compared to APC 841 and APC 850 (American Piezoceramics, PA). Though the details are described in Chapter 6, we will introduce the electrical impedance matching briefly here in advance. Figure 5.37 shows a rectification circuit with a full

wave rectifier and a capacitor for storing generated electrical energy of the Cymbal transducer in the case of off-
resonance. The EC parameters for D210 were determined
to be $C_p = 3.32\,$nF and loss $R_S = 9.59\,$kΩ at a frequency
of 100 Hz (because of this low frequency, the inductance L
contribution was neglected). Figure 5.38(a) shows the variation of the output power as a function of resistive external
load and frequency computed using the power formula on
the load:

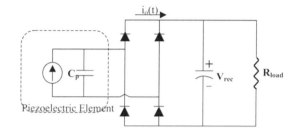

Fig. 5.37: A full-bridge rectifier with a resistive load
for piezoelectric energy harvesting.

$$P = \frac{V_{\text{load}}^2}{2R_{\text{load}}}. \tag{5.253}$$

Power is maximum at specific magnitude of load called "matching impedance load". Figure 5.38(b)
shows the variation of the matching load and maximum power as a function of frequency. In the range of
10–200 Hz, the magnitude of matching load decreases rapidly and at 100 Hz the magnitude of matching
load is 480 kΩ. The magnitude of power increases linearly with frequency and at 100 Hz the magnitude of
harvested power is around 48 mW. It should be noted here that all the calculations were done assuming
negligible damping, though in practical situations there is always some positive damping.

5.4.3.2.4 Case study — Cymbal energy harvesting

Cymbal design was optimized from the generated electrical energy level.[22] Using the circuit in Fig. 5.37, the
output voltage and the power were initially measured across the resistive load directly without any amplification circuit to characterize the performance of different transducers. The maximum rectified voltage V_{rec}
of a capacitor $C_{\text{rec}}(10\,\mu\text{F})$ was charged up to 248 V after saturation. Figure 5.39 shows the output electrical
power from various Cymbal transducers under AC and DC mechanical loads as a function of external load
resistance. The endcap thickness made of steel was changed from 0.3 mm to 0.5 mm. Prestress (DC bias
load) to the Cymbal was 66 N and applied AC (100 Hz) force was varied experimentally from 44 to 70 N.
For a small force drive (40 and 55 N), the power level increased with decreasing the endcap thickness
(from 0.5, then 0.4, and finally 0.3 mm) because of the input mechanical energy increase. With increasing
the force level up to 70 N, the maximum power of 53 mW was obtained at 400 kΩ with a 0.4-mm steel
endcap, because the Cymbal sample with a 0.3-mm-thick endcap could not endure under this high-force
drive (i.e., the cavity depth is collapsed, though device failure does not occur), which degrades the output
power. Note that the maximum power was obtained merely due to large input mechanical vibration level.

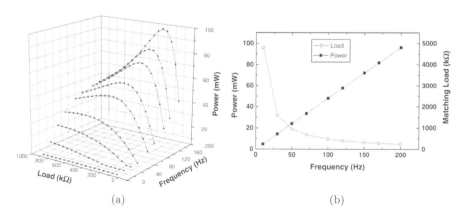

(a)   (b)

Fig. 5.38: (a) Variation of the power as a function of resistive load and frequency calculated using the pure resistive load model.
(b) Variation of the matching load and maximum power as a function of frequency calculated from the model.

The 0.3-mm-thick samples were better from the energy efficiency viewpoint (i.e., higher k_v^2), though the practically generative energy is limited, which is discussed again later in Table 5.7.[23]

5.4.3.3 *Multilayer and hinge lever mechanism*

We studied so far the k_{31}-based bimorphs, and k_p-based cymbals, among which we learned that $k_{\text{bimorph}} < k_{31} < k_{\text{cymbal}} < k_p$, but that the composite structures with the PZT and elastic components are essential to adapt the mechanical matching for the targeted vibration source, that is, mechanical flexibility (acoustic impedance) or the resonance frequency tuning for the targeting vibration source. In this direction, it is reasonable to consider the k_{33}-based structure, because k_{33} is the largest electromechanical coupling factor in PZT transducers. For the reader's information, PZT 5AH (soft PZT) shows $k_{31} = 37\%$, $k_p = 62\%$ and $k_{33} = 73\%$ (Example Prob-

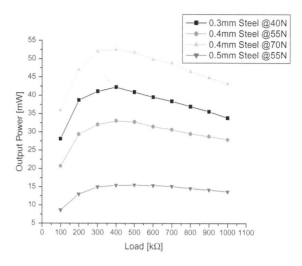

Fig. 5.39: Change in output electrical power from the various Cymbal transducers under different 100 Hz AC mechanical loads (shown as @ xx N under 66 N constant DC bias) with external electrical load resistance.

lem 5.2). In comparison with unimorph designs with $k_{\text{eff}}^2 = 1\%$, the multilayer k_{33} devices can convert the energy significantly higher by the factor of 50 times. In order to tune the mechanical impedance or flexibility to the vibration source, a sort of "mechanical transformer" is usually adopted. We consider in this section on hinge-lever mechanisms, which can enhance the input force by the lever length ratio in general under the energy harvesting system.

5.4.3.3.1 Multilayer (ML) and mass combination

Prior to discussing the complex hinge-lever, let us start from the mass load effect on the ML actuator. Knowing the fact that the ML structure is equivalent to a PZT rod in analytical viewpoints (except for the driving voltage/current and the capacitance/electrical impedance), the analysis here is merely on the PZT rod. We consider a model of a piezo rod and mass composite, as shown in Fig. 5.40(a). When only a mass is attached on the piezo rod, we can expect a monotonous decrease with the mass volume in "static" electromechanical coupling factor in comparison to the k_{33} of the pure piezo rod, in general, because the mechanical input energy (say, the force on the top of the mass) will split into the mass and piezo rod, then the decreased mechanical energy in the piezo rod is converted to electrical energy with the rate of k_{33}^2. However, the "dynamic" electromechanical coupling factor k_{vn} may change in a different way with the mass.

In order to analyze the situation, a 6-terminal EC is taken into account for the k_{33}-type piezo rod with one-end clamp and another end mass load (Fig. 5.40(b)). As learned in Example Problem 5.3 in the k_{33} type, the negative capacitance ($-C_d$) is installed in the electrical branch (or $-C_d/\Phi'^2$ in the mechanical branch) in the k_{33} type. Note the difference between Φ and Φ' in the 4- and 6-terminal models:

$$\Phi' = \Phi/2 = \frac{wbd_{33}}{Ls_{33}^D} \tag{5.254}$$

Z_1 and Z_2 in the 6-terminal model are described as follows:

$$C_d = \frac{wb\varepsilon_0\varepsilon_{33}^X(1 - k_{33}^2)}{L} \tag{5.255}$$

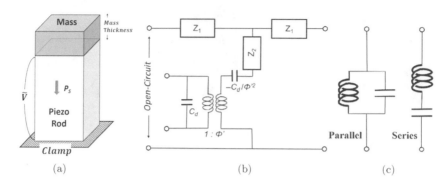

Fig. 5.40: (a) Model of a piezo rod and mass composite; (b) equivalent circuit for the k_{33}-type piezo-rod with one-end clamp and another end mass load. (c) mass load models.

$$Z_0 = wb\rho v = wb \left(\frac{\rho}{s_{33}^D} \right)^{1/2} = \frac{wb}{v_{33}^D s_{33}^D} \tag{5.256}$$

$$Z_1 = jZ_0 \tan\left(\frac{\omega L}{2v_{33}^D} \right), \quad Z_2 = \frac{Z_0}{j \sin\left(\frac{\omega L}{v_{33}^D} \right)} \tag{5.257a,b}$$

Here, $w = b = 5\,\text{mm}$, $L = 20\,\text{mm}$ in our model piezo rod (semi-soft PZT 4 is used). Regarding the terminal connection, the mechanical clamp on one end corresponds to the open terminal in the EC, and the mass on another end corresponds to the LC connection in the EC. As shown in Fig. 5.40(c), there are two ways for the LC connection for the mass load: parallel and series. We adopt here a parallel connection, because the resonance frequency change with the mass thickness is easily analyzed. The mass block with the same width, thickness ($5 \times 5\,\text{mm}^2$), height L (2–40 mm) is bonded on the top of the piezo rod in our simulation. When the load is modeled by the LC "parallel" connection in this case with the parameters, inductance l_{elast} and capacitance c_{elast} are provided by the following equations (without considering losses), where ρ, s_{metal} are the metal's density, elastic compliance, respectively:

$$l_{\text{elast}} = (\rho)(Lbw)$$
$$c_{\text{elast}} = (1/n^2\pi^2)(L/wb)s_{\text{metal}} \tag{5.258}$$

For the reader's reference, if the metal block is bonded symmetrically on both side surfaces of the piezo rod, the load is modeled by the above *LC* components in "series" connection, in order to keep the resonance frequency almost the same. You may understand this situation by taking into account the mechanical impedance series or parallel connection.

The general EC equation $\begin{bmatrix} F \\ I \end{bmatrix} = \begin{bmatrix} Z_1 & -A \\ A & Y_1 \end{bmatrix} \begin{bmatrix} \dot{u} \\ V \end{bmatrix}$ can be represented in practice by

$$\begin{bmatrix} 0 \\ I \end{bmatrix} = \begin{bmatrix} -\frac{\Phi'^2}{j\omega C_d} + Z_1 + Z_2 + \frac{1}{j\omega c_{\text{elast}} + \frac{1}{j\omega l_{\text{elast}}}} & -\Phi' \\ \Phi' & j\omega C_d \end{bmatrix} \begin{bmatrix} \dot{u} \\ V \end{bmatrix} \tag{5.259}$$

Then, the admittance $Y = I/V$ can be obtained as

$$Y = j\omega C_d + \frac{\Phi'^2}{-\frac{\Phi'^2}{j\omega C_d} + Z_1 + Z_2 + \frac{1}{j\omega c_{\text{elast}} + \frac{1}{j\omega l_{\text{elast}}}}} \tag{5.260}$$

When we initially consider the "static" electromechanical coupling factor, by taking $\omega \to 0$, c_{elast} and l_{elast} contributions (i.e., mass effect) disappear, and from

$$Y_{\text{low}\omega} \approx j\omega C_d + \cfrac{1}{-\cfrac{1}{j\omega C_d} + \cfrac{1}{\Phi'^2}\left[-jZ_0 \cot\left(\frac{\omega L}{v_{33}^D}\right)\right]} \tag{5.261}$$

we obtain the relation $k_{\text{eff}}^2 = k_{33}^2 = 47\%$, exactly the same formula as the one of the piezo rod (in this case, PZT 4) without mass.

Now, we simulate the "dynamic" electro-mechanical coupling factors, k_{vn}^2 with changing the mass length L from 2 to 40 mm, by keeping the cross-section area. Different from the Cymbal transducers, we cannot expect additional particular vibration modes (such as the endcap flex-tensional modes), related with the mass design. Different from the k_{31} or k_p (transverse piezoelectric effect), the k_{33} (longitudinal piezo effect) needs to use the formula $\frac{k_{vn}^2}{1-k_{vn}^2} = (f_{A,n}/f_{R,n})^2 - 1$ (Eq. (5.256)). Figure 5.41 summarizes mass length dependence of the resonance/antiresonance frequencies and dynamic electromechanical coupling factor k_{vn}^2 (for $n = 1$ and 2) in the k_{33}-type piezo rod. The resonance f_R and antiresonance frequency f_A monotonously decrease with the mass length, while very interestingly, k_{v1}^2 is found to increase actually up to 10-mm length. If we adopt an estimation

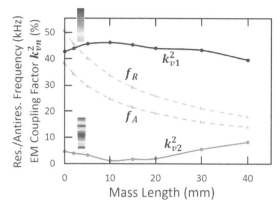

Fig. 5.41: Mass length dependence of the resonance/ antiresonance frequencies and electromechanical coupling factor k_{vn}^2 in the k_{33}-type piezo-rod.

$$\sum_n k_{vn}^2 = k_{33}^2 = 47\% \tag{5.262}$$

the increase in the fundamental mode k_{v1}^2 is interpreted by the decrease in the second mode k_{v2}^2. Refer to the inserted two figures in Fig. 5.41, which correspond to the fundamental and second extensional modes of the piezo rod. The fundamental mode shows simple extension/shrinkage with $(1/4)\lambda$ wavelength along the rod, while the second mode shows one nodal plane roughly 1/3 from the top. A reasonable mass on the piezo rod top may stabilize the fundamental mode, but impede the second mode; this is why increase in k_{v1}^2 and decrease in k_{v2}^2 are introduced for the initial mass addition. This k_{v1}^2 enhancement is essential to use this design under its first resonance mode operation.

5.4.3.3.2 Acoustic horn

When we change the mass design from a straight rod to a tapered shape, acoustic energy transfer can be modified. "Acoustic horn" is a displacement amplification mechanism at its mechanical resonance condition.

Example Problem 5.7.

At the tip of the *Langevin transducer* in Fig. 5.42, you find three "Horn (1:4)" structures for the ultrasonic machining tools. Describe the role of the "horn" in the ultrasonic transducers.

Solution:

Ultrasonic machining is based on scrubbing the surface of an object (ceramic, glass, etc., which are difficult to cut or polish with a metallic cutting edge) with polishing abrasives. Since the original ultrasonic displacement generated on the metal surface (typically micron meter) is not large enough to be used in practice, we need to increase the vibration displacement without losing the mechanical energy. A "horn" is an AC resonance displacement amplification mechanism at its resonance frequency. It effectively produces

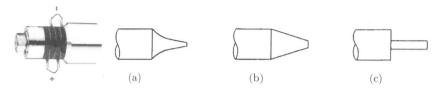

Fig. 5.42: Three commonly used horn types for the tip of a Langevin transducer (Left): (a) exponential cut horn, (b) linear taper horn, and (c) step contoured horn.

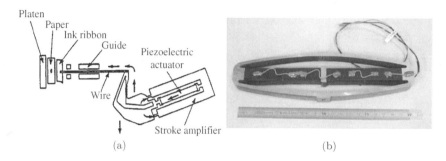

Fig. 5.43: (a) Hinge-lever displacement amplification mechanism (NEC)[24] and (b) a flextensional amplified piezoactuator (Cedrat).[25]

an amplification of the resonance displacement that is inversely proportional to the cross-section area of the vibrator. Because the acoustic energy density (product of force and displacement) for each cross-section slab should be maintained, with tapering the tip part, larger longitudinal displacement is expected. Three commonly used types are depicted in Fig. 5.42. The *exponential cut horn* (a) exhibits the highest energy transmission efficiency, but the fabrication procedure needed to produce a precisely cut horn of this shape is not simple, thus not cost efficient from a production point of view. The *linear taper horn* (b) exhibits an intermediate efficiency and is somewhat easier to fabricate. The *step contoured horn* (c), which is popularly used is the easiest to fabricate, but is the least efficient due to reflection of a portion of the vibration energy at the neck of the structure.[1]

5.4.3.3.3 Hinge lever structure designing

Hinge-lever and flextension-type displacement amplification mechanisms are adopted for the multilayer actuators as *mechanical transformers*. A famous hinge-type design is the one utilized in a dot-matrix printer developed by NEC, in Fig. 5.43(a).[24] Figure 5.43(b) shows a flextensional type developed by Cedrat.[25]

A *monolithic hinge* structure is made from a monolithic elastic body by cutting indented regions in the monolith as shown in Fig. 5.44(a). Because the hinge is made of monolith merely by changing the elasticity locally (no separation of two arms), the motion backlash can be minimized. This effectively creates a lever mechanism that may function to either amplify or reduce the displacement. As shown in Fig. 5.44(b), the displacement at the force point is amplified at the operation point in proportion to the length ratio from the fulcrum (i.e., *lever principle*). It was initially designed to reduce the displacement

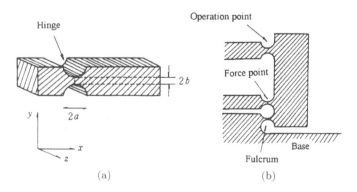

Fig. 5.44: Monolithic hinge lever mechanisms: (a) monolithic hinge structure and (b) hinge lever mechanism.

(including backlash) produced by a stepper motor, and thus increase its positioning accuracy. However, monolithic hinge levers have been combined with piezoelectric actuators to amplify their displacement. or even on MEMS structures.

If the indented region of the hinge can be made sufficiently thin to promote optimum bending while maintaining extensional rigidity (ideal case!). a mechanical amplification factor for the lever mechanism close to the apparent geometric lever length ratio is expected. The actual amplification. however. is generally less than this ideal value. If we consider the dynamic response of the device (even at an off-resonance ~ 1 kHz). a somewhat larger hinge thickness (identified as *2b* in Fig. 5.44(a)) producing an actual amplification of approximately half the apparent geometric ratio is found empirically to be optimum for achieving maximum generative force and response speed (mechanical energy transfer rate. *e*). The characteristic response (i.e., in terms of the displacement, generative force and response speed) of an actuator incorporating a hinge lever mechanism is generally intermediate between that of the multilayer and bimorph devices.

NEC developed a printer head element utilizing a longitudinal multilayer actuator. which provides much superior characteristics to the bimorph type in printing speed and durability.[26, 27] Since the longitudinal multilayer actuator does not exhibit a large displacement. a suitable displacement magnification mechanism is essential for this system. A printing wire stroke of $500\,\mu m$ should be produced from the $10\text{-}\mu m$ actuator displacement. High-energy transfer efficiency is also desired for this magnification mechanism. Let us consider initially a conceptual design of monolithic hinge lever design. as shown in Fig. 5.46(a). The simulated results are summarized in Fig. 5.45, where the admittance spectrum is shown for the low-frequency 6 resonance/antiresonance peaks. The structure is rather complex. the k_{vn}^2 are calculated on a trial basis from the longitudinal piezo effect relation: $\frac{k_{vn}^2}{1-k_{vn}^2} = \frac{f_{A,n}^2}{f_{R,n}^2} - 1$. where $f_{R,n}$ and $f_{A,n}$ are the *n*th resonance and antiresonance frequencies, respectively, though the vibration is not a simple longitudinal mode. The results for k_{vn}^2 are summarized as

Harmonics (n)	1	2	3	4	5	6
k_{vn}^2	6.3%	6.2%	0.5%	10.0%	19.8%	1.7%

It is interesting that the total k_{vn}^2 for the lower six modes ($\sum_{n=1}^{6} k_{vn}^2 = 44.5\%$) is already close to the static $k_{33}^2 = 46.8\%$ of the piezo ceramic PZT 4. We can conclude that as long as the idealistic hinge-lever is used

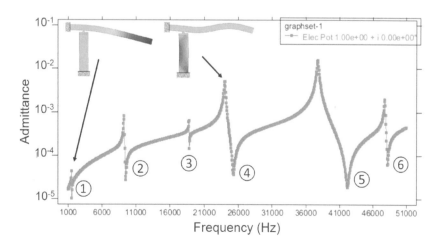

Fig. 5.45: Admittance spectrum of an ML-hinge lever mechanism for the low-frequency six pairs of resonance/antiresonance peaks. Vibration modes are also inserted.

at pseudo-DC condition, the input mechanical energy will be converted almost perfectly with the highest mechanical coupling factor k_{33}^2.

However, as long as we use the hinge-lever mechanism in a dynamic condition, say at 1 kHz (of course, lower than the structure resonance frequency), the mechanical structure of the hinge-lever stores some energy and not all the mechanical energy is transferred to the tip printing wire.

Figure 5.46 shows the development history of monolithic hinge lever mechanisms for dot-matrix printers:

(a) conceptual design, (b) prototype to be used for a printing wire array and (c) improved design in terms of energy transfer rate.

The off-resonance "mechanical energy transfer rate" e is defined and measured as

$$e = \frac{\begin{array}{c}\text{Output Energy from the}\\\text{Magnification Mechanism}\end{array}}{\begin{array}{c}\text{Stored Energy}\\\text{in the Piezoactuator}\end{array}}$$

$$= \frac{(\zeta_m^2/2C_m)}{(\zeta_c^2/2C_c)} \qquad (5.263)$$

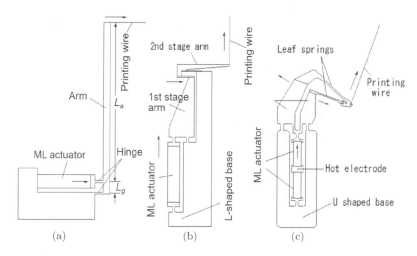

Fig. 5.46: Monolithic hinge lever mechanisms for printers: (a) conceptual design, (b) prototype for a wire array, and (c) improved design.

where ζ_c is the free displacement (m) of the piezo-ceramic device, C_c is the elastic compliance of the ceramic (m/N), ζ_m is the free displacement (m) of the displacement magnification mechanism and C_m is the elastic compliance of the lever mechanism (m/N).

Since the initial conceptual design in Fig. 5.46(a) gave $e < 10\%$ at 3 kHz, because of a too long arm problem, the design was shifted to the 2-stage amplification structure in Fig. 5.46(b), which gave $e < 25\%$, much better but not sufficient. A new problem was the L-shaped "asymmetric" base portion, which loses the energy via canting or bending motion of the ML actuator. The improved design in Fig. 5.46(c) was called "two-stage differential motion mechanical amplifier" with using a symmetrical U-shaped base portion (no-bending of the ML), which increased the energy transfer rate significantly $e > 60\%$.

A differential two-stage magnification mechanism is redrawn in Fig. 5.47(a).[26] The displacement induced in a multilayer actuator pushes up the force point, and rotates lever 1 and lever 2 around the fulcrum counter-clockwise and clockwise, respectively, so that the tip displacements of levers 1 and 2 are amplified five times. This is the primary displacement magnification mechanism occurring through a *monolithic hinge lever*. These opposing displacements are transferred to lever 3 in the secondary

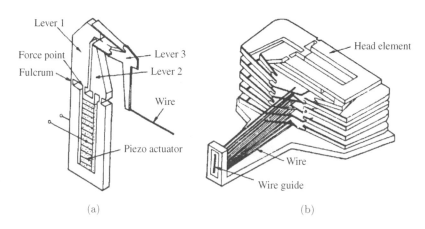

Fig. 5.47: Inkjet printer head developed by NEC: (a) the differential two-stage magnification mechanism used in a single element and (b) the entire printer head assembly.

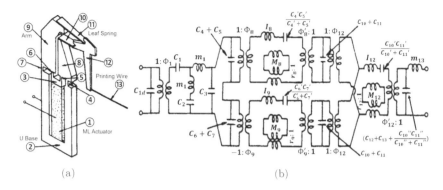

Fig. 5.48: Equivalent circuit for the dot matrix printer head elements: (a) an individual differential-type piezoelectric printer head element and (b) and equivalent circuit for the element.[26]

amplification stage, generating a large wire stroke. This is the *differential magnification mechanism*. Typical characteristics of this design (in a static condition) are an 8-μm displacement of the piezo actuator, a 240-μm wire stroke and overall displacement magnification of thirty. Note also that the wire displacement direction is orthogonal to the ML actuator displacement, which facilitates to make an array structure. The off-resonance energy transmission coefficient for this design can be as large as 60%. The printer head was constructed from a stack of these elements arranged in an alternating configuration as shown in Fig. 5.47(b).[27] The nine printing wires in the prototype, each of diameter 0.25 mm, were brought together along the wire guide.

The EC analysis was introduced for the printing wire action. The components appearing in the schematic for the circuit shown in Fig. 5.48(b) are designated by subscripts "i"s, which correspond to the numbered features of the individual element pictured in Fig. 5.48(a). The lever's *moment of inertia* is represented by inductance I_i and the lever *mass* by mutual inductance M_i. Moment of inertia and translation motion of a lever are represented by parallel connection of components. The compliances of each part are associated with the capacitances designated by C_i, C_i' and C_i''. The quantity designated by Φ_1, at the first transformer (left-hand side of the diagram), represents the electromechanical transduction rate (i.e., "force factor") and the quantities Φ_i and Φ_i', appearing near the other transformers in the circuit, represent the transduction of the linear movement into an angular displacement, a sort of "mechanical transformer", including the displacement amplification rate. The transduction rate associated with the movement of the center of mass is represented by the quantities, r_i. The damped capacitance of the ML actuator is designated by the initial C_{1d}. The equivalent mass of the actuator is represented by m_1 and the equivalent mass of the wire by m_{13}.

A model printer head, with 0.45 mm between the wire tip and the printing media, was fabricated and the experimental data related to the wire head displacement and the impact force collected for the prototype were compared with those generated in a computer simulation carried out on the EC. Simulation results on nonlinear stiffness associated with the printer head load and the response characteristics of the wire head are shown in Figs. 5.49(a) and 5.49(b), respectively.[28] The load of the printer head, consisting of the paper, ink ribbon and platen, has a significant nonlinear stiffness (third to sixth order) with respect to the impact force, as well as a distinct hysteresis. An approximate representation of this nonlinear relationship in Fig. 5.49(a) indicates the decrease in the repulsion coefficient, that occurs due to the hysteresis. This can be represented by an equivalent resistance, R_L, connected in parallel to an equivalent capacitance, C_L, where the nonlinear stiffness is associated with the quantity $(1/C_L)$. Though the EC is not a perfect tool for the transient response, the wire head action was roughly simulated by combining higher-order harmonics, as shown in Fig. 5.49(b).

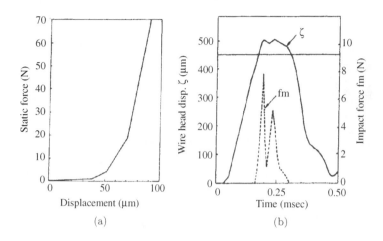

Fig. 5.49: (a) Nonlinear stiffness associated with the printer head load. (b) Response characteristics of the wire head.

5.5 Energy Flow Analysis

In order to evaluate the mechanical-to-electrical converted energy in the cymbal, the rectified voltage was used for charging the capacitor C_{rec} of $10\,\mu\text{F}$ in open condition during the charging time (t), for which the capacitor was charged up to 200 V. Note that the measured results in Table 5.8 include the energy Joule (not the power W, in this case) during a certain charging period. Table 5.8 summarized the energy flow analysis on three types of Cymbal transducers: endcap thickness of 0.3 mm and 0.4 mm with and without bias force under various cyclic vibration levels and drive durations.[22, 23] Note first that the mechanical-to-mechanical energy transfer rate is good for the 0.3-mm-thick cymbals ($83 \sim 87\%$), while it is rather low for the 0.4-mm-thick cymbal (46%). This is related with the mechanical impedance matching; the 0.4-mm endcap seems to be too rigid (effective stiffness is too high) to match the vibration source shaker (Toyota Prius engine for the then-final target).

Second, the received mechanical energy to electrical energy transduction rate can be evaluated from the effective electromechanical coupling factor. Because the value of k_{eff} of the cymbal is around 25–30% (0.3 mm exhibits larger), the energy transduction rate k_{eff}^2 can be evaluated around 6.25–9%, which agrees very well with the experimental results in Table 5.8. Thus, we obtained the conversion rate from the

Table 5.8: Energy flow analysis on three types of Cymbal transducers.[21]

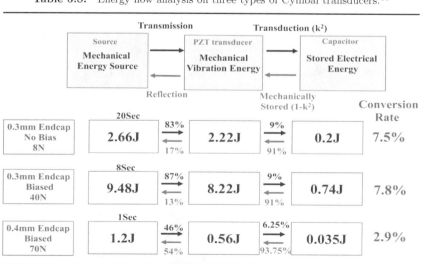

Device Design		K_{eff} (%)	$(K_{eff})^2$ (%)	Response
Unimorph/ Bimorph	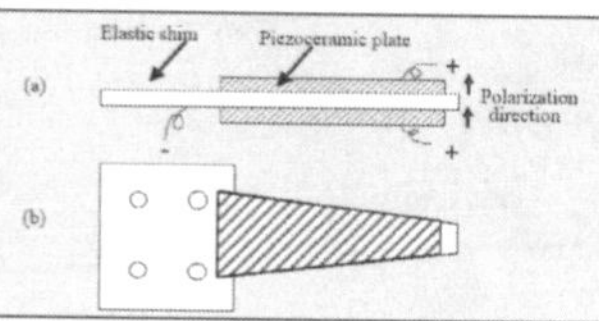	10%	1 %	0.5 – 2 kHz
Moonie/ Cymbal	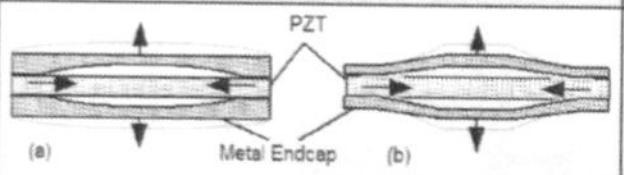	30%	9 %	10 – 40 kHz
Multilayer	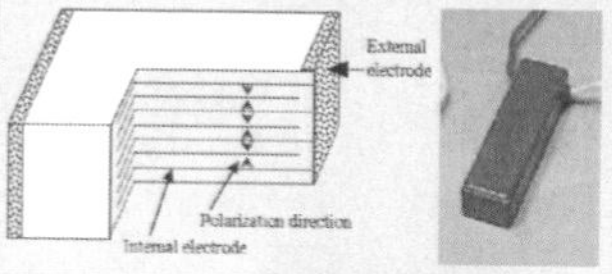	70%	49 %	50 – 300 kHz
Multilayer + Hinge Lever	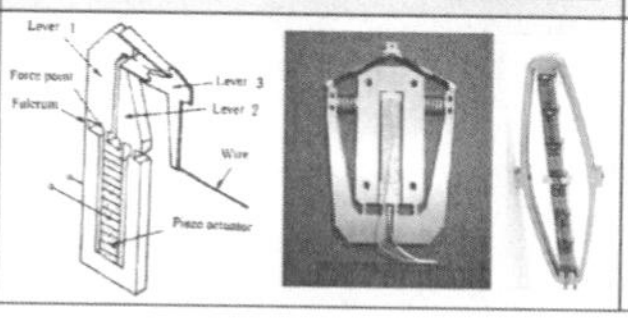	70%?	49 %?	1 – 20 kHz

Fig. 5.50: Promising piezoelectric device designs for energy harvesting applications.

vibration source energy to the stored electric energy in the cymbal transducer as 7.5–7.8% for the 0.3-mm endcap cymbals, and 2.9% for the 0.4-mm endcap cymbal. The reduction of the conversion rate for the 0.4-mm endcap type originated from the mechanical/acoustic impedance mismatch and low k_v.

As indicated in Table 5.8, the key to dramatic enhancement in the efficiency is to use a high electromechanical coupling factor k_{eff} mode, such as k_{33}, k_t or k_{15}, rather than the smallest k_{31}-based bimorph modes. Figure 5.50 summarizes promising piezoelectric device designs for energy harvesting applications. The hinge-lever mechanism is an ideal "mechanical transformer" without losing mechanical energy under an off-resonance condition by keeping a high electromechanical coupling factor k_{eff}. However, how to integrate an ML actuator and hinge-lever structure in a flat and compact system is the current problem to be solved. Figure 5.51 shows an idea on the out-of-plane displacement amplification with a hinge-lever structure, originally invented for a video magnetic head positioner. Based on the ML actuator, the first amplification is made by the arm length ratio $\bar{BC}/\bar{AB}$, then the second

Fig. 5.51: Out-of-plane displacement amplification with a hinge-lever structure.

stage changes the displacement direction to out-of-plane with a flextensional mode; the flat portion D will move down, as illustrated in the figure. This design can easily be manufactured using *Micro Machining* technologies with a very thin design in MEMSs (Micro Electro Mechanical Systems). This design may expand the tunability of the mechanical impedance still by keeping a high electromechanical coupling factor k_{eff} in the range of $k_{33} \approx 70\%$ for the off-resonance usage.

Chapter Essentials

1. FOM for "piezoelectric energy harvesting":

    (a) FOM for stress input — $g \cdot d$ (product of piezoelectric voltage and strain constants)

    (b) FOM for mechanical energy input — k_{eff}^2 (effective electromechanical coupling factor)

2. Piezoelectric materials for "energy harvesting"

    (a) Soft PZT — High g is preferred.

    (b) Piezo single crystal — PMN-PT (high d and high k).

    (c) Piezo polymers/composites — PVDF, PZT-polymer composites (good acoustic impedance matching).

3. Five definitions of electromechanical coupling factor:

    (a) *Mason's definition*

$$k^2 = (\text{Stored mechanical energy/Input electrical energy})$$

$$k^2 = (\text{Stored electrical energy/Input mechanical energy})$$

    (b) *Material's definition* (under static condition)

$$U = U_{MM} + 2U_{ME} + U_{EE}$$

$$= (1/2) \sum_{i,j} s_{ij}^E X_j X_i + 2 \cdot (1/2) \sum_{m,i} d_{mi} E_m X_i + (1/2) \sum_{k,m} \varepsilon_0 \varepsilon_{mk}^X E_k E_m$$

$$k^2 = U_{ME}^2 / U_{MM} U_{EE}.$$

    (c) *Device definition* (under static condition)

    When the primary constitutive equations are defined in a certain piezo component, as

$$\begin{bmatrix} x \\ D \end{bmatrix} = \begin{bmatrix} s^E & d \\ d & \varepsilon_0 \varepsilon^X \end{bmatrix} \begin{bmatrix} X \\ E \end{bmatrix},$$

$$k^2 = \frac{(\text{Coupling factor})^2}{(\text{Product of the diagonal parameters})} = \frac{(d)^2}{(s^E \varepsilon_0 \varepsilon^X)}.$$

    (d) *Constraint condition method*:

    Between E-constant, E-constant elastic compliances, s^E, s^D, stiffness c^E, c^D; and stress-free, strain-free permittivity $\varepsilon_0 \varepsilon^X$, $\varepsilon_0 \varepsilon^x$, inverse permittivity $\kappa_0 \kappa^X$, $\kappa_0 \kappa^x$:

$$1 - k^2 = \frac{s^D}{s^E} = \frac{c^E}{c^D} = \frac{\varepsilon^x}{\varepsilon^X} = \frac{\kappa^X}{\kappa^x}; \quad \kappa k^2 = \frac{(d)^2}{(s^E \varepsilon_0 \varepsilon^X)} = \frac{h^2}{c^D (\kappa_0{}^x)}.$$

    (e) *Dynamic definition*: 4-terminal EC – Voltage V and current I, mechanical terminal parameters force F and vibration velocity $\dot{u}$:

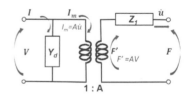

4-terminal equivalent circuit for k_{31} piezo-plate.

$$\begin{bmatrix} F \\ I \end{bmatrix} = \begin{bmatrix} Z_1 & -A \\ A & Y_1 \end{bmatrix} \begin{bmatrix} \dot{u} \\ V \end{bmatrix},$$

$$k_v^2 = \left| \frac{\left(\frac{A^2}{Z_1 Y_1}\right)}{1 + \left(\frac{A^2}{Z_1 Y_1}\right)} \right|.$$

$$k_v = k \text{ for } \omega \to 0$$

$$k_{v,n} = P_n k_v^2 \text{: } k \text{ for } nth \text{ resonance mode}$$

4. Comparison among various electromechanical coupling factors (Example PZT 5AH):

   (a) Transversal effect $k_{31} = \dfrac{d_{31}}{\sqrt{s_{11}^E \varepsilon_0 \varepsilon_{33}{}^X}} - \sim 37\%$

   - Bimorph with d_{31} – less than $\sqrt{\frac{3}{4}} k_{31} - \sim 13\%$ (smallest)

   (b) Planar mode $k_p = \sqrt{\dfrac{2}{1-\sigma}} k_{31} = \dfrac{d_{31}}{\sqrt{s_{11}^E \varepsilon_0 \varepsilon_{33}{}^X}} \cdot \sqrt{\dfrac{2}{1-\sigma}} - \sim 62\%$

   - Cymbal transducer based on $k_p - k_{\text{eff}} \sim 30\%$

   (c) Longitudinal effect $k_{33} = \dfrac{d_{33}}{\sqrt{s_{33}^E \varepsilon_0 \varepsilon_{33}{}^X}} - \sim 73\%$

   - Hinge-lever with an ML – $k_{\text{eff}} \sim 60\%$

5. Comparison among transverse and longitudinal effect modes:

	Transverse effect (k_{31})	Longitudinal effect (k_{33}, k_t)
Electric condition ($\boldsymbol{k}//x$)	$\frac{\partial E}{\partial x} = 0$	$\frac{\partial D}{\partial x} = 0$
Elastic constant	s_{11}^E	$c_{33}^D = 1/s_{33}^D$
Admittance	$Y = j\omega C_d \left[1 + \frac{k_{31}^2}{1-k_{31}^2} \frac{\tan(\Omega_{11})}{\Omega_{11}} \right]$	$Y = \dfrac{j\omega C_d}{1 - k_{33}^2 \frac{\tan(\Omega_{33})}{\Omega_{33}}}$
Resonance	$\tan(\Omega_{11}) = \infty$	$1 - k_{33}^2 \frac{\tan(\Omega_{33})}{\Omega_{33}} = 0$
Half-wave frequency ($\omega_{\lambda/2}$)	ω_R	ω_A
EC		

6. From the resonance $f_{R,n}$ and antiresonance $f_{A,n}$ frequencies, the dynamic electromechanical factor k_{vn} for the n-th harmonic vibration mode can be obtained:

$$\begin{cases} k_{vn}^2 = 1 - \dfrac{f_{R,n}^2}{f_{A,n}^2} & \text{(for the mode strain/stress} \perp P_S) \\[2mm] \dfrac{k_{vn}^2}{1-k_{vn}^2} = \dfrac{f_{A,n}^2}{f_{R,n}^2} - 1 & \text{(for the mode strain/stress} // P_S) \end{cases}$$

The dynamic k_{vn} is also related with the static k_v as

$$k_{vn}^2 = P_n k_v^2,$$

where P_n is the *capacitance factor*, which is given by

$$\begin{cases} P_n = \dfrac{8}{\pi^2 n^2} & \text{(for the longitudinal mode)} \\[2mm] P_n = \dfrac{2(1+\sigma)}{\alpha_n^2 - (1-\sigma^2)} & \text{(for the in-plane mode)} \end{cases}$$

7. (a) Impulse mechanical input generates linear displacement change in a piezoelectric component, while it generates a sinusoidal reaction in an EC, a discrepancy from the experimental result (see Fig. 3.3). This is the limitation of the EC usage for the impulse drive.

   (b) Sinusoidal mechanical input generates sinusoidal displacement change in both piezoelectric component and in an EC.

8. Piezoelectric resonance and antiresonance are both natural mechanical resonance modes. When the piezo component is electrically short-circuited, the resonance mode is realized, and while open-circuited, the antiresonance mode is realized. The shunt condition with electrical impedance $(1/\omega C)$ exhibits another resonance mode in between the resonance and antiresonance frequencies.

9. Total piezoelectric harvested energy for an "impulse force" with the matched external impedance $Z = 1/\omega C$:

$$\frac{1}{2}k^2 U_M \sum_{n=0}^{\infty} \left[\left(1 - \frac{1}{2}k^2\right) e^{-\frac{\pi}{2Q_M}}\right]^n = \frac{1}{2}k^2 U_M \frac{1}{1 - \left(1 - \frac{1}{2}k^2\right) e^{-\frac{\pi}{2Q_M}}}.$$

For reasonable $Q_M > 50$, this approaches U_M (input mechanical impact energy) after many "vibration ringings", irrelevant to the k value.

10. Piezoelectric harvested power for a "sinusoidal cyclic force" with the matched external impedance $Z = 1/\omega C_d$ (k_{31} case):

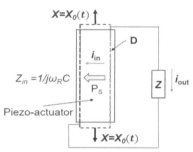

$$|P_{\text{out}}| = \frac{1}{2}\omega(Lwb)\frac{(d_{31}X_0)^2}{\varepsilon_0\varepsilon_{33}^X}\frac{1}{\left(1 - k_{31}^2\right) + \frac{1}{\left(1-k_{31}^2\right)}}.$$

The stored electric energy $\frac{1}{2}\frac{P^2}{\varepsilon_0\varepsilon_{33}^X}$ multiplied by $\frac{1}{\left(1-k_{31}^2\right)+\frac{1}{\left(1-k_{31}^2\right)}}$, which is slightly smaller than 1/2. A half of the stored electric energy can be harvested theoretically under the impedance matching condition.

Piezoelectric energy harvesting model under the external impedance Z.

11. In a unimorph design with a piezo-plate and a metal shim, the maximum electromechanical coupling k_{eff} is obtained when the metal thickness is adjusted in a similar level of the piezo-plate.

12. The metal endcap thickness of a cymbal transducer should be selected to an optimum range: (1) acoustic impedance matching with a vibration source, and (2) electromechanical coupling factor is larger for the thinner specimen, but (3) a too thin specimen cannot endure the vibration source pressure.

13. In a multilayer actuator with a hinge-lever mechanism, though the electromechanical coupling k_{eff} is close to k_{33} with $\omega \to 0$, the dynamic k_v (even at an off-resonance frequency) changes significantly according to the hinge-lever structure design.

14. Off-resonance mechanical energy transfer rate e in a hinge-lever mechanism is defined as

$$e \equiv \frac{\text{Output Energy from the Magnification Mechanism}}{\text{Stored Energy in the Piezoactuator}} = \frac{(\zeta_m^2/2C_m)}{(\zeta_c^2/2C_c)}$$

where ζ_c: free displacement (m) of the piezo-ceramic device, C_c: elastic compliance of the ceramic (m/N), ζ_m: free displacement (m) of the displacement magnification mechanism, and C_m: elastic compliance of the lever mechanism (m/N).

15. EC for the printing wire action in Fig. 5.48: Lever's *moment of inertia*: inductance I_i; lever *mass*: mutual inductance M_i. Moment of inertia and translation motion of a lever: parallel connection of components. Compliances of each part: capacitances C_i, C_i' and C_i''. Quantity Φ_1 at the first transformer: "force factor"; Φ_i and Φ_i' at other transformers in the circuit: transduction of the linear movement into an angular displacement. r_i: transduction rate associated with the movement of the center of mass. C_{1d}: damped capacitance of ML actuator. m_1, m_{13}: equivalent mass of the actuator and wire.

Check Point

1. What is the major problem of the piezo-MEMS devices for electric energy harvesting by using the piezo film of 1-μm PZT thickness? Answer simply.

2. (T/F) PZT: polymer composites with the 1:3 connectivity enhance piezoelectric g constant significantly by sustaining d constant rather constant, which enhances the piezoelectric energy harvesting amount under constant stress condition. True or False.

3. (T/F) In order to develop compact energy harvesting devices, we had better develop the piezo component which can generate the high output electric energy at its resonance frequency. True or False?

4. (T/F) The unimorph piezoelectric structure is most popularly used, because it exhibits the highest energy harvesting rate among various piezo component designs. True or False?

5. Calculate the electrical impedance $1/j\omega C$ of a piezoelectric component with capacitance 1 nF at the off-resonance frequency 100 Hz, which is larger than the internal impedance of rechargeable batteries ($\sim$50 Ω).

6. Describe the "figure of merit" for piezo energy harvesting material under a constant force/stress condition.

7. Describe the "figure of merit" for piezo energy harvesting material under a constant mechanical energy condition.

8. (T/F) When we excite a piezoelectric k_{31}-type plate mechanically with pulse force along the length direction, it shows the sinusoidal total displacement change with time. True or False?

9. When we analyze an EC for a piezoelectric k_{31}-type plate excited mechanically with pulse force along the length direction, how is the total displacement change (curve shape) with time? Answer simply on curve shape and overshoot rate.

10. A k_{31} piezo-plate was excited mechanically by an impulse force from the both ends, and the total displacement change was monitored under short-circuit and open-circuit conditions, as below. Which curve is for the "short-circuit" condition, (a) or (b)?

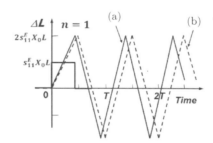

11. (T/F) Total piezoelectric harvested energy for an impulse force (impact energy $= U_M$) with the matched external impedance $Z = 1/\omega C$ approaches $(1/2)U_M$ after many vibration ringings. True or False?

12. (T/F) When we excite a piezoelectric k_{33}-type rod mechanically with sinusoidal force along the length direction, it shows the mechanical excitation (resonance) at its "antiresonance" frequency, when the rod is electrically short-circuited. True or False?

13. Which EC model is for the k_{33}-type piezo rod, the left-side or right-side figure?

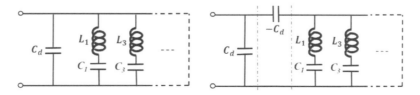

14. A semi-soft PZT 4 possesses the following physical parameters:

$$d_{31} = -123 \times 10^{-12}\,(\text{m/V}), \quad s_{11}^E = 12.3 \times 10^{-12}\,(\text{m}^2/\text{N}),$$

$$s_{12}^E = -4.05 \times 10^{-12}\,(\text{m}^2/\text{N}), \quad \text{and} \quad \varepsilon_{33}^X = 1300.$$

Calculate the electromechanical coupling factors for k_{31} and k_p for PZT 4.

15. (T/F) An acoustic horn produces effectively an amplification of the resonance longitudinal displacement that is inversely proportional to the radius of the vibrator. True or False?

16. (T/F) There is a piezoelectric bimorph plate in a cantilever support. The second resonance frequency is 3 times the fundamental resonance frequency. True or False?

17. (T/F) Poisson's ratio σ ranges from 0 to 1, and the most popular value in PZT ceramics is 0.5. True or False?

18. Which is higher among the electromechanical coupling factors k_{33} and k_{31} in the same PZT ceramic?

19. Which is higher among the electromechanical coupling factors k_{31} and k_p in the same PZT ceramic?

20. (T/F) The displacement amplification factor in a cymbal transducer is roughly proportional to the ratio (cavity depth/cavity diameter) of the metal endcap. True or False?

21. (T/F) A 4-terminal EC is expressed by $\begin{bmatrix} F \\ I \end{bmatrix} = \begin{bmatrix} Z_1 & -A \\ A & Y_1 \end{bmatrix} \begin{bmatrix} \dot{u} \\ V \end{bmatrix}$, where voltage V and current I, mechanical terminal parameters force F and vibration velocity $\dot{u}$ are used. In this case, the dynamic electromechanical coupling factor is expressed by $k_v^2 = \left| \frac{(\frac{A^2}{Z_1 Y_1})}{1 + (\frac{A^2}{Z_1 Y_1})} \right|$. True or False?

Chapter Problems

5.1 Obtain the resonance mode displacement profile shown in Fig. 5.22(b) in a k_{31}-type cantilever-type bimorph (Fig. 5.20) by considering the boundary condition, i.e., one-end clamp.

5.2 We consider a multilayer and mass combination, as illustrated in Fig. 5.40(a). The admittance is expressed by the following formula using mass parameters, c_{elast} and l_{elast}:

$$Y = j\omega C_d + \frac{\Phi'^2}{-\frac{\Phi'^2}{j\omega C_d} + Z_1 + Z_2 + \frac{1}{j\omega c_{\text{elast}} + \frac{1}{j\omega l_{\text{elast}}}}}.$$

Calculate the k_{v1}^2 and k_{v2}^2 from the formula

$$\frac{k_{vn}^2}{1 - k_{vn}^2} = (f_{A,n}/f_{R,n})^2 - 1$$

for different mass length by changing c_{elast} and l_{elast} values. Then, discuss the maximum of k_{v1}^2 around math length $L \sim 10\,\text{mm}$.

5.3 Derive the dynamic electromechanical factor k_{vn} for the n-th harmonic vibration mode from the resonance $f_{R,n}$ and antiresonance $f_{A,n}$ frequencies, for the modes (a) strain/stress $\perp P_S$ (left-side figure) and (b) strain/stress $//\ P_S$ (right-side figure).

$$\begin{cases} k_{vn}^2 = 1 - \frac{f_{R,n}^2}{f_{A,n}^2} & \text{(for the mode strain/stress } \perp P_S) \\ \frac{k_{vn}^2}{1 - k_{vn}^2} = \frac{f_{A,n}^2}{f_{R,n}^2} - 1 & \text{(for the mode strain/stress } //\ P_S) \end{cases}.$$

References

1. K. Uchino, *Micromechatronics*, 2nd Edition, Boca Raton, FL: CRC Press (2020), ISBN-13: 978-0-367-20231-6.
2. J. Kuwata, K. Uchino and S. Nomura, *Ferroelectrics*, 37, 579 (1981).

3. J. Kuwata, K. Uchino and S. Nomura, *Jpn. J. Appl. Phys.*, 21, 1298 (1982).

4. K. Yanagiwawa, H. Kanai and Y. Yamashita, *Jpn. J. Appl. Phys.*, 34, 536 (1995).

5. S. E. Park and T. R. Shrout, *Mat. Res. Innovt.*, 1, 20 (1997).

6. X. H. Du, J. Zheng, U. Belegundu and K. Uchino, *J. Appl. Phys. Lett.*, 72, 2421 (1998).

7. X. H. Du, U. Belegundu and K. Uchino, *Jpn. J. Appl. Phys.*, 36(9A), 5580 (1997).

8. S. Kalpat and K. Uchino, *J. Appl. Phys.*, 90(6), 2703–2710 (2001).

9. D. Damjanovic, D. V. Taylor and N. Setter, *Proc. Mater. Res. Soc., Symp.*, VIII (1999).

10. X. H. Du, Q. M. Wang, U. Belegundu and K. Uchino, *J. Ceram. Soc. Japan.*, 107(2), 190 (1999).

11. K. Uchino, *Micromechatronics*, New York, NY: Marcel Dekker (2003), ISBN: 0-8247-4109-9

12. K. Uchino, *Ferroelectric Devices*, 2nd Edition, New York, NY: CRC Press (2009).

13. W. P. Mason, *Physical Acoustics and the Properties of Solids*, New York: Van Nostrand (1958).

14. M. Onoe (ed.), *Fundamentals of Solid State Vibration for Electrics and Electronics*, Ohm Publ., Tokyo, Japan (1982).

15. M. Onoe and H. Jumonji, *Jpn. J. Comm. Soc.*, 50, 5 (1967).

16. H.-W. Kim, K. Uchino and T. Daue, *Proc. CD 9th Japan Int'l. SAMPE Symp. & Exhibit.*, SIT Session 05, November 29–December 2 (2005).

17. P. Muralt, *Integrated Ferroelectrics*, 17, 297 (1997).

18. *IEEE Standard on Piezoelectricity*, IEEE Standard 176 (1987).

19. R. N. N. and N. N. Rogacheva, *J. Appl. Math. Mech.*, 65(2), 317 (2001).

20. H.-W. Kim, S. Priya and K. Uchino, *Jpn. J. Appl. Phys.*, 45(7), 5836 (2006).

21. G. A. Wempner, *Proc. 3rd Natl. Congr. Applied Mechanics*, ASME, 473 (1958).

22. H.-W. Kim, S. Priya, K. Uchino and R. E. Newnham, *J. Electroceramics*, 15, 27 (2005).

23. K. Uchino, *Proc. 5th Int'l Workshop on Piezoelectric Mater. Appl.*, State College, PA, October 6–10 (2008).

24. T. Yano, E. Sato, I. Fukui and S. Hori, *Proc. Int'l Symp. Soc. Information Display*, p. 180 (1989), p. 180.

25. http://www.cedrat.com/en/mechatronic-products.html.

26. K. Yano, T. Hamatsuki, I. Fukui and E. Sato, *Proc. Japan. Electr. Commun. Soc.*, 1–157 (Spring, 1984).

27. K. Yano, I. Fukui, E. Sato, O. Inui and Y. Miyazaki, *Proc. Japan. Electr. Commun. Soc.* (Spring, 1984), pp. 1–156.

28. K. Yano, T. Inoue, S. Takahashi and I. Fukui, *Proc. Japan. Electr. Commun. Soc.* (Spring, 1984), 1–159.

Chapter 6

Electrical-to-Electrical Energy Transfer

Piezoelectric materials generally convert mechanical energy to electrical energy with relatively high voltage which means output impedance is relatively high (several $100\,\text{k}\Omega$) at an off-resonance frequency. On the other hand, energy storage devices such as a rechargeable battery have low input impedance (10–$100\,\Omega$). Thus, a large portion of the excited electrical energy is reflected back without charging, if we connect the battery immediately after the rectified voltage. In order to improve energy transfer efficiency, electrical impedance matching is required using a suitable DC–DC converter.

The electrical energy generated from the environmental mechanical energy on the piezoelectric component is usually low frequency (1–$100\,\text{Hz}$) sinusoidal, or sudden impulse force (i.e., snap-action type). Thus, we need two process steps prior to the rechargeable battery: (1) AC–DC rectification and (2) DC–DC converter for reducing the electrical impedance.

6.1 Principle of Electric Impedance Matching

6.1.1 *Electric Impedance Matching*

Let us start from a prerequisite knowledge check (you must have done it already!) on this issue.[1,2]

Example Problem 6.1

Given a power supply (such as a DC battery) with an internal impedance, Z_0, what is the optimum circuit impedance, Z_1, required for maximum power transfer? Refer to Fig. 6.1.

Solution:

Referring to Fig. 6.1, the current and voltage associated with an external impedance Z_1 are expressed by $V/(Z_0 + Z_1)$ and $[Z_1/(Z_0 + Z_1)]V$, respectively. The product of these yields the power spent in this external load Z_1:

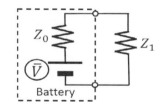

Fig. 6.1: Impedance matching with a power supply.

$$P = \left[\frac{V}{Z_0 + Z_1}\right] \cdot \frac{Z_1 V}{Z_0 + Z_1} = \frac{Z_1}{(Z_0 + Z_1)^2} V^2. \qquad (P6.1.1)$$

To maximize the power (or most effectively spend the battery energy), the following maximization relation

$$\frac{\partial P}{\partial Z_1} = \frac{Z_0 - Z_1}{(Z_0 + Z_1)^3} V^2 = 0 \qquad (P6.1.2)$$

should be satisfied. When impedance is resistive, the power will be maximum at $Z_1 = Z_0$ with $P_{\max} = (\frac{1}{4})\frac{V^2}{Z_0}$. Note that the same amount of power $P = (\frac{1}{4})\frac{V^2}{Z_0}$ is spent inside the battery (usually, this is converted to the heat generation in the battery), leading to the total power $P = (\frac{1}{2})\frac{V^2}{Z_0}$. When the impedance is complex, $Z_1 = Z_0^*$, which can be derived from the power definition $P = Re[(1/2)(V\ I^*)]$. Derive this by yourself.

6.1.2 *Electric Impedance Matching in Piezo Energy Harvesting*

A piezoelectric energy harvesting system under sinusoidal stress application on a piezoelectric component connected with the external electrical impedance Z is shown in Fig. 6.2. When we assume sinusoidal input stress $X = X_0 e^{j\omega t}$ and output electric displacement $D = dX_0 e^{j\omega t}$ via direct piezoelectric effect (d constant), we can derive the following current and voltage relationships from Fig. 6.2. We can understand that the piezo power supply has the internal impedance $1/j\omega C$ under an off-resonance frequency (by neglecting the dielectric loss or effective conducting loss $\sigma = 0$), and this piezoelectric "current supply" generates the total current (we consider the unit area for simplicity)

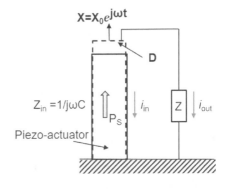

Fig. 6.2: Piezoelectric energy harvesting model.

$$i = \frac{\partial D}{\partial t} = j\omega dX_0; \tag{6.1}$$

This current is split into internal "displacement current" i_{in} and external current i_{out},

$$i = i_{\text{in}} + i_{\text{out}} \tag{6.2}$$

Then, because the potential/voltage should be the same on the top electrode of the piezo component, we get

$$Z_{\text{in}} i_{\text{in}} = Z i_{\text{out}} \tag{6.3}$$

Inserting the relation $i_{\text{in}} = \left(\frac{Z}{Z_{\text{in}}}\right) i_{\text{out}} = j\omega CZ \cdot i_{\text{out}}$ into Eq. (6.2),

$$i_{\text{out}}(1 + j\omega CZ) = j\omega dX_0. \tag{6.4}$$

Thus, we can obtain the output electric energy as

$$P = \left(\frac{1}{2}\right) \text{Re}(i_{\text{out}}\ V^*) = \frac{1}{2} Z \frac{(\omega dX_0)^2}{(1 + (\omega CZ)^2)}. \tag{6.5}$$

Figure 6.3 shows the electric load (resistive) dependence of the output electric energy, which concludes that the maximum electric energy $|P| = \frac{1}{4}\frac{\omega d^2 X_0^2}{C}$ can be obtained at $Z = 1/\omega C$, when we consider Z resistive, which is the situation for charging up a rechargeable battery. In other words, the "generated" electric energy in a piezo component can be spent maximum when the external load impedance matches exactly with the internal impedance. Note the internal impedance is capacitive with phase lag of $-j$(or $-90°$).

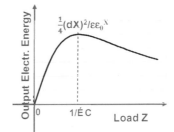

Fig. 6.3: Output electric energy vs. external electrical load Z.

The reader is reminded of the discussion in Section 5.3.2.5, calculated from the piezoelectric constitutive equations for the k_{31} mode. The differences from Eq. (6.5), in which we did not consider the dependence of permittivity on the mechanical constraint (i.e., ε_{33}^X or $\varepsilon_{33}^{x_1}$) from the exact solution in Eq. (5.118), are as follows:

(1) Matching impedance should be replaced by the "damped" capacitance C_d rather just a capacitance C.

(2) The calibration factor $\frac{1}{(1-k_{31}^2)+\frac{1}{(1-k_{31}^2)}}$ is more precise, rather than 1/2 in $|P| = \frac{1}{2}\frac{\omega d^2 X_0^2}{C}\frac{1}{(1-k_{31}^2)+\frac{1}{(1-k_{31}^2)}}$.

We consider further two additional impedance matchings: $Z = \left(\frac{1}{j\omega C}\right)^*$ and $Z = \left(\frac{1}{j\omega C}\right)$. When we consider Z "complex", $Z = Z_{\text{in}}^* = \left(\frac{1}{j\omega C}\right)^* = j\omega\left(\frac{1}{\omega^2 C}\right)$ provides the original electrical impedance matching. This condition corresponds to LC series connection (i.e., $\left(\frac{1}{j\omega C}\right)$ and $j\omega L$), where $L = 1/\omega^2 C$ is satisfied, leading to the LC resonance frequency exactly equal to the stress application frequency ω. The energy generated by a piezo component will be exchanged between the internal capacitance and external inductance, like a "catch-ball", without losing energy or providing work externally. The effective elastic compliance s_{eff}^E in Eq. (6.16) approaches infinite. To the contrary, when we consider $Z = \left(\frac{1}{j\omega C}\right)$, the same capacitance as the internal one is connected to the external load. Since converted energy is split to two equal capacitance, $s_{\text{eff}}^E = s^E(1 - \frac{1}{2}k^2)$, in-between the short- and open-circuit conditions. In order to take the energy out into a rechargeable battery, resistive load is essential to connect.

6.1.3 *Energy Transmission Coefficient*

Since we need to use or accumulate energy externally, we consider "resistive shunt" for further discussions. As we have already discussed, in the case of the impulse force application, the input total mechanical energy U_M is converted to the stored electrical energy by the factor of electromechanical coupling factor k, $U_E = k^2 U_M$. By accumulating $\frac{1}{2}k^2 U_M$ every half cycle into a rechargeable battery, we can collect the whole U_M during all vibration ring-down periods (typically 100–1000 cycles). However, when the vibration source generates sinusoidal continuous force, not all the electrically stored energy (in this case, energy per second, "power") can be actually used under the mechanical drive, and the actual work done in the piezoelectric component depends on the electrical load, as we derive the electrical impedance matching above. Let us reconsider the power expendable on the external electrical load Z from the *energy transmission coefficient* viewpoint. Figure 6.4 summarizes the calculation processes of the input mechanical and output electric energy under various impedance Z's: (a) the stress vs. electric displacement relation, the area on this domain does not mean the energy; thus, we need to translate this plot into (b) stress vs. strain relation to calculate the input mechanical energy; and (c) electric displacement vs. electric field to calculate the output electric energy. With zero mechanical load or a complete clamp (no strain), no output work ("Pushing a curtain, and pushing a wall") is done in an electrical driven piezo actuator. Vice versa, no electrically converted energy can be actu-

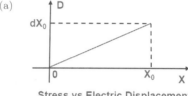
(a) Stress vs Electric Displacement

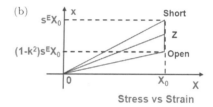
(b) Stress vs Strain

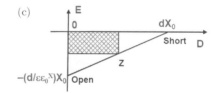
(c) Electric Displacement vs Field

Fig. 6.4: Calculation models of the input mechanical and output electric energy.

ally spent under the short-circuit (zero impedance) or open-circuit (infinite impedance) condition in the energy harvesting case. The *energy transmission coefficient* is defined by

$$\lambda_{\max} = (\text{Output mechanical energy/Input electrical energy})_{\max} \qquad (6.6a)$$

in the actuator application (refer to Section 2.6.2.2), or equivalently,

$$\lambda_{\max} = (\text{Output electrical energy/Input mechanical energy})_{\max} \qquad (6.6b)$$

in the energy harvest application. The difference of the above from the definition of electromechanical coupling factor k^2 is "stored/converted energy" or "output/spent energy".

Let us consider the formula derivation for the latter case. First, the electrical energy output from Fig. 6.4(c) can be calculated from the constitutive equations

$$\begin{pmatrix} D \\ x \end{pmatrix} = \begin{pmatrix} \varepsilon_0 \varepsilon & d \\ d & s^E \end{pmatrix} \begin{pmatrix} E \\ X \end{pmatrix}. \tag{6.7}$$

The short-circuit condition ($E = 0$) gives $D_0 = dX_0$, while the open-circuit condition ($D = 0$) gives $E_0 = -dX_0/\varepsilon_0\varepsilon$ (i.e., "depolarization field"). The triangular areas of $OD_0E_0(\frac{1}{2}dX_0 \cdot \left(\frac{dX_0}{\varepsilon_0\varepsilon}\right) = \frac{1}{2}\frac{(dX_0)^2}{\varepsilon_0\varepsilon})$ mean the total electric energy converted from the mechanical vibration. *Thus*, under an impedance Z shunt condition, we can expect a point (E, D) on the line between the terminals of the above D_0 and E_0:

$$E = \frac{1}{\varepsilon_0\varepsilon}(D - dX_0) \tag{6.8}$$

The output electrical energy can be calculated as

$$U = -DE = -\frac{D}{\varepsilon_0\varepsilon}(D - dX_0) = -\frac{1}{\varepsilon_0\varepsilon}\left(D - \frac{1}{2}dX_0\right)^2 + \frac{1}{4}\frac{(dX_0)^2}{\varepsilon_0\varepsilon}. \tag{6.9}$$

Note that the rectangular area in Fig. 6.4(c) generated by the point (E, D) stands for the output electrical energy on the load Z. It is easy to maximize this output energy by taking the point at the middle of maximum D_0 and E_0; that is, the maximum energy (per unit volume) is given by

$$|U|_{\max} = \left|\frac{D_0}{2} \cdot \frac{E_0}{2}\right| = \frac{1}{4}\frac{(dX_0)^2}{\varepsilon_0\varepsilon} \tag{6.10}$$

Taking into account the device size, accumulating the energy per second, Eq. (6.10) is equivalent to the power $|P| = \frac{1}{4}\frac{\omega d^2 X_0^2}{C}$, obtained previously. Another derivative approach (first derivative in terms of time) is introduced: Equation (6.7) provides

$$\dot{D} = \varepsilon_0\varepsilon\dot{E} + d\dot{X}. \tag{6.11}$$

Knowing $\dot{D}S = I$, $\dot{E} = j\omega E$, $\dot{X} = j\omega X$, and $E = -V/t$,

$$I = -j\omega\varepsilon_0\varepsilon\left(\frac{S}{t}\right)V + j\omega dXS \tag{6.12}$$

Integrating $I = V/Z$, $C = \varepsilon_0\varepsilon\left(\frac{S}{t}\right)$, and $XS = X_0$, we obtain

$$V = j\omega dX_0 / \left(\frac{1}{Z} + j\omega C\right) \tag{6.13}$$

Then, the output power via the load Z is calculated as

$$P_{\text{out}} = \text{Re}\left[\left(\frac{1}{2}\right)(I\,V^*)\right] = \frac{1}{2}Z\frac{(\omega dX_0)^2}{(1 + (\omega CZ)^2)} \tag{6.14}$$

which is exactly the same result as Eq. (6.5). When the "ambient vibration energy is unlimited", the above maximum output energy condition is the final targeted status.

However, you should notice that the input mechanical energy differs significantly also depending on the external electrical load Z. If you recall the tunable elasticity according to the electric constraint, that is, s^E (short-circuit) or s^D (open-circuit), and further $s^D = s^E(1 - k^2)$ in particular, in the transverse effect k_{31} case, you can understand that the input mechanical energy (e.g., triangular area made by OX_0x_0 for a short-circuit condition) differs largely depending on the electrical load Z, as illustrated in Fig. 6.4(b) (the slope, elastic compliance changes according to Z). Now, second, let us calculate the load Z dependence of the input mechanical energy. We calculate now the "input mechanical energy" from the second constitutive equation in Eq. (6.7):

$$x = dE + s^E X = -d\left(\frac{V}{t}\right) + s^E X = -\left(\frac{d}{t}\right)\left[\frac{j\omega d X_0}{\frac{1}{Z} + j\omega C}\right] + s^E X. \tag{6.15}$$

The last transformation used Eq. (6.13). We obtained effective complex elastic compliance as

$$s_{\text{eff}}^E = \frac{x}{X} = s^E\left[1 - \left(\frac{S}{t}\right)\frac{j\omega Z d^2}{s^E(1 + j\omega C Z)}\right]. \tag{6.16}$$

You can verify the above "effective elastic compliance" is equal to s^E or $s^D = s^E[1 - k^2]$, when $Z = 0$ or ∞, respectively. The total input mechanical power under Z-shunt condition is derived by multiplying the volume (St) as

$$P_{\text{in}} = \text{Re}\left[(St)\left(\frac{\omega}{2}\right)X_0\left(s_{\text{eff}}^E X_0\right)^*\right] = (St)\left(\frac{\omega}{2}\right)s^E\left[1 - \left(\frac{S}{t}\right)\frac{(C\omega Z d)^2}{s^E\left[1 + (\omega C Z)^2\right]}\right]X_0^2 \tag{6.17}$$

The *energy transmission coefficient* defined by

$$\lambda_{\text{max}} = (\text{Output electrical energy/input mechanical energy})_{\text{max}}$$

in the energy harvest application can be calculated from

$$\lambda = \frac{|P|_{\text{out}}}{|P|_{\text{in}}} = S^2\frac{1}{2}Z\frac{(\omega d X_0)^2}{(1 + (\omega C Z)^2)}/(St)\frac{\omega}{2}s^E\left[1 - \left(\frac{S}{t}\right)\frac{C(\omega Z d)^2}{s^E[1 + (\omega C Z)^2]}\right]X_0^2$$

$$= \omega C Z k^2/[1 + (\omega C Z)^2(1 - k^2)]. \tag{6.18}$$

Taking the maximization process in terms of Z, we can obtain the optimized impedance Z as $Z = 1/[\omega C(\sqrt{(1 - k^2)}]$. This impedance value is in-between the free capacitance C and the damped capacitance $(1 - k^2)C$. Then, the energy transmission coefficient is expressed as

$$\lambda_{\text{max}} = [(1/k) - \sqrt{(1/k^2) - 1}]^2$$

$$= [(1/k) + \sqrt{(1/k^2) - 1}]^{-2}. \tag{6.19}$$

However, we need to be aware that since the input mechanical energy is changed (even if we keep the stress/force constant) due the elastic compliance change with the external electrical impedance (see Eq. (6.16)), the condition for realizing the "maximum transmission coefficient" is slightly off from the electrical impedance matching point. When we take the matched electrical impedance $Z = 1/\omega C$, we obtain

$$\lambda = \frac{k^2}{2}\frac{1}{(2 - k^2)} \tag{6.20}$$

which is slightly smaller than λ_{max} of Eq. (6.19). We can also notice that from Eq. (6.19) (monotonous increase with respect to k increase)

$$k^2/4 < \lambda_{\text{max}} < k^2/2,$$

depending on the k value. For a small $k(< 0.3)$, $\lambda_{\text{max}} \approx k^2/4$, and for a large k (up to 0.9), $\lambda_{\text{max}} \approx k^2/3 \sim k^2/2$. When k is unrealistically high (~ 0.98), λ_{max} approaches to 1.

6.2 AC–DC Rectification

Electric power generated in a piezoelectric component by the ambient or machinery mechanical vibration is usually low-frequency (less than 100 Hz) AC (mostly off-resonance, but sometimes resonance). Except for the direct usage of this low-frequency electric power, we need an AC–DC rectifier in order to charge the energy into a battery or an energy storage capacitor. We review the electric rectification technology in this section.

6.2.1 *Diode*

Silicon can be modified to p-type or n-type semi-conductor materials by doping acceptor (e.g., B^{3+}) or donor (e.g., P^{5+}) doping. In comparison with the Si^{4+}, higher valence P^{5+} ion acts as $+1$ which traps the electron, while lower valence B^{3+} acts as -1 which traps the hole, intuitively speaking. A p–n junction is a boundary or interface between these two p- and n-types of semiconductor materials, inside a Silicon single crystal. As illustrated in Fig. 6.5(a), the "p" (positive) side contains an excess of holes, while the "n" (negative) side contains an excess of electrons in the outer shells of the electrically neutral atoms. This allows electrical current to pass through the junction only in one direction (i.e., *rectification*). When the p- and n-types are contacted (Fig. 6.5(a)), the (electron-base) energy band decreases by eV_d, in order to adjust the E_F (*Fermi level*) constant.

Bias field is applied across a p–n junction; "forward bias" is in the direction of p- to n-type, and reverse bias is in the direction of n- to p-type. When the forward-bias is applied (Fig. 6.5(b)), the band level in the n-type increases by $+eE$, facilitating the electron flow to the p-type area; vice versa to the hole. When the reverse-bias is applied (Fig. 6.5(c)), the Fermi level in the p-type increases by $-eE$, leading to large barrier height at the junction. The flows of the electron or holes are suppressed significantly. Thus, the p–n junction allows electric charges to flow in one direction; thus, this electronic component is called "diode". Figure 6.6 shows realistic voltage vs. current relation; when forward-biased, electric charge flows freely, while reverse-biased, however, the junc-

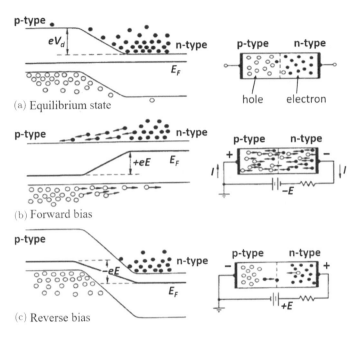

(a) Equilibrium state

(b) Forward bias

(c) Reverse bias

Fig. 6.5: Band structure of p-n junction: (a) zero bias, (b) forward bias, and (c) reverse bias state.

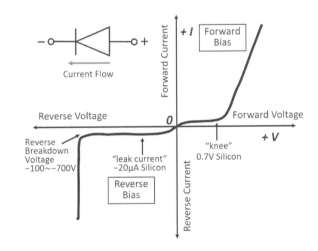

Fig. 6.6: Voltage vs. current relation in an actual p-n junction diode.

tion barrier (and therefore resistance) becomes greater and charge flow is minimal (i.e., "electric rectification"). Precisely speaking, note the leak current under reverse bias, which provides the normal diode energy consumption sub mW. Piezo energy harvesting systems, whose energy level is less than mW, are useless because the cultivating energy is spent in the circuit.

Another issue is the "knee", that is, the forward current requires some threshold voltage, which does not exhibit a serious problem in the piezo energy harvesting system, since the generated voltage is usually several tens to hundreds V. Since we can find the diode with the "reverse breakdown voltage" in the range of $-100\,V$ to $-1\,kV$, we can adopt the diode for the electric power rectification in the piezo energy harvesting system.

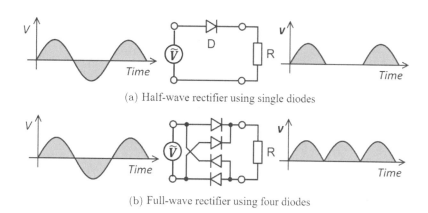

(a) Half-wave rectifier using single diodes

(b) Full-wave rectifier using four diodes

Fig. 6.7: Rectifier with diodes: (a) a half-wave rectifier and (b) a full-wave rectifier using four diodes.

6.2.2 Half-Wave Rectification

Figure 6.7(a) shows a half-wave rectifier using a single diode. In half-wave rectification of a single-phase supply, either the positive or negative half of the AC wave is passed, while the other half is blocked. Mathematically, it is a step function (for positive pass, negative block): passing positive corresponds to the ramp function being the identity on positive inputs, blocking negative corresponds to being zero on negative inputs. The no-load output DC voltage of an ideal half-wave rectifier for a sinusoidal input voltage is

$$\begin{cases} V_{\text{rms}} = \dfrac{V_{\text{peak}}}{2} \\ V_{\text{DC}} = \dfrac{V_{\text{peak}}}{\pi} \end{cases}. \tag{6.21}$$

6.2.3 Full-Wave Rectification

A full-wave rectifier converts the whole of the input waveform to one of constant polarity (positive or negative) at its output. Mathematically, this corresponds to the absolute value function. Full-wave rectification converts both polarities of the input waveform to pulsating DC (direct current), and yields a higher average output voltage. Four diodes in a bridge configuration and any AC source are needed, as illustrated in Fig. 6.7(b). The average and RMS no-load output voltages of an ideal single-phase full-wave rectifier are

$$\begin{cases} V_{\text{rms}} = \dfrac{V_{\text{peak}}}{\sqrt{2}} \\ V_{\text{DC}} = V_{\text{av}} = 2 \cdot \dfrac{V_{\text{peak}}}{\pi} \end{cases}. \tag{6.22}$$

Figure 6.8 shows a full-bridge rectifier with a resistive load and a storage capacitor for our cymbal piezoelectric energy harvesting. A piezoelectric component is composed of an AC current supply (converted from the external AC force via "direct piezoelectric effect" with driving condition at 100 Hz) and the internal impedance $1/j\omega C_p$, where C_p is the free capacitance of the piezo component (by neglecting dielectric loss). Using $C_p = 4nF$ and 100 Hz, we obtain $|Z_{\text{in}}| = 400 \, \text{k}\Omega$. Note this large impedance (high voltage and low current) of the

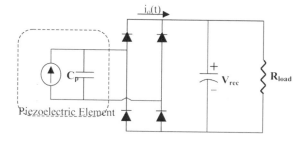

Fig. 6.8: A full-bridge rectifier with a resistive load for piezoelectric energy harvesting.

piezo component in general. Figure 6.9 shows the output electrical power from various Cymbal transducers under AC and DC mechanical loads as a function of external load resistance. Endcap thickness made of steel was changed from 0.3 mm to 0.5 mm.[3] Prestress (DC bias load) to the Cymbal was 66 N and applied AC (100 Hz) force was varied experimentally from 44 N to 70 N. For a small force drive (40 and 55 N), the power level increased with decreasing the endcap thickness (from 0.5, then 0.4, and finally 0.3 mm) because of the input mechanical energy increase (endcap softening). With increasing the force level up to 70 N, the maximum power of 53 mW was obtained at 400 kΩ with a 0.4-mm steel endcap, because the Cymbal sample with a 0.3-mm-thick endcap could not endure under this high force drive (i.e., the cavity depth is collapsed, though device (ceramic) failure does not occur), which degrades the output power. Note that the maximum power was obtained at around 400 kΩ of the external load, which is exactly the impedance matching value with $1/\omega C$ of the piezo component internal impedance.[3,4]

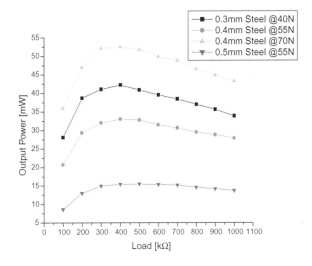

Fig. 6.9: Change in output electrical power from the various Cymbal transducers under different 100 Hz AC mechanical loads (shown as @ xx N under 66 N constant DC bias) with external electrical load resistance.[3]

6.3 DC–DC Converter

Piezoelectric "Cymbal" design was optimized from the generated electrical energy level in Section 5.4.3.2. Using a "full-bridge rectifier" circuit shown in Fig. 6.8, the output voltage and the power were measured across the resistive load directly without any amplification circuit to characterize the performance of different cymbal transducers. The maximum rectified voltage V_{rec} of a storage capacitor C_{rec} (10 μF) was charged up to 248 V after saturation. In Fig. 6.9, the output electrical power from Cymbal transducers exhibits the maximum under the external resistance of 300–400 kΩ, which corresponds to the internal impedance value $1/\omega C$, and is the "electrical matching impedance". Taking into account the input impedance of rechargeable batteries, such as 50–100 Ω, it is difficult to accumulate the electric energy from the piezo component to the rechargeable battery merely through the rectifier. We need to integrate the DC–DC converter to adjust or match the supply impedance to the battery.

The author's group at Penn State University explored various topologies of the DC–DC converters: buck converter, buck-boost converter, flyback converter, etc. The section below discusses these converters from the piezo energy harvesting application viewpoint, after reviewing the basic principles of MOSFET.

6.3.1 *MOSFET*

6.3.1.1 *MOS junction*

Example Problem 6.1

Explain the generation process of the depletion and inversion layers in a metal-oxide-semiconductor (MOS) structure (p-type Si) using a simple energy band model as pictured in Fig. 6.10, when a positive voltage is applied on the metal (reverse-biased). Describe the *Fermi level* in the

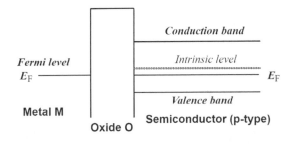

Fig. 6.10: Energy band model for a MOS.

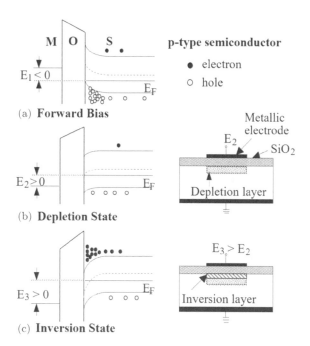

Fig. 6.11: Energy band change of a MOS structure (with a p-type semiconductor) under an applied voltage: (a) forward bias, (b) depletion state, and (c) inversion state.[1]

metal region and the hole and electron concentrations using + and − symbols in the energy band model. For simplicity, you can use the assumption that the flat band voltage is close to zero.

Solution:

Figure 6.11 illustrates the energy band change of a MOS structure (with a p-type semiconductor) under an applied voltage. Negative voltage application on the metal exhibits the "forward-bias" condition. We explain the depletion and inversion layer generation according to the bias voltage.

(a) When the metal is negatively biased (i.e., the gate is *forward-biased*), the Fermi level is increased in the metal, and holes in the p-type semiconductor will accumulate in the semiconductor region near the oxide interface (Fig. 6.11(a)). Since holes cannot drift into the oxide (SiO_2 in this case) insulator, no particular phenomenon occurs in the semiconductor region.

(b) On the contrary, when the metal is positively biased (i.e., the gate is *reverse-biased*), the Fermi level is decreased in the metal, and holes in the p-type semiconductor will be expelled in the semiconductor region near the oxide. Thus, this region will lose both holes and electron, and a *depletion layer* (that is, a region depleted in holes) is generated (see Fig. 6.11(b) Right). Note this depletion layer exhibits rather high resistivity.

(c) When we further increase the positively applied voltage above a certain threshold voltage (V_T), large voltage attracts the electrons in the p-type semiconductor (though the electron density is not large in the p-type region), and an *inversion layer* (an electron-rich region) is generated. This condition is given by semiconductor surface voltage $\psi_s > \psi_B$, where ψ_B is the difference between the intrinsic Fermi level (E_i, dashed line) and the Fermi level (E_F, solid straight line) ($\psi_B = (kT/q)\ln(N_a/n_i)$). This inversion layer is highly conductive due to high electron density. As illustrated in Fig. 6.11(c) Right, under high positive voltage on the metal, the p-type semiconductor region generates two layers adjacent to the interface with the oxide (SiO_2 in this case) insulative layer: (1) inversion layer — conducting channel and (2) depletion layer — insulative cover, which interfaces the p-type semiconductor.

6.3.1.2 *MOSFET*

Based on Example Problem 6.1, we learned how to generate the *inversion layer*; now we learn MOSFET. The *Field-Effect Transistor* (FET) is a type of transistor which uses an electric field to control the flow of current. FETs are devices with three terminals: *source*, *gate* and *drain*. FETs control the flow of current by the application of a voltage to the gate, which in turn alters the conductivity between the drain and source. Learn how we can make a function, only when both the gate and drain are biased, the drain current flows, from the following example, which is the basic principle of on/off switching function.

Example Problem 6.2

Let us consider an n-channel enhancement mode MOSFET (based on p-type semiconductor) as illustrated in Fig. 6.12. A positive gate voltage induces the *electron inversion layer*, which then connects the n-type source and the n-type drain regions. Discuss the drain current behavior as a function of the drain/source voltage.[1]

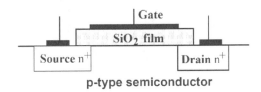

Fig. 6.12: MOSFET fabricated with a p-type semiconductor (n-channel enhancement mode).

Solution:

A large positive gate voltage induces the *electron inversion layer* (conducting channel), covered by the *depletion layer* (resistive layer). The inversion layer connects the n-type source and drain regions as illustrated in Fig. 6.13(a). The source terminal is the source of carriers that flow through the channel (i.e., inversion layer) to the drain terminal. In such an n-channel device, electrons travel from the source to the drain so that the conventional current flows from the drain to the source.

Remember that the *depletion layer* has low conductivity, which is analogous to a shielded cable, a lead wire (the inversion layer) covered by an insulating coat (the depletion layer). This is also analogous to water flowing in a tube, where the water (the electron) flows in a tube (the inversion layer) surrounded by rubber material (the depletion layer). When the tube is pinched off, the water flow is restricted.

Making the assumption that the flat band voltage is close to zero, the application of the gate voltage E_G easily creates the inversion layer. When a small drain voltage (by keeping $E_{DS} < E_G$) is applied, the electrons in the inversion layer will flow from the source to the positive drain terminal. Since for small E_{DS}, the channel region has the characteristics of a conducting lead, we expect

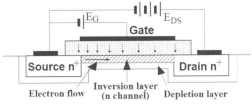

(a) Drain voltage $E_{DS} <$ Gate voltage E_G

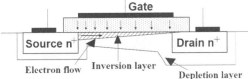

(b) Drain voltage $E_{DS} =$ Gate voltage E_G

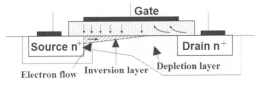

(c) Drain voltage $E_{DS} >$ Gate voltage E_G

Fig. 6.13: Change in the n-channel with the drain/source voltage for an n-channel enhancement mode MOSFET.

$$I_D = g_d E_{DS}. \tag{P.6.2.1}$$

When E_{DS} is increased to the point where the potential drop across the oxide at the drain terminal is equal to zero (precisely speaking, equal to the *threshold voltage* V_T), the induced inversion charge density is zero at the drain terminal. This effect is schematically shown in Fig. 6.13(b). At this point ($E_{DS} = E_G$), the incremental conductance at the drain becomes zero. The slope of the I_D vs. E_{DS} curve becomes zero.

When E_{DS} becomes larger than the above value (E_G), the point in the channel at which the inversion charge is just zero shifts toward the source terminal (Fig. 6.13(c)). In this case, electrons enter the channel at the source, travel through the channel toward the drain, and then, at the pinch-off point, the electrons are injected into the space charge region (depletion layer) where they are swept by the E-field to the drain contact. If we assume that the change in channel length is small compared to the original length, the drain current becomes constant for $E_{DS} > E_G$. This region is referred to as the *saturation region*. Figure 6.14 shows the I_D vs. E_{DS} curves for various gate voltages E_G. In practical applications, we usually keep E_{DS} constant at above the threshold in the saturation region,

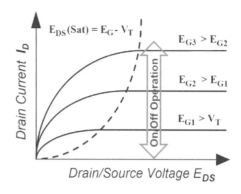

Fig. 6.14: I_D vs. E_{DS} curves for various.

then, only the gate voltage E_G is changed below to above the V_T (almost zero in practice) so that the drain current I_D can be changed zero to large, which is the basic principle of the on/off switching with MOSFET. That is, by switching E_G from zero to some voltage, drain current is switched from zero to high current like an On-Off switch.

6.3.2 *Switching Regulator*

Let us review a *Power MOSFET*, which is popularly used in the piezo actuator drive.[2] The symbol of the MOSFET is shown in Fig. 6.15(a), and its drain current vs. drain-source voltage characteristics are plotted in Fig. 6.15(b), which corresponds to 500 V, 50 A Class *Power MOSFET*. Under keeping the drain-source voltage (V_{DS}) at 5 V, the drain current can be significantly changed from 0 to 60 A with changing the gate-source voltage (V_{GS}) from 0 to 5.5 V, which creates the current on-off switch by controlling the gate-source voltage.

On the basis of the current switching function of the MOSFET, we can design *switching regulators*, which are also called *step-down buck choppers*. The switching regulator is occasionally used as a step-down voltage/step-up current converter (popular DC–DC converter), significant output impedance modulation, without using an electromagnetic transformer, which kills the size and weight in a power system. Figure 6.16 shows the simplest switching regulator composed of only single MOSFET, inductor L, diode and the resistive load R (capacitor C is occasionally added to stabilize the ripple wave form) under

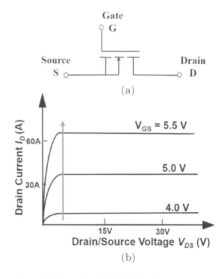

Fig. 6.15: (a) MOSFET, (b) its output characteristics.

a DC constant voltage (battery). Figs. 6.16(a) and 6.16(b) illustrate the operations for the MOSFET on and off stages, respectively. When the gate-source voltage with a rectangular wave form (reasonably high carrier frequency with the *duty ratio* $d = \left(\frac{T_{on}}{T_{on}+T_{off}}\right)$) is applied, the MOSFET behaves as an ON-and-OFF switch. During the ON stage (Fig. 6.16(a)), $E - FET - L - R$ is the current flow route, so that $v_0 = E$, but v_R does not reach E so quickly because of the inductor (which needs to accumulate the electrical energy first). On the contrary, during the OFF state (Fig. 6.16(b)), $v_0 = 0$, but v_R does not reach to 0 so quickly; moreover, because the inductor will release the electrical energy, or generate the *reverse electromotive force*, the current still continues to flow in the route $L - R - D$ now. Note first that the average voltage of a rectangular wave form (0–E) with the duty ratio (d) is estimated as $d \cdot E$. Voltage wave forms of v_0, v_L and v_R of the step-down buck chopper are illustrated as a function of time in Fig. 6.16(c), where you can find first v_0 as exactly the "similar" rectangular wave form to the MOSFET gate voltage with 0 to E voltage

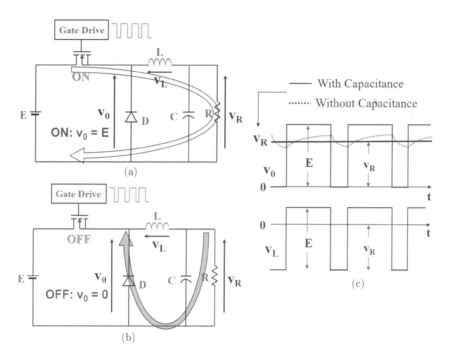

Fig. 6.16: Switching regulator: (a) On, and (b) Off operation. (c) Voltage wave forms of v_0, v_L and v_R of the step-down buck chopper.

height (unipolar). On the contrary, v_L shows the same wave form, but a negative bias voltage equal to v_R. Since v_R is obtained as the subtraction $v_0 - v_L$, we can conclude that the v_R is almost constant around the average voltage $d \cdot E$. More details on the v_R behavior are described using Fig. 6.16(c). Without using a capacitance, the v_R exhibits a ripple mode of an exponential curve with time constant $\tau = (L/R)$ in a L–R circuit. In order to minimize the ripple level, the carrier time period (inverse of the carrier frequency) should be chosen much less than the circuit time constant (L/R) first. An additional smoothing capacitance C helps more with realizing almost constant output voltage $d \cdot E$. Carrier frequency is set at 1 kHz in our Cymbal energy harvesting system, which is high enough compared with the operating frequency 100 Hz, but should be as low as possible not to increase the switching losses.

6.3.3 *On–Off Signal Generator and Bipolar Switching Regulator*

In order to control the Gate voltage of a switching regulator, we need to introduce a circuit to synthesize the desired rectangular signal with a certain duty ratio. Figure 6.17 illustrates the principle of *pulse width modulation* (PWM) based on a triangular carrier wave. When we use a triangular carrier wave v_C, a certain input signal level v_S is easily converted into an ON–OFF signal with a certain duty ratio by subtracting these two voltage values (Fig. 6.17(c)). The subtraction operational amplifier is called "Comparator" (Fig. 6.17(a)), and its practical device example, a low-power CMOS (complementary metal–oxide–semiconductor) clocked comparator, is shown in Fig. 6.17(b), where you can find complementary and symmetrical pairs of p-type and n-type MOSFETs for logic functions.

So far, we demonstrated only a mono-polar drive (0 to E V) switching regulator. However, we occasionally use a bi-polar drive ($-E$ to $+E$ V) power supply, in particular, in DC to AC converters, such as for driving piezoelectric transformers described in Section 6.3.5. The key is to use a *bridge circuit* illustrated in Fig. 6.18(a). In order to obtain "positive" E, we control Tr_1^+ (on and off) by keeping Tr_2^- (on), while to obtain negative E, we control Tr_1^- (on and off) by keeping Tr_2^+ (on). Figure 6.18(b) shows a DC–DC voltage converter from a small signal voltage v_S to amplified voltage v_{ave} utilizing $\pm$ triangular carrier. Note here that when we denote the duty ratio as d, the average output voltage $v_{\mathrm{ave}} = k \cdot E$, where $k = 2d - 1$.

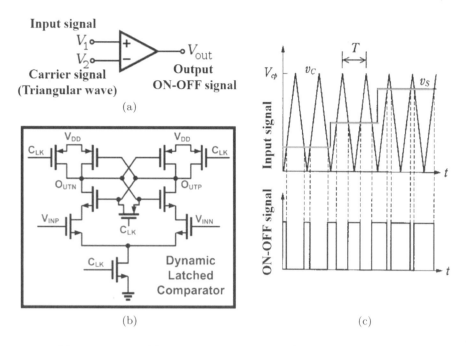

Fig. 6.17: (a) Comparator, (b) low-power CMOS clocked comparator, and (c) Duty ratio realization with a triangular carrier signal.

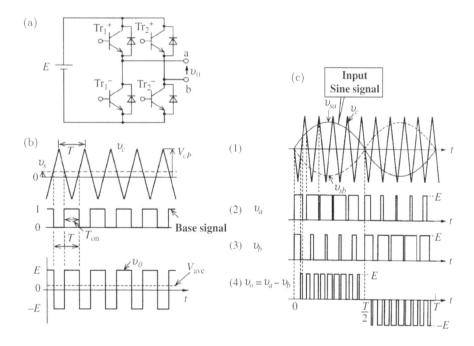

Fig. 6.18: (a) Bridge circuit, (b) DC voltage converter, and (c) Principle of AC voltage pulse width modulation.

Figure 6.18(c) shows the principle of AC voltage pulse width modulation. We will start from a triangular carrier signal v_C with $\pm v_C$. We now consider two sine input signals, v_{Sa} and v_{Sb} (Fig. 6.18(c)(1)), each of which generates a pulse-width modulated wave shown in v_a or v_b at the terminal a or b (Fig. 6.18(c)(2) and (3)), respectively. Since the final output voltage v_0 is provided by the subtraction $v_a - v_b$, we obtain the pulse-width modulated $\pm$ signals corresponding to the input sine signal (Fig. 6.18(c)(4)).

6.3.4 *Review of DC–DC Converters*

6.3.4.1 *Switching power supply vs. linear voltage regulator*

High-frequency switching converters are power conditioning circuits whose semiconductor devices operate at a frequency that is "fast compared to the variation of input and output waveforms".[5] They are used most often over linear regulators as a more efficient interface between DC systems operating at disparate voltage levels, and are the work horse of computer and consumer electronic power supply circuitry today. Switching power supplies offers two distinct advantages over linear voltage regulators:

(1) Because they are truly power converters and not simply voltage regulators, switching converters are quite efficient even when the difference between input and output voltages is large. On the other hand, the average values of the input and output currents in a linear regulator must be the same. Therefore, the power lost through a linear regulator is the product of the input current with the difference between the input and regulated output voltages. Because piezoelectric voltage signals typically have a relatively large domain, and it is disadvantageous to clamp this signal, linear regulators encounter some obvious drawbacks which switching converters are well suited to meet.

(2) Switching converters conserve input to output power regulating either voltage or current at the output stage. In contrast, a linear regulator is an active resistive divider in series with the load. They regulate the output voltage by dissipating excess power for given input current. The switching converter can be thought of as "impedance converters" because the average dc output current can be smaller or larger step-up or step-down, respectively, than the average dc input current.[6] This impedance conversion property is the second notable advantage to use a switching converter over a linear regulator with direct discharge because of the large impedance mismatch inherent in this design.

6.3.4.2 *Analysis of DC–DC converters*

Three kinds of DC–DC converters were analyzed by our group for the piezoelectric energy harvesting circuit application.[7]

- Buck PWM DC–DC converter.
- Buck-Boost PWM DC–DC Converter.
- Transformer-isolated Buck-Boost PWM Converter or Flyback Converter.

Their circuit designs are shown in Fig. 6.19. A *Buck converter* (a) is identical to the switching regulator explained in Section 6.3.2. A *Buck–Boost converter* (b) is a type of DC-to-DC converter that has an output voltage magnitude that is either greater than or less than the input voltage magnitude. Note the position exchange between the diode and inductance from the Buck converter. A *Flyback converter* (c) is a kind of Buck-Boost converter with slight difference in the usage of a transformer, instead of a single inductor.

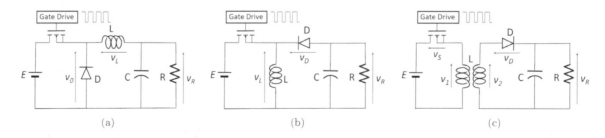

Fig. 6.19: DC–DC converters: (a) Buck converter, (b) Buck-Boost converter, and (c) Flyback converter.

Each converter's analysis is divided into four main categories:

(a) Converter description (see Fig. 6.19).
(b) Analysis of ideal converter in discontinuous conduction mode.
(c) Determination of optimal duty cycle of the converter for energy harvesting.
(d) Loss analysis of a non-ideal converter in discontinuous conduction mode.

For the sake of simplicity, all the converters considered here are fixed frequency. Given below is the origin of losses in a conventional Fixed-Frequency DC–DC converter:

Load-Dependent (Conduction) Losses

- MOSFET "ON" Resistance
- Diode Forward drop
- Inductor/transformer winding resistance
- Capacitor ESR

Frequency-Dependent (Switching) Losses

- MOSFET output capacitance
- MOSFET gate capacitance
- Diode Capacitance
- Diode stored minority charge
- Inductor and transformer core loss
- Snubber loss
- Gate driver loss

Other Fixed Losses

- Controller circuitry losses.
- MOSFET, diode and capacitor leakage currents.

Reference[7] contains detailed derivation of the optimal duty cycle, operation and estimated power loss of the converters in the "Discontinuous Conduction" for the buck-boost and flyback converter topologies. Table 6.1 summarizes the conduction losses in Buck, Buck-Boost and Flyback converters operating in discontinuous conduction mode.

Losses in almost all the cases are as follows:

- Proportional to the cube of input voltage: Losses increase with high voltage generating devices.
- Inversely proportional to the square root of the inductance value and switching frequency.
- Inversely proportional to the output voltage.

Assuming input voltage $\sim 150\,\text{V}(\sim 310/2)$, output voltage $\sim 12\,\text{V}$, the losses in buck converter are 10% less than buck-boost converter under optimal operating conditions. More precisely, we obtained the following results, where the loss is compared with the rate (Buck converter/Buck-Boost converter):

Conduction loss		Switching loss
Inductor loss	0.83	0.83
MOSFET loss	0.91	Assuming same t_{off} and L
Diode loss	0.90	

Based on the loss analysis for the DC–DC converters above, the following results can be drawn:

- Strong arguments can be made to show that buck converter is better than buck-boost in terms of conduction and switching losses.

Table 6.1: (a) Bridge circuit, (b) DC voltage converter and (c) Principle of AC voltage pulse width modulation.

Converter	Diode loss	Inductor loss	MOSFET loss
Buck	$\dfrac{8R_F\left(\frac{\omega C_{cymbal}}{\pi}\right)^{3/2}(V_g-V_o)^{3/2}V_g^{3/2}}{3\sqrt{f_s L}\cdot V_o}$ $+\dfrac{\left(\frac{2\omega C_{cymbal}}{\pi}\right)V_g(V_g-V_o)V_f}{V_o}$	$\dfrac{8R_L\left(\frac{\omega C_{cymbal_s}}{\pi}\right)^{3/2}(V_g-V_o)^{1/2}V_g^{5/2}}{3\sqrt{Lf_s}\cdot V_o}$	$\dfrac{8R_{DS}\left(\frac{\omega C_{cymbal_s}}{\pi}\right)^{3/2}(V_g-V_o)^{1/2}V_g^{3/2}}{3\sqrt{f_s L}}$
Buck-Boost	$\dfrac{8R_F\left(\frac{\omega C_{cymbal}}{\pi}\right)^{3/2}V_g^3}{3\sqrt{Lf_s}\cdot V_o}$ $+\dfrac{V_g^2\left(\frac{2\omega C_{cymbal}}{\pi}\right)}{V_o}\cdot V_f$	$\dfrac{8R_L\left(\frac{\omega C_{cymbal}}{\pi}\right)^{3/2}V_g^2(V_g+V_o)}{3\sqrt{Lf_s}\cdot V_o}$	$\dfrac{8R_{DS}\left(\frac{\omega C_{cymbal}}{\pi}\right)^{3/2}V_g^2}{3\sqrt{Lf_s}}$
Flyback	$\dfrac{8R_F V_g^3\left(\frac{NP}{NS}\right)\left(\frac{\omega C_{cymbal}}{\pi}\right)^{3/2}}{3\sqrt{Mf_s}\cdot V_o}$ $+\dfrac{V_g^2\frac{2\omega C_{cymbal}}{\pi}}{V_o}V_f$	$\dfrac{8R_P\left(V_g+\left(\frac{N_p}{N_s}\right)V_o\right)^2\left(\frac{\omega C_{cymbal}}{\pi}\right)^{3/2}}{3\sqrt{Mf_s}}$ $+\dfrac{8V_g^3 R_s\left(\frac{\omega C_{cymbal}}{\pi}\right)^{3/2}}{3\sqrt{Mf_s}\cdot V_o}\left(\frac{N_p}{N_s}\right)$	$\dfrac{8R_{DS}V_g^2\left(\frac{\omega C_{cymbal}}{\pi}\right)^{3/2}}{3\sqrt{Mf_s}}$

- Buck converter is an even better choice for the devices which do not generate high voltages.
- For Flyback converter, however, the design is not optimized, so a direct comparison is not possible. But, strong intuitive arguments can be made to show higher losses associated with this converter.

6.3.5 *Cymbal Energy Harvesting into Rechargeable Battery*

A Buck DC–DC Converter was chosen for the Cymbal energy harvesting systems in order to store the energy in a rechargeable battery.[8] The beneficial reasons include the following:

- Low power loss: $\sim 3\,\text{mW}$ for the gate drive circuitry.
- Inexpensive and small: component count for the entire circuit = 8.
- Operates at a fixed "optimal duty cycle" of 2% for high excitation: no complicated control of frequency generation is required.
- Very reliable and durable: the circuit had been tested for several hours and it works without any performance degradation.
- Low stress on the MOSFET switch and other components: better efficiency and cheap components.

The basic theory for the converter is summarized: (1) The purpose of the external circuit is to maximize the *Average Power* transferred to the electrical load. (2) The condition for maximum average power transfer requires that the load impedance must be equal to the complex conjugate of the Thevenin impedance. (3) The optimal duty cycle of the converter is calculated to be as the following equation, using the mechanical excitation frequency ω, inductance value of Buck filter L and the switching frequency of the converter f_S:

$$D_{optimal} = \sqrt{\frac{4\omega L V_g C_{cymbal}f_s}{\pi(V_g - V_o)}}. \qquad (6.23)$$

Figure 6.20 shows a DC–DC Buck-converter designed on the bases of the "maximum power transfer theorem" at an "optimal duty cycle". We have chosen the switching frequency $\sim 1\,\text{kHz}$, high enough in comparison with the operation frequency of the piezo component ($\sim 100\,\text{Hz}$), but low enough to reduce the switching losses. The *duty cycle* $\sim 2\%$

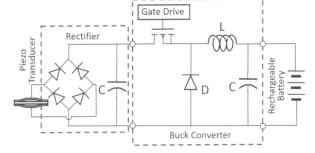

Fig. 6.20: DC–DC Buck-converter designed for converting the original impedance $300\,\text{k}\Omega$ down to $5\,\text{k}\Omega$ with a 2% duty cycle.

can be calculated by substituting the values in D_{optimal} in Eq. (6.23). We may expect a 50-time voltage step-down and a 50-time current step-up, leading to the 2500-time impedance reduction ideally, if we neglect any losses in the circuit. Since the matching impedance to the Cymbal component is $400\,\text{k}\Omega$, we may expect the matching impedance to come down to $200\,\Omega$. Figure 6.21 plots the duty cycle (1–5%) dependence of the obtained power through the Buck converter. Irrelevant to the electrical load, 2% provides the maximum, as Eq. (6.23) indicates. However, from Fig. 6.21, the maximum output power can be obtained at $5\,\text{k}\Omega$ (i.e., matching impedance), which is an order of magnitude higher than ideally expected above. The size of the inductor and filter capacitor were set to $33\,\text{mH}$ and $100\,\mu\text{F}$, which are high enough induc-

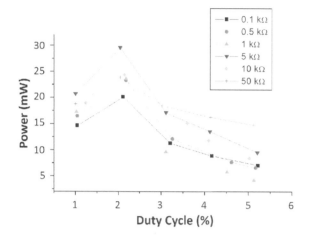

Fig. 6.21: Duty cycle dependence of the obtained power through the Buck converter.

tance to reduce the output current ripple and high enough capacitance to keep the output voltage ripple-free. A low power consumption PWM generator "LM555" was used then for the "Duty Cycle" generation. Amplified output current was applied for charging the battery, with the power increase by approximately 10 times as compared to direct charging. Power supply was continued to the load irrespective of the mechanical input fluctuations. A DC–DC Buck-converter was designed to allow transfer of $43\,\text{mW}$ power out of $53\,\text{mW}$ from the Cymbal (81% efficiency) by converting the original impedance $300\,\text{k}\Omega$ down to $5\,\text{k}\Omega$ with a 2% duty cycle and at a switching frequency of $1\,\text{kHz}$. The power loss in the gate drive circuit of Buck converter was estimated at $\sim 3\,\text{mW}$, which is much smaller than the cultivated energy of $43\,\text{mW}$ into a rechargeable battery.

6.3.6 *Multilayered Cymbal*

Since the *Buck-Converter* introduced in the previous section cannot reduce the output impedance sufficiently to match with the rechargeable battery ($\sim 50\,\Omega$), we should further reduce the output impedance. However, reducing the duty ratio much less than 2% is not practically feasible. Thus, the alternative strategy is to reduce the output impedance of the piezo transducer by changing the piezo transducer structure. Multilayered (ML) cymbal structures have lower impedance,[9] different from a popular reason for the ML usage to reduce the driving voltage for actuator applications. Figures 6.22(a) and 6.22(b) show cross-sectional views of single- and ML Cymbal transducers. With increasing the number n of the layers

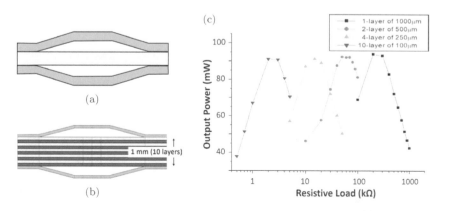

Fig. 6.22: (a) Single-layer Cymbal, (b) Multilayered (ML) Cymbal transducer. (c) Output power characteristics of the ML Cymbal transducers.[9]

by keeping the total thickness constant, output impedance decreases by a factor of $(1/n^2)$ owing to their capacitance increase in an opposite way, as you are familiar with. As shown in Fig. 6.22(c), the maximum output power shifts to the lower resistance direction with an increase of the layer number, which means output impedance can be controlled by the number of the layers in the piezo transducer structure. You can find that the matching impedance $300\,k\Omega$ in the single-layer cymbal is reduced down to $3\,k\Omega$ by using a 10-layer cymbal. Note that the total power is the same for all ML piezo components, as long as we keep the same total PZT volume.

It is worth noting that the performance becomes much more superior by combining the ML cymbal structure and the DC–DC Buck converter, as shown in Fig. 6.23. When we compare first the output voltage from the 10-layer cymbal with or without the DC–DC converter, no big difference can be found, just a continuous increase in the output voltage for both cases with increasing the resistive load. To the contrary, the output power of the 10-layer cymbal with or without the DC–DC converter presents a significant difference; using the Buck converter reduces the maximum power level from $100\,mW$ to $80\,mW$ with the matching load shift from 5 to $2\,k\Omega$. However, the output power around $50\,\Omega$ (*matching impedance* to the rechargeable battery) differs significantly: $50\,mW$ with the converter and less than $10\,mW$ without the converter. This load-insensitive broadening effect is essential to the usage of the DC–DC Back converter, but the reason needs to be clarified in the future.

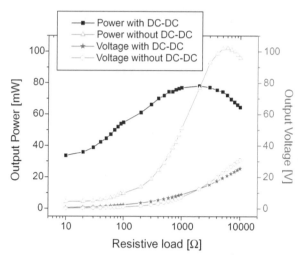

Fig. 6.23: Load-resistance third resonance mode type harvesting energy from the ML piezo component.

6.4 Usage of A Piezoelectric Transformer

Another unique circuit design is with a piezoelectric transformer.[2] The reason for piezoelectric transformer usage in the DC–DC converter is its low output impedance, which may help with matching the impedance to rechargeable battery.

6.4.1 *Piezoelectric Transformer — Principle*

Because conventional inductive coil transformers kill the size/weight of analogue power systems significantly, a piezoelectric transformer is one of the alternative components. The operating principle and some application examples are introduced.

One of the bulkiest and most expensive components in solid-state actuator systems is the power supply with an electromagnetic transformer. Electromagnetic transformer losses occurring through the thin wire loss, core loss and skin effect increase dramatically as the size of the transformer is reduced. Therefore, it is difficult to realize miniature low-profile electromagnetic transformers with high efficiency. The piezoelectric transformer (PT) is an attractive alternative for such systems due to its high efficiency, small size and lack of electromagnetic noise. They are highly suitable as miniaturized power inverter components, which may have found applications in lighting up the cold cathode fluorescent lamp behind a color liquid crystal display or in generating the high voltage needed for air cleaners.[1]

The original design to step-up or step-down an input AC voltage using the converse and direct piezoelectric effects of ceramic materials was proposed by Rosen.[10] This type of transformer operates by exciting a piezoelectric element like the one pictured in Fig. 6.24(a) at its mechanical resonance frequency. An electrical input is applied to one-half part of the piezo-electric element (at the top left electrode), which produces the fundamental mechanical resonance (i.e., a half wave length, around 40–60 kHz), as shown in the middle of the figure. This mechanical vibration is then converted back into an electrical voltage at another half end (right edge electrode) of the piezoelectric plate. Assuming that the electric field level is similar among the input and output parts, the voltage step-up ratio (r) without load (i.e., open-circuit conditions) is primarily given by the electrode gap ratio (recall the voltage vs. electric field relation with electrode gap, $E = V/Gap$):

$$r \propto k_{31}k_{33}Q_m(l/t), \qquad (6.24)$$

where l and t are the electrode gap distances for the input and output portions of the transformer, respectively (Fig. 6.24(a)). Note from this relationship how the length to thickness ratio, the electromechanical coupling factors, and/or the mechanical quality factor Q_m are the primary means of increasing the step-up ratio.

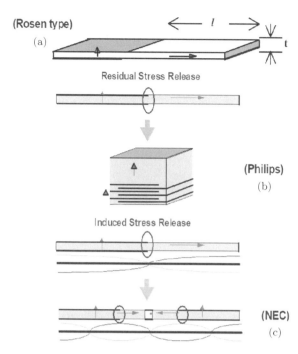

Fig. 6.24: Piezoelectric transformer designs: (a) Rosen design,[10] (b) multilayer design by NEC[11] and (c) third resonant mode type developed by NEC.[12]

This transformer was utilized on a trial basis in some color televisions during the 1970s.

In spite of its many attractive features, the original Rosen transformer design had a serious reliability problem. Mechanical failure tends to occur at the center of the device where the residual stress originating from the poling process is most highly concentrated. This happens to also be coincident with the nodal point of the vibration where the highest induced AC stress occurs. Two recently developed transformers pictured in Figs. 6.24(b) and 6.24(c) are designed to avoid this problem and are commercially produced for use as back-light inverters in liquid crystal displays. Both of the newer designs make use of more mechanically tough ceramic materials. The NEC and Philips Components' transformer shown in Fig. 6.24(b) further alleviates the problem by using a multilayer structure to avoid the development of residual poling stress in the device.[11] Another NEC design pictured in Fig. 6.24(c) makes use of an alternative electrode configuration to excite a third resonance excitation (longitudinal) in the rectangular plate to further redistribute the stress concentrations in a more favorable manner.[12]

6.4.2 DC–DC Converter with a Piezoelectric Transformer

We developed a ring-dot-type multilayer transformer shown in Fig. 6.25(b). A single disk ring-dot transformer (Hard PZT) with electrode shapes is shown in Fig. 6.25(a).[13] Figure 6.25(c) plots load resistance dependence of the efficiency of the ML Ring-Dot piezoelectric transformer,[14,15] which indicates low-output impedance around 50 Ω (i.e., resistive at the transformer resonance frequency), and the efficiency of the piezoelectric transformer at its resonance 78 kHz can reach above 97%.[14,15] This low-output impedance is suitable for an impedance matching to the energy storage devices. Figure 6.26(a) illustrates a block diagram of a piezoelectric energy harvesting circuit with a PT, and Fig. 6.26(b) shows an actual piezoelectric energy harvesting circuit with a *ring-dot-type ML transformer*.[14]

As discussed in Section 6.3, the basic principle of a switching regulator using a MOSFET is the key to develop a suitable *DC–DC converter*. The output voltage v_0 through a load R switches the maximum E

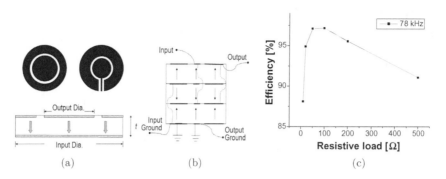

Fig. 6.25: (a) Single layer ring-dot transformer, (b) ML ring-dot transformer, and (c) load resistance dependence of the efficiency of the ML ring-dot piezoelectric transformer.[14,15]

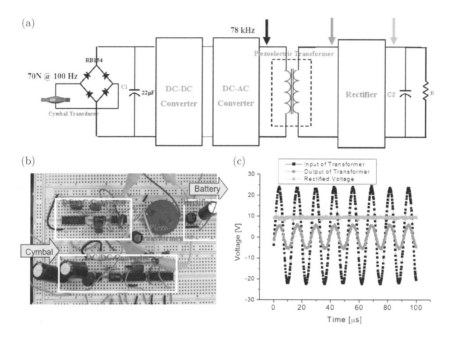

Fig. 6.26: (a) Block diagram of a piezoelectric energy harvesting circuit with a piezoelectric transformer. (b) Actual piezo-energy harvesting circuit with a ring-dot-type multilayer transformer. (c) Input and output voltage signals of the piezo-transformer and the final rectified voltage, which correspond to the three arrow parts in (a).

(some 100 s V) and minimum 0 according to the Gate/Source voltage (rectangular voltage with the duty ratio of D), so that the average voltage is estimated by

$$v_0 = \frac{T_{\mathrm{on}}}{T_{\mathrm{on}} + T_{\mathrm{off}}} E = dE. \quad (D: \text{duty ratio}). \tag{6.25}$$

If the loss of the electronic component is negligibly small, ideally Input power = Output power. Thus, using the duty ratio d, the voltage step-up ratio and current step-down ratio should be d, leading to the impedance change by the factor of d^2. If we take $d = 2\%$, the impedance can be reduced by 2,500 theoretically. Among Forward converter, Buck Converter, Buck-Boost Converter and Flyback Converter, we have chosen a *DC–DC Buck-Converter* shown in Fig. 6.20, because of the simplest topology using a single MOSFET and the lowest losses in this level of low power, which allows transfer of 43 mW power out of 53 mW from the Cymbal (81% efficiency) in practice by converting the original impedance 400 kΩ down to 5 kΩ with a 2% duty cycle and at a switching frequency of 1 kHz.[9,16] The impedance modification of 80 times differs from the above expectation (2.500 times), which may be explained by the components' losses in the circuit.

Because of insufficient impedance reduction of DC–DC converter, we added (1) DC-AC converter, (2) piezoelectric transformer and (3) another rectifier, for further impedance reduction (Fig. 6.26(a)). Input and output voltage signals of the PT and the final rectified voltage, which correspond to the three arrow parts in Fig. 6.26(a), are plotted in Fig. 6.26(c). Large and small AC voltages correspond to the input and output voltage through the piezo transformer (i.e., step-down), and DC voltage corresponds to the rectified voltage after the full-wave rectifier. You can find that the further step-down voltage ratio seems to be 5. In conclusion, the PT may be an alternative device to be used for modifying the impedance matching.

Chapter Essentials

1. The internal impedance of a piezoelectric component is "capacitive" under an off-resonance operation, $1/j\omega C$, with an absolute value of $100s\,\mathrm{k\Omega}$ at $100\,\mathrm{Hz}$ operation.
2. Electric Impedance Matching: External impedance Z_1 is adjusted to "conjugate" of the internal impedance Z_0^*, in general. However, in this impedance matching, energy will make catch-ball between the internal capacitance and external inductance. Resistive load is essential to harvest the energy, and is adjusted to the "absolute internal impedance" of the piezo component ($1/\omega C$) for obtaining the maximum output power.
3. When the input mechanical energy is unlimited (such as environmental water or wind flow), the impedance matching for obtaining the maximum output power is the primary target.
4. Under a constant force X_0 applied, the input mechanical energy into a piezoelectric transducer changes with the shunted external impedance Z:

$$|P|_{\mathrm{in}} = \frac{\omega}{2}\left| s^E\left[1 - \left(\frac{S}{t}\right)\frac{j\omega Z d^2}{s^E(1+j\omega CZ)}\right]X_0^2\right| \quad (S: \text{area}, \ t: \text{electrode gap of piezo component})$$

This originates from the effective complex elastic compliance under Z shunt expressed as

$$s_{\mathrm{eff}}^E = \frac{x}{X} = s^E\left[1 - \left(\frac{S}{t}\right)\frac{j\omega Z d^2}{s^E(1+j\omega CZ)}\right]$$

Recall that s^E or $s^D = s^E(1-k^2)$, when $Z = 0$ or ∞, respectively.
5. When the input mechanical energy is limited, the energy transmission coefficient is the Figure of Merit:

$$\lambda_{\max} = (\text{Output electrical energy/Input mechanical energy})_{\max}$$
$$= [(1/k) - \sqrt{(1/k^2)-1}]^2 = [(1/k) + \sqrt{(1/k^2)-1}]^{-2}.$$

6. AC–DC rectification uses p–n junction diodes — half-wave and full-wave rectifiers.
7. MOSFET is popularly used for an ON–OFF switching regulator.
8. Because the output impedance from the piezo energy harvesting device is too high to accumulate the energy directly into a rechargeable battery, we should adopt a DC–DC converter for the impedance matching purpose.
9. DC–DC converters: Among Forward converter, Buck converter, Buck-Boost converter and Flyback converter, the Buck converter is best suited for piezo energy harvesting, from the simplest design and high efficiency viewpoints.
10. The electronic components of energy harvesting circuit spend mW level continuously (diode $\sim$mW, MOSFET $\sim 2\,\mathrm{mW}$, DC–DC converter $\sim 5\,\mathrm{mW}$). Thus, piezo energy harvesting systems reporting less than mW are useless for practical energy storage purposes.

11. A piezoelectric transformer may be an alternative device to be used for modifying the impedance matching in the energy harvesting circuit.

Check Point

1. (T/F) External load resistance should be adjusted to the internal impedance of the piezo energy harvesting component for obtaining the maximum output power. True or False?
2. There is a battery (total energy $= 1\,\mathrm{kJ}$) with an internal impedance $75\,\Omega$. How much energy can be spent for the external work roughly by matching the external load impedance?
3. Calculate the electrical impedance $1/j\omega C$ of a piezoelectric component with capacitance $1\,\mathrm{nF}$ at the off-resonance frequency $100\,\mathrm{Hz}$.
4. The above impedance is significantly larger than the internal impedance of rechargeable batteries ($\sim\!50\,\Omega$). What kind of circuit is required to match the impedance between the piezo component and the battery?
5. Among three DC–DC converters, Buck, Buck-Boost and Flyback, which is the best suitable converter for the piezoelectric energy harvesting system?
6. Fill in the blank: A DC–DC converter is composed of (a) MOSFET switching device, (b) inductor, (c) capacitor and (d) ⬚ .
7. (T/F) A piezoelectric transformer is designed merely for an efficient step-up voltage purpose. It is difficult for the step-down voltage purpose theoretically. True or False?
8. (T/F) A full-wave rectifier is represented by the following circuit. True or False?

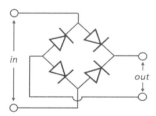

9. (T/F) When the losses of the electronic components are neglected, input power $=$ output power is ideally expected in a Buck converter. When we take the duty ratio (on/off time period ratio) as 2%, the impedance can be reduced by 50 theoretically. True or False?
10. The effective complex elastic compliance under Z shunt is expressed as

$$s_{\mathrm{eff}}^{E} = \frac{x}{X} = s^{E}\left[1 - \left(\frac{S}{t}\right)\frac{j\omega Z d^2}{s^{E}(1 + j\omega C Z)}\right]$$

When we choose $Z = 1/j\omega C$ (capacitive load), calculate s_{eff}^{E}, which should be between s^{E} and $s^{D} = s^{E}(1 - k^2)$.

Chapter Problems

6.1 The total energy loss in the piezoelectric energy harvesting circuit as shown in Fig. 6.20 was about 3–5 mW in the early 2000s. You are requested to check the up-to-date minimum loss level for recent electronic components, in particular, (a) diode and rectifier, (b) power MOSFET, (c) storage capacitor, (d) inductor, rechargeable battery (leak loss, etc.)

6.2 Power MOSFET has the drain-source-gate structure shown in the right-hand side. Similar to Fig. 6.13, explain the switching function of this MOSFET by considering the change in the n-channel with the drain/source voltage.

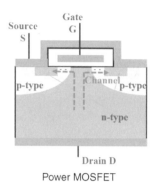

Power MOSFET

References

1. K. Uchino, *Ferroelectric Devices 2nd Edition*, CRC Press, Boca Raton, FL (2010).
2. K. Uchino, *Micromechatronics Second Edition*, CRC Press, Boca Raton, FL (2020), ISBN-13: 978-0-367-20231-6.
3. H. -W. Kim, S. Priya, K. Uchino and R. E. Newnham, *J. Electroceramics*, 15, 27 (2005).
4. K. Uchino, *Proc. 5th Int'l Workshop on Piezoelectric Mater. Appl.*, State College, PA, October 6–10 2008.
5. J. Kassakian, M. Schlecht, and G. Verghese, *Principles of Power Electronics*, Addison-Wesley Publishing Company, Inc., Reading, Massachusetts, (1991), p. 103.
6. P. Horowitz and W. Hill, *The Art of Electronics*, Cambridge University Press, Cambridge, UK, (1980), p. 360.
7. A. Batra, MS (EE) Thesis, The Pennsylvania State University, "Energy Harvesting Using a Piezoelectric "Cymbal" Transducer in a Dynamic Environment" (August 2004).
8. H.-W. Kim, A. Batra, S. Priya, K. Uchino, D. Markley, R. E. Newnham and H. F. Hofmann, *Japan. J. Appl. Phys.*, 43(9A) 6178 (2004).
9. H. -W. Kim, S. Priya, H. Stephanau and K. Uchino, *IEEE Trans. — UFFC*, 54(9), 1851 (2007).
10. C. A. Rosen, *Proc. Electronic Component Symp.*, 205 (1956).
11. NEC, "Thickness Mode Piezoelectric Transformer", US Patent No. 5,118,982 (1992).
12. S. Kawashima, O. Ohnishi, H. Hakamata, S. Tagami, A. Fukuoka, T. Inoue and S. Hirose, *IEEE Int'l Ultrasonic Symp. Proc.* (November 1994).
13. S. Priya, H.-W. Kim, S. Ural and K. Uchino, *IEEE Trans. — UFFC*, 53(4), 810 (2006).
14. H.-W. Kim, S. Priya, and K. Uchino, *Proc. 10th Int'l Conf. New Actuators*, Bremen, Germany, June 14–16, 2006, A5.4, (2006), pp. 189–192.
15. K. Uchino and A. Vazquez Carazo, *Proc. 11th Int'l Conf. New Actuators*, Bremen, Germany, June 9–11, A3.7, (2008), pp. 137.
16. H.-W. Kim, S. Priya, and K. Uchino, *Japan. J. Appl. Phys.*, 45, 5836 (2006).

Chapter 7

Case Studies on Energy Flow Analysis

Energy flow analysis is essential to identify problematic steps to be resolved in order to improve the energy harvesting efficiency. This chapter introduces three typical cases: (1) continuous AC mechanical input force, (2) impulse (one time) mechanical input force and (3) continuous pulse mechanical input force (human walk).

7.1 Cymbal Piezo Energy Harvesting — Case Study 1

A steady sinusoidal mechanical input case is described in this section.

7.1.1 *Three Phases in Energy Harvesting Process*

There are three major phases/steps associated with piezoelectric energy harvesting (see Fig. 7.1): (i) *mechanical–mechanical energy transfer*, including mechanical stability of the piezoelectric transducer under large stresses, and mechanical impedance matching (Chapter 4), (ii) *mechanical–electrical energy transduction*, relating with the electromechanical coupling factor in the composite transducer structure (Chapter 5), and (iii) *electrical–electrical energy transfer*, including electrical impedance matching. A suitable DC/DC converter is required to accumulate the electrical energy from a high-impedance piezo device into a rechargeable battery (low impedance) (Chapter 6).

The following sections mainly deal with detailed energy flow analysis in piezoelectric energy harvesting systems with typical stiff "Cymbals" (~100 mW) and under cyclic mechanical load (off-resonance) at 100 Hz (simulating the automobile engine vibration), in order to provide comprehensive strategies on how to improve the efficiency of the harvesting system. Energy transfer rates are practically evaluated for all three steps above. Our application target of the "Cymbal" was set to hybrid vehicles with both an engine and an electromagnetic motor, reducing the engine vibration and harvesting electric energy to car batteries to increase the mileage.

7.1.2 *Source to Transducer: Mechanical Impedance Matching*

The piezoelectric component was a cymbal with 0.3-mm-thick stainless steel endcaps, inserted below a 4-kg engine weight (40 N bias force). The electromagnetic shaker shook at 100 Hz for 8 s, and the accumulated energy during that time period was measured at each phase. The energy levels at all states are summarized in Fig. 7.2.

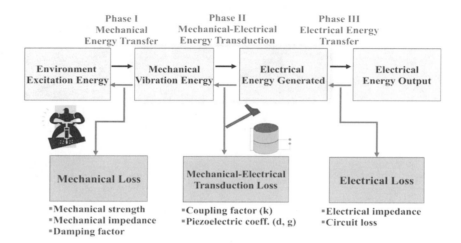

Fig. 7.1: Three major phases associated with piezoelectric energy harvesting: (i) mechanical-mechanical energy transfer, (ii) mechanical-electrical energy transduction, and (iii) electrical-electrical energy transfer.

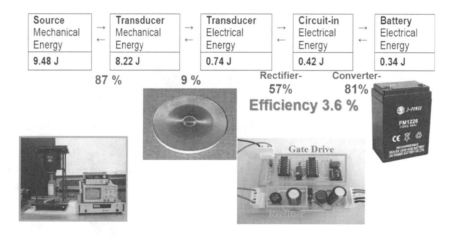

Fig. 7.2: Energy flow/conversion analysis in the cymbal energy harvesting process.

Let us discuss why the energy amount decreases with transmitting and transducing successively by numerical analyses:

(1) Initial engine vibration energy for 8 s: 9.48 J
(2) Mechanical energy transmitted into a Cymbal transducer: 8.22 J.

This ratio (= 87%) is the energy transfer rate from the vibration source to piezo cymbal transducer.

The rigidity of the Cymbal endcaps needs to be reasonable (not too elastically hard or soft) to match the mechanical/acoustic impedance to the engine chamber stiffness. If the mechanical impedance is not seriously considered in the transducer (too mechanically soft or hard), the mechanical energy will not transfer efficiently, and a large amount will be reflected at the contact/interface plane.

7.1.3 *Transduction in the Transducer: Electromechanical Coupling*

Electromechanical coupling factor square k^2 is the energy transduction rate:

$$k^2 = \text{(Stored electrical energy/Input mechanical energy)}. \tag{7.1}$$

Since the Cymbal is based on a piezoelectric disk of the k_p type, the static electromechanical coupling factor is estimated by

$$k_p = \sqrt{\frac{2}{(1-\sigma)}} k_{31} \ (\sim 60\% \text{ in soft PZT}). \tag{7.2}$$

The dynamic (off-resonance) electromechanical coupling factor is usually lower than the k_p. Practically, our samples show $k_v = k_{\text{eff}} = 0.25$ (0.4-mm endcap) ~ 0.30 (0.3-mm endcap) in the cymbal, leading to the energy conversion rate 9% maximum, which was equal to 0.74 J. The remaining 90% portion will remain as the original mechanical vibration energy. The converted electrical energy 0.74 J/mechanical energy transmitted into a Cymbal transducer 8.22 J = 9.0%. Since k_{eff} of the bimorph or unimorph is much smaller (10–15%), the energy conversion rate is smaller than 2%. Using transducer modes with higher k values is highly recommended.

7.1.4 *Transducer to Harvesting Circuit: Rectifier*

Under the continuous AC mechanical force, a half of the converted electric energy $U_E = k^2 U_M$ can be spent out using the impedance matched resistor, while for the impulse mechanical force, almost all input mechanical energy U_M (one time) can be taken out. In this particular burst (resonance) mode measurement, we expect $\frac{1}{2}k^2 U_M$ or higher energy harvesting. We measured the DC energy 0.42 J after the full-wave rectifier. In comparison with the converted electrical energy 0.74 J, 0.42 J/0.74 J = 57% was obtained, higher than 50%. This reduction is thus expected even we use a full-wave (not a half-wave) rectifier; only a little more than half of the stored electrical power in the piezo device can be spent under the impedance matching condition. A part of the reduction also includes the electrical component (diode) loss (2–3 mW loss per diode).

7.1.5 *Harvesting Circuit to Rechargeable Battery: DC–DC Converter*

We used a "Buck converter" to convert the original matching impedance 400 kΩ to a rechargeable battery level, less than 100 Ω. However, in practice, 400 kΩ became 5 kΩ, with 80-time reduction of the matching impedance even we used the duty ratio of 2% in this switching converter. We obtained the energy 0.34 J in a rechargeable battery. Since the input DC energy was 0.42 J, 0.34/0.42 = 81% is the converter efficiency. This reduction partially originated from the energy consumption in the circuit, and partially originated from the electrical impedance mismatch between the circuit output (still around 5 kΩ) and the rechargeable battery impedance (around 10–50 Ω). The circuit losses include (1) conduction losses (A MOSFET consumes 1 mW level during operation inevitably, inductor loss and diode loss), and (2) switching loss, leading to the total loss around 3 mW. Thus, a piezoelectric energy harvesting component, which can generate less than 2 mW, is not an energy "harvesting" device, but a "losing" device in practice. The PZT thin film energy harvesting systems (PZT–MEMS) report just μW or lower-level harvested energy, for which the author feels "frustration", because the energy level "μW" is called "sensing" capability, far from an energy "harvesting" system.

7.1.6 *Total Energy Efficiency of the Piezo Energy Harvesting*

In summary, 0.34 J/9.48 J = 3.6% is the energy harvesting rate from the "vibration source energy" to the "storage battery" in the current system. Taking into account the efficiency of popular amorphous silicon solar cells of around 5–9%, this prototype piezoelectric energy harvesting system with the 3.6% efficiency seems to be rather promising.

7.2 Programmable Air-Burst Munition — Case Study 2

This section deals with the impact/impulse mechanical input case. Total energy efficiency is not important, but the harvesting energy level is critical. Micromechatronics Inc., PA developed Programmable Air-Burst Munition (PABM) under a Small Business Innovation Research Program from the US Army. The author introduces its development procedure as a case study of the piezoelectric energy harvesting system for the reader's sake. Note that the used design and evaluated values are intentionally fictitious, because of the confidentiality agreement restriction.

7.2.1 *Research Background of PABM*

Until the 1960s, the development of weapons of mass destruction (WMD) was the primary focus, including nuclear bombs and chemical weapons. However, based on the global trend for "Jus in Bello (Justice in War)", environment-friendly "green" weapons became the mainstream in the 21st century, that is, minimal destructive weapons with a pin-point target such as laser guns and rail guns. In this direction, programmable air-bust munition (PABM) was developed from 2003.

After the World Trade Center was attacked by Al Qaeda on September 11, 2001, the US military started the "revenge" war against Afghanistan. The US troops initially destroyed all the buildings by bombs, which dramatically increased the war cost for restructuring new buildings. Thus, the US Army changed the war strategy: without collapsing the building, just kill the Al Qaeda soldiers inside by using a sort of micro missile. A micro missile passes through the window by making a small hole on a window glass and explodes in the air 3–5 m (programmable!) inside the window, so that the building structure damage becomes minimized. For this purpose, each bullet needs to be installed with a micro-processor chip, which navigates the bullet to a certain programmed point. ATK Integrated Weapon Systems, AZ started to produce button-battery operated PABM first. Though they worked beautifully for an initial couple of months, due to severe weather conditions (incredibly hot in the daytime in Afghanistan), most of the batteries wasted out in three months. No soldier was willing to open a dangerous bullet to exchange a battery!

Under the circumstances, the Army Research Office contacted Micromechatronics Inc., State College, PA for asking the development of a compact electric energy source to be embedded in the PABM bullet. Refer to Fig. 7.3. The initial shooting impact from the bottom by the explosive can be converted to electricity via a multilayer (ML) piezoelectric device, which should fulfill energy to be spent in the micro-processor for 5 s while maneuvering the bullet. Micromechatronics Inc. successfully developed an energy source for a 25-mm caliber-"Programmable Ammunition". Instead of a battery, an ML PZT piezo actuator is used for generating electric energy under shot mechanical impact to activate the operational amplifiers which ignite the burst according to the command program (Fig. 7.3). This is one of the successful million-selling products with piezoelectric energy harvesting at present.

Fig. 7.3: Programmable air-burst munition (PABM) developed by ATK and MMech (25 mm caliber).

7.2.2 *25-mm-Caliber Capability — Literature Survey*

We initially searched the 25-mm-caliber capability — design (size and weight), reach/distance, speed and acceleration — from the literature and the Army Research Office (program officer at this project contact). In particular, the initial acceleration ü of the bullet is essential to design the energy harvesting component.

Table 7.1: Capabilities of 25-mm ammunitions.

Case Length (mm)	Cartridge Length (mm)	Cartridge Weight (g)	Projectile Weight (g)	Muzzle Velocity (m/s)	Max Effective Range (m)	Tracer Burn Range (m)	Acceleration (m/s^2)
136.5	220	500	180	1300	2000	>2000	4×10^4

Figure 7.4 shows an example ammunition, M791 Armor-Piercing Discarding Sabot with Tracer (APDS-T).[1] The capabilities of 25-mm ammunitions are summarized in Table 7.1.[2] The ammunition weighs 680 g in total, with the initial ignition explosive 500 g in cartridge. The top projectile will reach 2–3 km in 2–3 s. Instead of the fuse at the tip of the nose cap, which senses and ignites the explosion, the explosion can be programmed in the PABM.

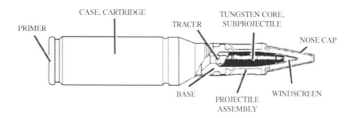

Fig. 7.4: M791 Armor-Piercing Discarding Sabot with Tracer (APDS-T).

We assume $\ddot{u} = 5 \times 10^4 \, \text{m/s}^2$ in the following analysis, so that the recoiling force F on the piezoelectric ML by the active mass M can be estimated by $F = M\ddot{u}$, which generates the electrical energy via the ML piezo actuator.

7.2.3 *Step-by-Step Design Processes*

7.2.3.1 *Working model*

Figure 7.5 shows a piezoelectric energy harvesting device model for PABM with a PZT ML sandwiched by steel masses. When the bullet is fired, the metal material above the ML piezoelectric ceramic piece will give the piezo ceramic a force. The piezoelectric ceramic generates electric energy due to the piezoelectric effect and is sent to the microprocessor for programming work. Thus, we need to know the required electrical energy, the parameters of the ML PZT, the amount of energy generated by the PZT, the choice of the PZT material and the specific number of layers. We provide step-by-step designing processes in the following sections.

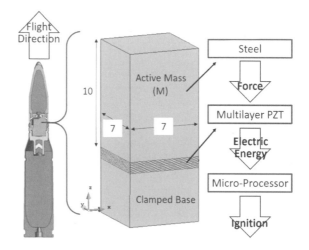

Fig. 7.5: Piezo energy harvesting device model for PABM; PZT multilayer is sandwiched by steel masses.

7.2.3.2 *Required energy estimation*

A microprocessor is a computer processor that incorporates the functions of a central processing unit on a single integrated circuit (IC),[3] or at most a few integrated circuits.[4] The application of the microprocessor is different, and its energy consumption is different. For a 25-mm bullet's microprocessor, we assume that its power consumption is "1mW", and for a "6 sec flight" (maximum), the energy we need is $U_{\text{processor}} = $ "6 mJ". If we design the circuit impedance of the microprocessor as "1 MΩ", we can calculate the current I as

$$I^2 = \frac{\text{Power}}{R} = \frac{1 \times 10^{-3}}{1 \times 10^6} A^2. \tag{7.3}$$

Thus, the voltage and current should be "32 V" and "32 μA" from $1 \, \text{mW} = 1 \, \text{M}\Omega \times I^2 = V^2/1 \, \text{M}\Omega$.

7.2.3.3 *Active mass (M) and force on the multilayer PZT*

The size of the steel material on the multilayer PZT stack in this system is $7\,\text{mm} \times 7\,\text{mm} \times 10\,\text{mm}$, the density is $7,800\,\text{kg/m}^3$, its mass is M and the force generated at a given acceleration a is F.

$$M = \rho \times V = 7,800 \times 7 \times 7 \times 10 \times 10^{-9}\,\text{kg} = 3.82 \times 10^{-3}\,\text{kg}. \tag{7.4}$$

This mass was employed as a counter mass to provide sufficient compressive force on the PZT ML. However, taking into account the base mass (or *cartridge mass*) 500 g as the impulse force starting point by explosive, the force to the ML PZT may be estimated by

$$F = \text{Ma} = 500 \times 10^{-3} \times 5 \times 10^4\,\text{N} = 2.5 \times 10^4\,\text{N}, \tag{7.5}$$

rather than the steel active mass usage for this estimation, because the impulse force comes from the base in practice.

Note that the compressive stress $2.5 \times 10^4\,\text{N}/(7\,\text{mm})^2 = 510\,\text{MPa}$ on the PZT is a rather critical range for its destruction (recall that 100 MPa is the critical range for the tensile stress). Thus, Micromechatronics Inc. received some feedback from the Army on the collapse of the PZT ML in the first year, originating from a slight misalignment of the PZT stack between the base and active steel mass.

7.2.3.4 *Piezoelectric device design*

7.2.3.4.1 PZT material's choice

From the manufacturer's catalog, PZT-5A, PZT-5H, PZT-4 and PZT-8 exhibit the properties listed in Table 7.2. Which material is to be employed for the energy harvesting application is described in the following, in terms of permittivity, piezoelectric constant d (larger charge generation).

Table 7.2: Physical parameters in piezoelectric PZT-5A, PZT-5H, PZT-4, and PZT-8.[5]

Piezo Material	PZT-5A	PZT-5H	PZT-4	PZT-8
Coupling coefficients				
k_{33}	0.71	0.75	0.70	0.64
k_{31}	0.34	0.39	0.33	0.30
Piezoelectric coefficient				
$d_{33}(\times 10^{-12}\,\text{m/V})$	374	593	285	225
$d_{31}(\times 10^{-12}\,\text{m/V})$	−171	−274	−122	−97
Permittivity				
$\varepsilon_{33}^{X}/\varepsilon_0$	1700	3400	1300	1000
$\varepsilon_{11}^{X}/\varepsilon_0$	1730	3130	1475	1290
Dielectric loss (%)	2.00	2.00	0.40	0.40
Elastic compliance				
$s_{33}^{E}(\times 10^{-12}\,\text{m}^2/\text{N})$	18.8	20.8	15.2	13.5
$s_{11}^{E}(\times 10^{-12}\,\text{m}^2/\text{N})$	15.0	16.4	12.2	11.5
Density (kg/m^3)	7500	7500	7500	7600
Curie Point (°C)	350	193	325	300

7.2.3.4.2 Capacitance of ML

Because of the outer diameter of a bullet, 25 mm, the multilayer (ML) PZT total size is designed as $7\,\text{mm}\,n \times 7\,\text{mm} \times 1\,\text{mm}$ thick, which consists of n layers (n is determined from the required capacitance discussed later) (see Fig. 7.6). Capacitance of the ML configuration consisting of n capacitors connected in parallel, with area S ($7\,\text{mm} \times 7\,\text{mm}$) and thickness t (t_{total}/n mm), is calculated from

$$C_{\text{total}} = \varepsilon_0 \varepsilon_{33}{}^X \times (S/t_{\text{total}}) \times n^2. \quad (7.6)$$

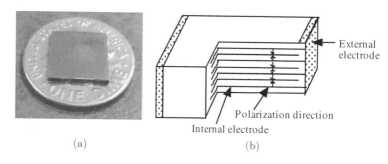

(a)   (b)

Fig. 7.6: PZT multilayer: (a) picture and (b) schematic illustration of the internal electrode structure.

Table 7.3 shows the total capacitance using the actual permittivity ε_{33}^X for $n = 1$, 10 and 100. $\varepsilon_0 = 8.854 \times 10^{-12}$.

7.2.3.4.3 Energy generated in the ML PZT

Since the impact force on the PZT stack by the active mass ($F = M\ddot{u}$) and the induced polarization is provided by $P = d_{33}X_3$, as long as the force application period is longer than the piezo responsivity (the resonance period of a 1-mm-thick ML is about $1\,\mu\text{s}$, which is inverse of the thickness resonance frequency around 1 MHz), the total charge Q generated by the PZT multilayer is

$$Q = nS \cdot P = nS(d_{33}X_3) = nS\,d_{33}(F/S) = n\,d_{33}F. \quad (7.7)$$

Energy generated by impact force can be calculated by

$$U_{\text{gen}} = (Q^2/C_{\text{total}}) = (nd_{33}F)^2/[\varepsilon_0\varepsilon_{33}{}^X(S/t_{\text{total}})n^2] = (t_{\text{total}}/S)F^2(d_{33}^2/\varepsilon_0\varepsilon_{33}{}^T) \quad (7.8)$$

(*Note*: When the charge is gradually accumulated by increasing the applied voltage, energy U is given by $\int_0^V Q\,dV = (\frac{1}{2})\frac{Q^2}{C}$. However, when the Q is suddenly provided, $U = \frac{Q^2}{C}$.)

Note that U_{gen} is irrelevant to the layer number n, and that the *energy harvesting figure of merit* is

$$d_{33}^2/\varepsilon_0\varepsilon_{33}{}^T = d_{33} \times g_{33} = k_{33}^2 \times s_{33}^E. \quad (7.9)$$

We calculate U_{gen} for the various ML actuators under $F = 2.5 \times 10^4\,\text{N}$ (for M = 500 g), and results are summarized in the first row of Table 7.4. Any PZT material can generate the minimum 6 mJ (even 60 mJ for a larger microprocessor) even after 6 seconds of the ammunition flight.

7.2.3.4.4 Layer number determination

The harvested energy by the impulse U_{gen} is irrelevant to the layer number. So, why are we using an ML structure? The layer number is determined from the time constant requirement via total capacitance. The

Table 7.3: Capacitance change with layer number in piezoelectric PZT-5A, PZT-5H, PZT-4, and PZT-8.

Piezo Material	PZT-4	PZT-5H	PZT-5A	PZT-8
$\varepsilon_{33}^X/\varepsilon_0$	1300	3400	1700	1000
$n = 1$	0.56 nF	1.48 nF	0.74 nF	0.43 nF
$n = 10$	56 nF	148 nF	74 nF	43 nF
$n = 100$	5,600 nF	14,800 nF	7,400 nF	4,300 nF

Table 7.4: Energy generated by different types of multilayer PZT. S = 7 mm × 7 mm, t total = 1 mm.

Piezo Material	PZT-4	PZT-5H	PZT-5A	PZT-8
Energy Generated	87.5 mJ	145 mJ	115 mJ	70.9 mJ
Min Layer No.	103	64	90	118

time of flight is at most 6 s, so the time constant of $Z_1 C$ should be longer than this time period:

$$Z_1 C_{\text{total}} > 6 \text{ s}, \tag{7.10}$$

where $Z_1 = 1 \text{ M}\Omega$ (internal impedance of a microprocessor), then we obtain the condition that capacitance $C_{\text{total}} > 6 \times 10^{-6}$ F. Note that the total capacitance is given by

$$C_{\text{total}} = \varepsilon_0 \varepsilon_{33}^X \times \left(\frac{S}{t_{\text{total}}}\right) \times n^2. \tag{7.11}$$

Table 7.3 provides example data for C_{total}, in which from the value for n = 1, and we can calculate the minimum number of the layers for each PZT composition, as shown in the second row of Table 7.4.

7.2.3.5 *Concluding remark*

First, Soft PZT-5H seems to be the best material for this application due to the highest figure of merit $d_{33} \cdot g_{33}$. Expected generative energy from the ammunition impact is 145 mJ, high enough above minimum required level of 6 mJ for operating a microprocessor during the bullet flight period (6 s). ML layer thickness of a little less than 20 μm seems to be satisfactory for keeping enough capacitance.

7.3 Piezo Tile Energy Harvesting — Case Study 3

This development, piezoelectric "tile" for obtaining electric energy from the human walk, is for a sort of impulse operation, but repeated impulse to accumulate the electric energy into rechargeable batteries. It is technologically neat, but a failure from the industrial viewpoint. The challenge by Smart Material Corp., Sarasota, FL is briefly introduced.[6]

7.3.1 *Energy Estimation from Human Walk*

The potential Energy of a person of 75 kg stepping on a tile giving in about 10 mm to convert into a bending/strain motion can be estimated as

$$75 \text{ kg} \times 9.8 \text{ m/s}^2 \times 10^{-2} \text{ mm} \sim 7.4 \text{ J (per step)}. \tag{7.12}$$

7.3.2 *Piezo Tile Structure*

Figure 7.7 shows the structure of "Piezo Tile" developed by Smart Material: (a) structure (cross-section), (b) deformation after force application (1 mm sinking) and (c) the product picture. The macro fiber composite (MFC) (Fig. 7.8(a)) was bonded on a metal plate to create a unimorph structure. When a human stamps the tile, the top plate sinks by 10 mm and provides force to the piezo unimorph, leading to electric energy harvesting. Typical efficiency of harvester, minus energy being stored in the tile spring to bend back the tile, was actually measured as about 2%, which corresponds roughly to k_{eff}^2 (after whole vibration ringings are accumulated).

7.3.3 *Piezo Tile Application Test Location*

Smart Material Corporation tested piezo tiles in "Grand Central Station" in New York. They have 750,000 visitors per day, or average of 31,250 visitors per hour. As estimated in Section 7.3.1, potential energy of a person of 75 kg (165 lbs) stepping on a tile gives ~7.4 J per step in about 10 mm to convert into a bending/strain motion. Tiles of 50 cm by 50 cm size cover all station floors (~48 acres), using more than 100,000 tiles! Supposing a person does on average 500 steps in one hour in the station, and one step per tile builds the same energy for all tiles, we can convert all visitors' contributions to about 19.2 kWh raw potential energy. The efficiency of the piezo tile, about 2% (actually measured), estimates the total harvested electrical energy at ~384 Wh; it would power just 4 bulbs of the ~4,000 total bulbs at Grand Station but at what cost (i.e., 100,000 tiles)!

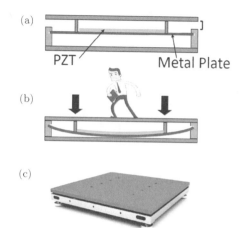

Fig. 7.7: Piezo Tile by Smart Material: (a) structure (cross-section), (b) deformation after force application, and (c) picture.

Another mistake on this project was found in "human behavior". They tested the piezo tile in a limited 10 m × 10 m square for a week. Though they obtained relatively comparable total harvested electrical energy ~384 Wh (by multiplying the total 48 acres) in the first day, the energy level exponentially dropped after the middle of the week. Because 10 mm sinking on the floor gives discomfort to a person, they may pass on the first day, but they will never try second or third times. They tried to escape from the walking on the discomfort area!

7.3.4 *Possible Business Success of MFC*

If the "Piezo Tiles" are arranged on the floor, from where nobody can escape, it may increase the effectiveness of the project. Japan Railway Company and Keio University, Japan, developed a similar piezo tile in 2007 at the ticket gate corridor, where nobody can escape from passing to go to the platform despite of discomfort. They obtained ~0.1 W level energy constantly.[7] However, this project has been terminated due to the piezo-component failure. With increasing the piezo component reliability and lifetime, this application may be revived in the future. Another business strategy can be found in energy harvesting shoes.[8] A piezoelectric "Thunder" actuator was embedded in the sole of shoe. Shenck reported 8.4 mW energy from the regular walk and impulse energy generation at 1-s walking cycle. Note that the energy harvesting from the shoes chases two "hares", that is, "shock absorbing" and "energy harvesting". As long as the sole mass increase by the energy harvesting system is minimized, both the comfort from shock absorbance and convenience from the portable equipment charging can be expected simultaneously.

Smart Material Corporation reported an interesting "fish tag" project, in collaboration with Pacific Northwest National Laboratory (PNNL), USA,[9] using MFC (Fig. 7.8(a)). This is a successful business project beneficial to both industrial and academic communities. Self-powered fish tags are essential to study the behaviors, dam passage and migration patterns of fish for longer periods of time. Implantable tags with transmitter electronic and an MFC harvest energy from fish's swimming motion with no battery. The tag is surgically inserted just under the skin. MFC harvester powers an ultrasound transmitter operating completely autonomously and sends ultrasound signals to track fish movements and locations as the fish swims. Figure 7.8(b) shows the developed self-powered fish tag (77 mm long), and Fig. 7.8(c) implant location in a fish.[9] PNNL already conducted live fish testing on rainbow trout and white sturgeons, which demonstrated successful signal transmissions using power solely from the MFC.

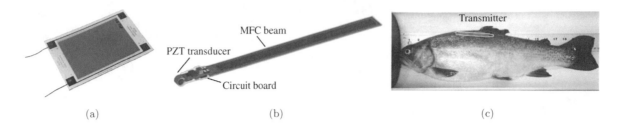

(a)   (b)   (c)

Fig. 7.8: (a) MFC (Smart Material), (b) self-powered fish tag (77 mm long), and (c) implant location in a fish (rainbow trout).[9]

The reader can now understand that even though the using technology is similar, the marketing application for floor tiles (piezo "tile") failed, but the "fish tag" project has successfully been launched. The business success depends significantly on how we find a niche marketing area, rather than the technological development.

Chapter Essentials

1. Three phases in the piezo energy harvesting process:
   (i) Mechanical–mechanical energy transfer, including mechanical stability of the piezoelectric transducer under large stresses, and mechanical impedance matching.
   (ii) Mechanical–electrical energy transduction, relating with the electromechanical coupling factor in the composite transducer structure.
   (iii) Electrical–electrical energy transfer, including electrical impedance matching. A suitable DC/DC converter is required to accumulate the electrical energy from a high-impedance piezo device into a rechargeable battery (low impedance).

2. Energy flow analysis example in the Cymbal harvesting process:

Source Mechanical Energy 9.48 J	→ ← 87%	Transducer Mechanical Energy 8.22 J	→ ← 9%	Transducer Mechanical Energy 0.74 J	→ ← 57%	Circuit-In Electric Energy 0.42 J	→ ← 81%	Battery Electric Energy 0.34 J

In summary, $0.34\,\text{J}/9.48\,\text{J} = 3.6\%$ is the energy harvesting rate from the "vibration source energy" to the "storage battery" in the current system. Taking into account the efficiency of popular amorphous silicon solar cells around 5–9%, the piezoelectric energy harvesting system with 3.6% efficiency is promising.

3. Energy flow analysis example in the Cymbal harvesting process indicates that the bottleneck of low-energy efficiency comes from the low mechanical-to-mechanical energy transfer rate, that is, the electromechanical coupling factor k^2. The higher k^2 design piezo component will increase the efficiency significantly.

4. One of the million-selling piezo energy harvesting devices includes PABM (Programmable Air-Burst Munition), because the competitive products (button batteries) cannot endure in the severe usage conditions (high temperature for a long period) in battlefields. Identifying the application target where the competitive products cannot be adopted is a good strategy for a piezoelectric product commercialization.

5. A "Piezo Tile" is an example of commercialization failure: technologically neat, but terrible cost/performance for the final application. Consideration of the "human behavior" is essential: because 10 mm sinking on the floor by stepping gives discomfort to a person, and they will skip stamping the "tiles" next time.

Check Point

1. Piezo-MEMS energy harvesting devices generate significantly high electric energy density, but the actual harvesting energy level is only μW or less. Is it possible to expect mW from the piezo-MEMS? What restricts the total energy level? Answer simply.

2. In order to develop compact energy harvesting devices, we had better develop a piezo component which can generate the high output electric energy, that is, high electromechanical coupling factor k. Among the k_{31}-type PZT plate and the k_{33}-type PZT plate, which is better suited for the energy harvesting purposes?

3. What is the major problem for the unimorph/bimorph piezoelectric structures to be used for the efficient piezo energy harvesting? Answer simply.

4. (T/F) The reason for the success of piezo energy harvesting for the PABM application exists in the situation where competitive products (such as button batteries) cannot endure in such a severe usage condition. True or False?

5. (T/F) The current energy harvesting efficiency from the "vibration source energy" to the "storage battery" is around 4%. Taking into account the efficiency of popular amorphous silicon solar cells around 5–9%, the piezoelectric energy harvesting system is rather promising. True or False?

6. (T/F) "Piezoelectric energy harvesting floor tiles" are niche and promising future products to be commercialized. True or False?

7. (T/F) The "piezoelectric energy harvesting fish tag" is a niche and promising future product to be commercialized. True or False?

8. Because of significant loss increase with increasing the handling energy density in PZT piezo ceramics, we have empirical limitation of the handling energy in the device. We used $1\,\text{cm}^3$ PZT for the piezo energy harvesting system. How much energy can we expect for the device to handle?

Chapter Problems

7.1 Picking up "million-selling" piezo energy harvesting devices, such as "PABM" by Micromechatronics Inc. PA, and "Lightning Switch" by Face International Corporation, VA, and failure example such as the "Piezo Tile" by Smart Material Corporation, consider the necessary factors for business success.

7.2 Business strategists occasionally use a "scoring table" to identify a development target. A sample of how to score is shown below. Refer to *Entrepreneurship for Engineers* (2010) by Kenji Uchino published by CRC Press, Boca Raton, FL [ISBN: 978-1-4398-0063-8] for the details. This table includes various factors which are significant, including financial factors (market and cost) and device performance. We compare the total scores, and select the higher priority for development (pecking order), which was used to persuade the executive of the sponsor, Samsung Electromechanics, to select the ultrasonic motor for Samsung mobile phone camera module in the author's development strategy. The table below demonstrates that the motor performance of the "PZT-tube" is better than that of the "Metal-tube", while the Metal-tube endures the shock test. The Metal-tube type is much cheaper in raw materials and manufacturing cost, because of the cheapest rectangular plate PZT usage. Regarding the marketing, the clients do not understand the inside of the black box (PZT-tube or Metal-tube does not matter, as long as the device specs are satisfied). Because the total score of the Metal-tube is higher than the PZT-tube, we chose the Metal-tube type.

Now, it is your turn: generate a Scoring Table for the "PZT piezo energy harvesting system" and "button battery" as a counterpart competitor for "PABM" application. Then, create a scenario to persuade the sponsor, Army Research Program Officer, to work with us on the PZT piezo energy harvesting system.

Scoring table for PZT-tube and Metal-tube motors for mobile phone applications.

	Device A PZT tube	Device B Metal tube
High Performance		
1) figure of merit	0 1 ②	0 ① 2
2) Lifetime (shock test)	0 ① 2	0 1 ②
Cheap Cost		
3) raw materials cost	⓪ 1 2	0 1 ②
4) preparation cost (machining, electroding)	⓪ 1 2	0 1 ②
5) labor cost (special skill)	0 ① 2	⓪ 1 2
Good Market		
6) design	0 1 ②	0 1 ②
7) production quantity	0 1 ②	0 1 ②
8) maintenance service	0 1 ②	0 1 ②
Total score	10	13

References

1. M242 25 mm Automatic Gun. http://www.inetres.com/gp/military/cv/weapon/M242.html.

2. 25 mm Ammunition-Orbital ATK. c2017. USA, ATK. https://www.orbitalatk.com/defense-systems/small-caliber-systems/25mm/docs/25mm_Fact_Sheet.pdf.

3. A. Osborne, *An Introduction to Microcomputers*, Volume 1: Basic Concepts, 2nd edn., Berkeley, California: Osborne-McGraw Hill (1980).

4. K. Kant, *Microprocessors And Microcontrollers: Architecture Programming And System Design*, PHI Learning Pvt. Ltd. (2007).

5. Ceramic Materials Properties. BOSTON PIEZO-OPTICS INC. https://www.bostonpiezooptics.com/assets/pdf/Ceramic_Materials.pdf.

6. T. Daue, Piezo Electric Vibration Harvester Using Macro Fiber Composites, *Proc. 70th ICAT International Smart Actuator Symposium*, State College, PA, October 3–4, 2017.

7. https://www.youtube.com/watch?v=RCOBA3Yfm1k.

8. N. S. Shenck, A demonstration of useful electric energy generation from piezoceramics in a shoe, PhD thesis, Department of electrical and computer science, Massachusetts Institute of Technology, May (1999).

9. H. Li *et al.*, *Scientific Reports*, 6, 33804 (2016).

Chapter 8

Hybrid Energy Harvesting Systems

8.1 Hybrid Energy Harvesting Background

8.1.1 *Difference between "Sensors" and "Energy Harvesting"*

The author introduces his "Functionality Matrix" concept here. Figure 8.1 lists the various effects in various materials/devices, relating the input (electric field, magnetic field, stress, heat and light) with the output (charge/current, magnetization, strain, temperature and light). Electrically conducting and elastic materials, which generate current and strain outputs, respectively, under the input, voltage or stress, are sometimes called "trivial" materials (well-known diagonal coupling phenomena, that is, Ohm's and Hooke's Laws). On the other hand, pyroelectric and piezoelectric materials, which generate an electric field with the input of heat and stress (unexpected phenomena!), respectively, are called "smart" materials. These off-diagonal couplings have corresponding *converse effects*, the electrocaloric and converse-piezoelectric effects, and both "sensing" and "actuating" functions can be realized in the same materials. Ferroelectric materials exhibit most of these effects with the exception of the magnetic phenomena. Thus, ferroelectrics are said to be very "smart" materials.

"Actuator" materials utilize an effect in which the output is "strain", such as converse piezoelectric effect, magnetostriction and photostriction, while "sensor" materials utilize an effect in which the output is "electric charge or current", such as piezoelectric effect, pyroelectric and photovoltaic effect. We usually spend external electrical energy for enhancing the monitoring signal. However, if the output sensing electric energy level is high (say, 10 mW), we can even harvest the remaining electric energy after subtracting the energy spent in the circuitry. The difference between "sensors" and "energy harvesting devices" comes from the generating electric energy level. A practically usable level of electrical energy for the following applications is as follows:

- Typical MOSFET — 1–3 mW
- Blue-tooth transmission device — mW (sub mW was developed recently)
- DC/DC converter — 2–4 mW
- Heart pacemaker — 5 mW
- Blood soaking syringe — 5–10 mW.

In this textbook, we use the "energy harvesting" when the obtained energy level is higher than 1 mW level.

8.1.2 *Functionality Matrix*

Taking into account (5×5) components of Fig. 8.1, we introduce a (5×5) *functionality matrix*. We invented "photostrictive" actuators, according to the following development procedure.[1] If one material

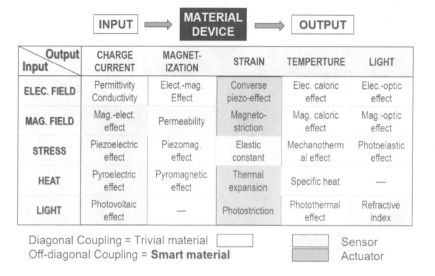

Fig. 8.1: Various effects in materials/devices.

has a "photovoltaic effect", the functionality matrix of this material can be expressed by

$$\begin{pmatrix} 0 & 0 & 0 & 0 & Light\ Emission \\ 0 & 0 & 0 & 0 & 0 \\ 0 & 0 & 0 & 0 & 0 \\ 0 & 0 & 0 & 0 & 0 \\ Photovoltaic & 0 & 0 & 0 & 0 \end{pmatrix}. \tag{8.1}$$

On the other hand, a piezoelectric has a functionality matrix of the following form:

$$\begin{pmatrix} 0 & 0 & Converse\ Piezo. & 0 & 0 \\ 0 & 0 & 0 & 0 & 0 \\ Piezo\ Ele. & 0 & 0 & 0 & 0 \\ 0 & 0 & 0 & 0 & 0 \\ 0 & 0 & 0 & 0 & 0 \end{pmatrix}. \tag{8.2}$$

As a composite effect of photovoltaic and piezoelectric, when the light illumination is input first, the expected phenomenon is expressed by the matrix product:

$$\begin{pmatrix} 0 & 0 & 0 & 0 & Light\ Emis. \\ 0 & 0 & 0 & 0 & 0 \\ 0 & 0 & 0 & 0 & 0 \\ 0 & 0 & 0 & 0 & 0 \\ Photovoltaic & 0 & 0 & 0 & 0 \end{pmatrix} \otimes \begin{pmatrix} 0 & 0 & Converse\ Piezo & 0 & 0 \\ 0 & 0 & 0 & 0 & 0 \\ Piezo\ Ele. & 0 & 0 & 0 & 0 \\ 0 & 0 & 0 & 0 & 0 \\ 0 & 0 & 0 & 0 & 0 \end{pmatrix}$$

$$= \begin{pmatrix} 0 & 0 & 0 & 0 & 0 \\ 0 & 0 & 0 & 0 & 0 \\ 0 & 0 & 0 & 0 & 0 \\ 0 & 0 & 0 & 0 & 0 \\ 0 & 0 & Photostriction & 0 & 0 \end{pmatrix}. \tag{8.3}$$

Note that only one component, "photostriction", is derived from this product calculation.

Now, let us consider the "magnetoelectric" (ME) effect. If one material has a "piezo magnetic effect" and its converse effect is "magnetostrictive effect", the functionality matrix of this material is expressed as follows:

$$
\begin{pmatrix}
0 & 0 & 0 & 0 & 0 \\
0 & 0 & Mag.Strictive & 0 & 0 \\
0 & Piezo.Mag & 0 & 0 & 0 \\
0 & 0 & 0 & 0 & 0 \\
0 & 0 & 0 & 0 & 0
\end{pmatrix}.
\tag{8.4}
$$

On the other hand, a piezoelectric has a functionality matrix as Eq. (8.2).

Thus, in a composite composed of the above materials, when the magnetic field is input first, the expected phenomenon is expressed by the matrix product:

$$
\begin{pmatrix}
0 & 0 & 0 & 0 & 0 \\
0 & 0 & Mag.str. & 0 & 0 \\
0 & Piezo.Mag & 0 & 0 & 0 \\
0 & 0 & 0 & 0 & 0 \\
0 & 0 & 0 & 0 & 0
\end{pmatrix}
\otimes
\begin{pmatrix}
0 & 0 & Converse\ Piezo & 0 & 0 \\
0 & 0 & 0 & 0 & 0 \\
Piezo\ Ele. & 0 & 0 & 0 & 0 \\
0 & 0 & 0 & 0 & 0 \\
0 & 0 & 0 & 0 & 0
\end{pmatrix}
$$

$$
=
\begin{pmatrix}
0 & 0 & 0 & 0 & 0 \\
Mag.Ele. & 0 & 0 & 0 & 0 \\
0 & 0 & 0 & 0 & 0 \\
0 & 0 & 0 & 0 & 0 \\
0 & 0 & 0 & 0 & 0
\end{pmatrix}.
\tag{8.5}
$$

Note that only one component, *magnetoelectric coupling*, is derived from this product calculation. To the contrary, if we start from the electric field input first, the expected phenomenon will be

$$
\begin{pmatrix}
0 & 0 & Converse\ Piezo & 0 & 0 \\
0 & 0 & 0 & 0 & 0 \\
Piezo\ Ele. & 0 & 0 & 0 & 0 \\
0 & 0 & 0 & 0 & 0 \\
0 & 0 & 0 & 0 & 0
\end{pmatrix}
\otimes
\begin{pmatrix}
0 & 0 & 0 & 0 & 0 \\
0 & 0 & Mag.str. & 0 & 0 \\
0 & Piezo.Mag & 0 & 0 & 0 \\
0 & 0 & 0 & 0 & 0 \\
0 & 0 & 0 & 0 & 0
\end{pmatrix}
$$

$$
=
\begin{pmatrix}
0 & Ele.Mag. & 0 & 0 & 0 \\
0 & 0 & 0 & 0 & 0 \\
0 & 0 & 0 & 0 & 0 \\
0 & 0 & 0 & 0 & 0 \\
0 & 0 & 0 & 0 & 0
\end{pmatrix}.
\tag{8.6}
$$

Note now that the resulting product matrix includes only one component, *electromagnetic effect*.

8.1.3 *Phase Connectivity in Composites*

Composites composed of two or more phases are promising materials because of their excellent tailorable properties. The geometry for two-phase composites can be classified according to the connectivity of each phase (1, 2 or 3 dimensionally) into 10 structures: 0-0, 0-1, 0-2, 0-3, 1-1, 1-2, 1-3, 2-2, 2-3 and 3-3.

Newnham *et al.* introduced the concept of "connectivity" for classifying the various two-phase composite structures.[2] When considering a two-phase composite, the connectivity of each phase is identified; for example, if a phase is self-connected in all x, y and z directions, it is called "3"; if a phase is self-connected only in z direction, it is called "1". A di-phasic composite is identified with this notation with two numbers "m–n", where m stands for the connectivity of an active phase (such as PZT) and n for another phase (such as a polymer). There are 10 types of diphasic composites: 0-0, 1-0, 2-0, ..., 3-2, 3-3, as illustrated in Fig. 8.2 (note that there are two different configurations for 3-2 and 3-3 among 12 figures).

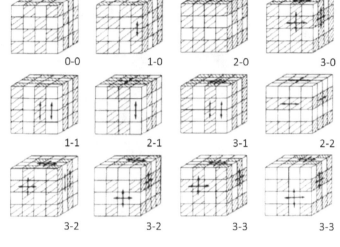

Fig. 8.2: Classification of two-phase composites with respect to spatial connectivity (10 types).[2] Two configurations are shown for 3-2 and 3-3.

A 0-0 composite, for example, is depicted as two alternating hatched and unhatched cubes, while a 1-0 composite has Phase 1 connected along the z direction. A 1-3 composite has a structure in which PZT rods (1D connected) are arranged in a 3D connected polymer matrix, and in a 3-1 composite, a honeycomb-shaped PZT contains the 1D connected polymer phase. A 2-2 indicates a structure in which ceramic and polymer sheets are stacked alternately, and a 3-3 is composed of a jungle-gym-like PZT frame embedded in a 3D connecting polymer.

8.1.4 *Composite Effects*

There are three types of composite effects (Fig. 8.3):[2,3] the sum effect, combination effect and product effect.

8.1.4.1 *Sum effects*

Let us discuss a composite function in a diphasic system to convert an input parameter X to an output parameter Y. Assuming Y_1 and Y_2 are the outputs from Phase 1 and 2, respectively, responding to the input X, the output Y^* of a composite of Phase 1 and 2 could be an intermediate value between Y_1 and Y_2. Figure 8.3(a) shows the Y^* variation with volume fraction of Phase 2 for a case of $Y_1 > Y_2$. The variation may exhibit a concave or a convex shape, but the averaged value in a composite does not exceed Y_1, nor is it less than Y_2. This effect is called a *sum effect*.

An example is a fishing rod, i.e., a lightweight and tough material, where carbon fibers are mixed in a polymer matrix (between

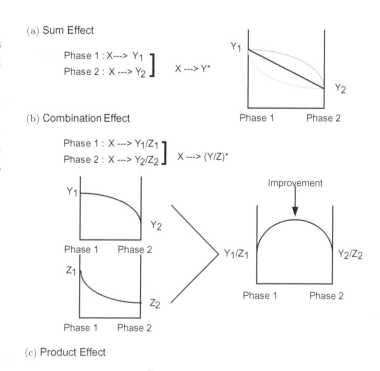

Fig. 8.3: Composite effects: (a) sum, (b) combination and (c) product effect.[3]

3-1 and 3-0). The density of a composite should be an average value with respect to volume fraction, if no chemical reaction occurs at the interface between the carbon fibers and the polymer, following the linear trend depicted in Fig. 8.3(a). A dramatic enhancement in the mechanical strength of the rod is achieved by adding carbon fibers in a special orientation, i.e., along a rod (showing a convex relation as depicted in Fig. 8.3(a)).

Another interesting example is an NTC-PTC material.[4] V_2O_3 powders are mixed in epoxy with a relatively high packing rate (3-3), as illustrated in Fig. 8.4. Since V_2O_3 exhibits a semiconductor-metal phase transition at 113°C, a drastic resistivity change is observed with increasing temperature. A further increase in temperature results in a larger thermal expansion for epoxy than for the ceramic, leading to a separation of each particle, and the structure becomes a 0-3 composite. The V_2O_3 particle separation increases the resistivity significantly at around 100°C. Thus, the conductivity of this composite is rather high only over a limited temperature range (around −110°C to 100°C), which is sometimes called the *conductivity window*.

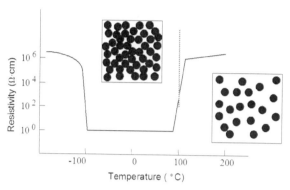

Fig. 8.4: NTC-PTC effect observed in a V_2O_3:epoxy composite.[4]

8.1.4.2 *Combination effects*

In certain cases, the averaged value of the output, Y*, of a composite does exceed Y_1 and Y_2. This enhanced output refers to an effect on a figure-of-merit Y/Z which depends on two parameters Y and Z. Suppose that the parameters Y and Z follow concave and convex type sum effects, respectively, as illustrated in Fig. 8.3(b), the combination value Y/Z will exhibit a maximum at an intermediate ratio of phases. This is called a *combination effect*.

Certain piezoelectric ceramic:polymer composites exhibit a combination property of g (the piezoelectric voltage constant) which is provided by $d/\varepsilon_0\varepsilon$ (d: piezoelectric strain constant, and ε: permittivity). In the 1-3 piezoelectric composite, where PZT rods are arranged in silicone rubber matrix,[5] the effective piezoelectric constant d_{33}^* of the composite is rather close to that of the PZT rod itself d_{33} down to 10 vol% of PZT, in while the effective permittivity is almost proportional to the volume % of PZT. Thus, we can realize 10 times higher piezoelectric constant g_{33}^* in a 10 vol% PZT composite, in comparison with the pure PZT specimen. Refer to Example Problem 2.1 for the derivation formula.

8.1.4.3 *Product effects*

When Phase 1 exhibits an output Y with an input X, and Phase 2 exhibits an output Z with an input Y, we can expect for the composite an output Z with an input X. A completely new function is created for the composite structure, called a *product effect* (see Fig. 8.3(c)). The phase selection is conducted on the basis of the "functionality matrix" concept (Section 8.1.2).

Philips Laboratories in the Netherlands developed ME materials based on the above concept.[3,6] The material was composed of magnetostrictive $CoFe_2O_4$ and piezoelectric $BaTiO_3$ mixed and sintered together. Figure 8.5(a) shows a micrograph of a transverse section of a unidirectionally solidified rod of the material with an excess of TiO_2 (1.5 wt.%). The four finned spinel dendrites $CoFe_2O_4$ and white cubical barium titanate grains are observed in cells (×100). Ideally, $CoFe_2O_4$ and $BaTiO_3$ grains should be 0-0 connectivity, without connecting each phase in a long range. Figure 8.5(b) shows the magnetic field dependence of

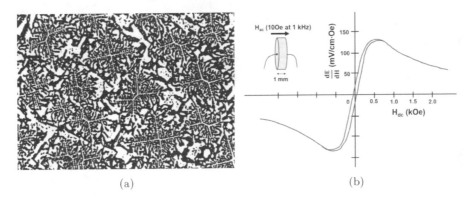

(a)   (b)

Fig. 8.5: (a) Micrograph of a mixture of magnetostrictive $CoFe_2O_4$ and piezoelectric $BaTiO_3$. (b) Magnetic field dependence of the magnetoelectric effect in a $CoFe_2O_4 : BaTiO_3$ composite.[3,6]

the ME effect in a $CoFe_2O_4 : BaTiO_3$ composite (at room temperature). Note maximum $(dE/dH) \approx$ 130 mV/cm · Oe, which is more than 10 times higher than the typical values observed in single-phase ME materials. Section 8.3 explains the product effect using the magnetostrictive and piezoelectric laminated composites.

8.1.5 *Concept of Hybrid Energy Harvesting Systems*

Most parts of this textbook are devoted to the piezoelectric energy harvesting. However, as the reader can imagine from Fig. 8.1, energy harvesting (theoretically equivalent to "sensor") can be obtained even from "ME", "pyroelectric" and "photovoltaic" effects, in addition to the "piezoelectric" effect. The key is found: how much energy level can we receive, or much higher than 1 mW? Though the primary research target should be put on each individual energy harvesting device development, it is also a general extension to seek hybrid energy harvesting systems by coupling two or more effects in order to increase the energy harvesting rate. If we couple "ME effect" with "piezoelectric effect", we can realize the energy harvesting device from either/both environmental magnetic noise and mechanical vibrations. Figure 8.6 summarizes conceptual ideas on possible hybrid energy harvesting devices based on piezoelectric plates. In addition to ME effect, we can couple with various multi-functional effects such as photovoltaic and pyroelectric effects to enhance the harvesting energy level.

Piezoelectric Plate
Elastic Plate

- Vibration Noise → Elastic material →
  Piezoelectric effect
- Magnetic Noise → Magnetostrictive material →
  Magnetoelectric effect
- Photo Illumination
  ○ Pure light → Elastic material →
    Photovoltaic effect
  ○ Photothermal heat → Elastic material →
    Pyroelectric effect

Fig. 8.6: Possible hybrid energy harvesting devices based on piezoelectrics.

8.2 Coupling Phenomenon Phenomenology

8.2.1 *Linear Handling of Free Energy*

We consider in this section various coupling phenomena comprehensively from phenomenological viewpoints. We adopt a practical formula of the Gibbs free energy $G(T, X, E)$ for the case of small value change in temperature $\theta = T - T_R$ (room temperature), external X and E (1D case). Magnetic properties are excluded for making the discussion simpler. If the change of parameters is small, we may adopt the three-parameter Taylor expansion approximation up to second derivatives in order to discuss just the linear relationships, based on the description by Mitsui *et al.*:[7]

$$G(T, X, E) = G_0 + \left(\frac{\partial G}{\partial T}\right)\theta + \left(\frac{\partial G}{\partial X}\right)X + \left(\frac{\partial G}{\partial E}\right)E$$

$$+ \frac{1}{2}\left(\frac{\partial^2 G}{\partial T^2}\right)\theta^2 + \frac{1}{2}\left(\frac{\partial^2 G}{\partial X^2}\right)X^2 + \frac{1}{2}\left(\frac{\partial^2 G}{\partial E^2}\right)E^2$$

$$+ \left(\frac{\partial^2 G}{\partial T \partial X}\right)\theta X + \left(\frac{\partial^2 G}{\partial T \partial E}\right)\theta E + \left(\frac{\partial^2 G}{\partial X \partial E}\right)XE. \tag{8.7}$$

Taking into account $dG = -SdT - xdX - DdE$, we obtain first the relations, $\left(\frac{\partial G}{\partial T}\right)_{\theta,X,E=0} = -S_0$, $\left(\frac{\partial G}{\partial X}\right)_{\theta,X,E=0} = -x_0$ and $\left(\frac{\partial G}{\partial E}\right)_{\theta,X,E=0} = -D_0$. Since S_0 is the entropy density at θ, X, $E = 0$, we take this as the original value and set it $S_0 = 0$. The values x_0 and $D_0(\approx P_0)$ are considered to be spontaneous strain and spontaneous polarization in the ferroelectric phase of this material, and we set them as zero in the discussion merely in the ferroelectric phase. Now, Eq. (8.7) can be transformed as

$$S = -\left(\frac{\partial G}{\partial T}\right) = -\left(\frac{\partial^2 G}{\partial T^2}\right)\theta - \left(\frac{\partial^2 G}{\partial T \partial X}\right)X - \left(\frac{\partial^2 G}{\partial T \partial E}\right)E, \tag{8.8a}$$

$$x = -\left(\frac{\partial G}{\partial X}\right) = -\left(\frac{\partial^2 G}{\partial T \partial X}\right)\theta - \left(\frac{\partial^2 G}{\partial X^2}\right)X - \left(\frac{\partial^2 G}{\partial X \partial E}\right)E, \tag{8.8b}$$

$$D = -\left(\frac{\partial G}{\partial E}\right) = -\left(\frac{\partial^2 G}{\partial T \partial E}\right)\theta - \left(\frac{\partial^2 G}{\partial X \partial E}\right)X - \left(\frac{\partial^2 G}{\partial E^2}\right)E. \tag{8.8c}$$

The diagonal terms, $\left(\frac{\partial^2 G}{\partial T^2}\right)$, $\left(\frac{\partial^2 G}{\partial X^2}\right)$ and $\left(\frac{\partial^2 G}{\partial E^2}\right)$, correspond to the trivial effect parameters in Fig. 8.1, while the off-diagonal expansion terms, $\left(\frac{\partial^2 G}{\partial T \partial X}\right)$, $\left(\frac{\partial^2 G}{\partial T \partial E}\right)$, $\left(\frac{\partial^2 G}{\partial T \partial X}\right)$, $\left(\frac{\partial^2 G}{\partial X \partial E}\right)$, $\left(\frac{\partial^2 G}{\partial T \partial E}\right)$ and $\left(\frac{\partial^2 G}{\partial X \partial E}\right)$, correspond to so-called "coupling coefficients" among temperature, stress and electric field.

8.2.2 *Isothermal Process — Piezoelectric Coupling*

8.2.2.1 *Thermodynamical Meaning of Piezoelectric Constant*

When the temperature is constant (i.e., isothermal), $\theta = 0$ in Eqs. (8.8b) and (8.8c), we can obtain the "intensive" parameter-based piezoelectric constitutive equations:

$$x = s^E X + dE, \tag{8.9a}$$

$$D = dX + \varepsilon_0 \varepsilon^X E, \tag{8.9b}$$

where the following notations are used, and denoted as s^E elastic compliance under constant E, $\varepsilon_0 \varepsilon^X$ dielectric permittivity under stress-free condition:

$$\begin{cases} s^E = -\left(\frac{\partial^2 G}{\partial X^2}\right) \\ \varepsilon_0 \varepsilon^X = -\left(\frac{\partial^2 G}{\partial E^2}\right). \\ d = -\left(\frac{\partial^2 G}{\partial X \partial E}\right) \end{cases} \tag{8.10}$$

The Maxwell relation $(\frac{\partial D}{\partial X})_{T,E} = (\frac{\partial x}{\partial E})_{T,X}$ verifies that the piezoelectric coefficient d in Eqs. (8.9a) and (8.9b) is thermodynamically the same.

When we start from the Helmholtz free energy A $(dA = -SdT + Xdx + EdD)$, by taking a similar Taylor expansion approach, we obtain another set of piezoelectric constitutive equations in terms of "extensive" parameters, x and D:

$$X = c^D x - hD, \tag{8.11a}$$

$$E = -hx + \kappa_0 \kappa^x D, \tag{8.11b}$$

where c^D is elastic stiffness under constant D, and $\kappa_0 \kappa^x$ is inverse permittivity $(\kappa_0 = 1/\varepsilon_0)$ under strain-free condition, and these coefficients are expressed by

$$\begin{cases} c^D = \left(\dfrac{\partial^2 A}{\partial x^2}\right) \\[2mm] \kappa_0 \kappa^x = \left(\dfrac{\partial^2 A}{\partial D^2}\right). \\[2mm] h = -\left(\dfrac{\partial^2 A}{\partial x \partial D}\right) \end{cases} \tag{8.12}$$

The Maxwell relation, $(\frac{\partial X}{\partial D})_{T,x} = (\frac{\partial E}{\partial x})_{T,D}$, verifies that the inverse piezoelectric coefficient h in Eqs. (8.12a) and (8.12b) is thermodynamically the same.

8.2.2.2 *Electromechanical coupling factor*

The term *electromechanical coupling factor* k is defined as the square value k^2 being the ratio of the converted energy over the input energy: when electric to mechanical

$$k^2 = (\text{stored mechanical energy/input electrical energy}), \tag{8.13a}$$

and when mechanical to electric,

$$k^2 = (\text{stored electrical energy/input mechanical energy}) \tag{8.13b}$$

Let us derive Eq. (26a) first practically, when an external electric field E_3 is applied to a piezoelectric material in a pseudo-static process. See Fig. 2.22(a), when we apply electric field on the top and bottom electrodes under stress-free condition $(X = 0)$. We can obtain

$$\begin{aligned} k_{33}^2 &= [(1/2)(d_{33}E_3)^2/s_{33}^E]/[(1/2)\varepsilon_0\varepsilon_3^X E_3^2] \\ &= d_{33}^2/\varepsilon_0\varepsilon_3^X \cdot s_{33}^E. \end{aligned} \tag{8.14a}$$

When an external stress X_3 is applied to a piezoelectric material in a pseudo-static process, refer to Fig. 2.22(b), under short-circuit condition $(E_3 = 0)$, we obtain

$$\begin{aligned} k_{33}^2 &= [(1/2\varepsilon_0\varepsilon_3^X)(d_{33}X_3)^2]/[(1/2)s_3^E X_3^2] \\ &= d_{33}^2/\varepsilon_0\varepsilon_3^X \cdot s_{33}^E. \end{aligned} \tag{8.14b}$$

It is essential to understand that the electromechanical coupling factor k (or k^2, which has a physical meaning of energy transduction/conversion rate) can be exactly the same for both converse Eq. (8.13b) and direct Eq. (8.14b) piezoelectric effects. The conditions under constant X (stress free) or constant E (short-circuit) are considered to be non-constrained.

8.2.2.3 *Constraint physical parameters*

It is important to consider the conditions under which a material will be operated when characterizing the dielectric constant and elastic compliance of that material. When a constant electric field is applied to a piezoelectric sample as illustrated in Fig. 2.20(a), the total input electric energy (*left*) should be equal to a combination of the energies associated with two distinct mechanical conditions that may be applied to the material: (1) stored electric energy under the *mechanically clamped state*, where a constant strain (*zero strain*) is maintained and the specimen cannot deform, and (2) converted mechanical energy under the *mechanically free state*, in which the material is not constrained and is free to deform. This situation can be expressed by

$$\left(\frac{1}{2}\right)\varepsilon^X\varepsilon_0 E_0^2 = \left(\frac{1}{2}\right)\varepsilon^x\varepsilon_0 E_0^2 + \left(\frac{1}{2s^E}\right)x^2 = \left(\frac{1}{2}\right)\varepsilon^x\varepsilon_0 E_0^2 + \left(\frac{1}{2s^E}\right)(dE_0)^2$$

such that

$$\varepsilon^X\varepsilon_0 = \varepsilon^x\varepsilon_0 + \left(\frac{d^2}{s^E}\right) \quad \text{or}$$

$$\varepsilon^x = \varepsilon^X(1 - k^2)\left[k^2 = \frac{d^2}{\varepsilon^X\varepsilon_0 s^E}\right]. \tag{8.15a}$$

When a constant stress is applied to the piezoelectric as illustrated in Fig. 2.20(b), the total input mechanical energy will be a combination of the energies associated with two distinct electrical conditions that may be applied to the material: (1) stored mechanical energy under the *open-circuit state*, where a constant electric displacement is maintained, and (2) converted electric energy (i.e., "depolarization" field) under the *short-circuit condition*, in which the material is subject to a constant electric field. This can be expressed as

$$\left(\frac{1}{2}\right)s^E X_0^2 = \left(\frac{1}{2}\right)s^D X_0^2 + \left(\frac{1}{2}\right)\varepsilon^X\varepsilon_0 E^2 = \left(\frac{1}{2}\right)s^D X_0^2 + \left(\frac{1}{2}\right)\varepsilon^X\varepsilon_0(d/\varepsilon_0\varepsilon^X)^2 X_0^2$$

which leads to

$$s^E = s^D + \left(\frac{d^2}{\varepsilon^X\varepsilon_0}\right) \quad \text{or} \quad s^D = s^E(1 - k^2)\left[k^2 = \frac{d^2}{\varepsilon^X\varepsilon_0 s^E}\right]. \tag{8.15b}$$

In principle, if we measure the permittivity in a piezoelectric specimen under stress-free and completely clamped conditions, we can obtain ε^X and ε^x, respectively. However, in practice, ε^x cannot be measured because of the experimental difficulty to maintain the ideal strain-free (clamped) condition for a long period. Similarly, if we measure the strain in a piezoelectric specimen as a function of applied stress pseudo-statically, under short-circuit and open-circuit conditions, we can obtain s^E and s^D, respectively. However, in practice, s^D cannot be measured because of the induced bound charge (or polarization) "screening" by migrating charge in the electrode, specimen or surrounding atmosphere in a couple of minutes. Thus, the clamped permittivity ε^x or open-circuit D-constant s^D can only be measured with high-frequency dynamical methods, such as an impedance analyzer around the resonance/antiresonance frequencies.

In conclusion, we obtain the following equations:

$$\varepsilon^x/\varepsilon^X = (1 - k^2), \tag{8.16a}$$

$$s^D/s^E = (1 - k^2), \tag{8.16b}$$

where

$$k^2 = \frac{d^2}{s^E \varepsilon^X \varepsilon_0}. \tag{8.17a}$$

We can also write equations of similar form for the corresponding reciprocal quantities:

$$\kappa^X / \kappa^x = (1 - k^2), \tag{8.18a}$$

$$c^E / c^D = (1 - k^2), \tag{8.18b}$$

where, in this context,

$$k^2 = \frac{h^2}{c^D (\kappa_0{}^x)} \tag{8.17b}$$

This new parameter k in Eq. (8.17b) is also the *electromechanical coupling factor* in the "extensive" parameter description, and identical to the k in Eq. (8.17a). Note the k expression derivation from the piezoelectric constitutive equations, Example Problem 2.7:

$$k^2 = \frac{(\text{Coupling factor})^2}{(\text{Product of the diagonal parameters})} = \frac{d^2}{s^E \varepsilon^X \varepsilon_0} = \frac{h^2}{c^D (\kappa^x / \varepsilon_0)}.$$

8.2.3 *Adiabatic Process 1 — Piezothermal Effect*

When we discuss the piezoelectric coupling phenomena, we assume the "isothermic condition"; that is, the specimen temperature is maintained constant, even the energy conversion is conducted between the electrical and mechanical energy. We consider next the "adiabatic process"; that is, the specimen is isolated from the external heat source, and temperature may be changed during the energy conversion process by the external input. This is the condition when we consider "thermal energy harvesting". When the electric field is constant, $E = 0$, in Eqs. (8.8a) and (8.8b), we can obtain the following equations:

$$S = -\left(\frac{\partial^2 G}{\partial T^2}\right)\theta - \left(\frac{\partial^2 G}{\partial T \partial X}\right) X, \tag{8.18a}$$

$$x = -\left(\frac{\partial^2 G}{\partial T \partial X}\right)\theta - \left(\frac{\partial^2 G}{\partial X^2}\right) X. \tag{8.18b}$$

Or

$$S = \frac{c_p}{T}\theta + \alpha_L X, \tag{8.19a}$$

$$x = \alpha_L \theta + s^E X, \tag{8.19b}$$

where the following notations are used, and denoted as c_p specific heat capacity under $X = 0$ and $E = 0$, s^E elastic compliance under constant E:

$$\begin{cases} c_p = -T \left(\dfrac{\partial^2 G}{\partial T^2}\right)_{X,E} \\ s^E = -\left(\dfrac{\partial^2 G}{\partial X^2}\right)_{E,T} \\ \alpha_L = -\left(\dfrac{\partial^2 G}{\partial T \partial X}\right)_E \end{cases} \tag{8.20}$$

Let us discuss here the diagonal expansion coefficient $-\left(\frac{\partial^2 G}{\partial T^2}\right)$ in terms of specific heat capacity. Recall the relation $dq = TdS$ in the "reversible" thermal process, where dq is the thermal energy flow per unit "mass", given by the total energy flow $dQ = \rho V dq$ (ρ: mass density, V: volume). The "specific heat capacity" c_p is defined by

$$c_p = \frac{\partial q}{\partial T} = T\left(\frac{\partial S}{\partial T}\right)_{X,E} = -T\left(\frac{\partial^2 G}{\partial T^2}\right)_{X,E}. \qquad (8.21)$$

It is noteworthy that the piezothermal coefficient α_L is usually called "linear thermal expansion coefficient", because $-\left(\frac{\partial^2 G}{\partial T \partial X}\right)_E = \left(\frac{\partial x}{\partial T}\right)$. This piezothermal coefficient α_L contributes to the converse effect, that is, temperature change under the stress application. The piezothermal coefficient α_L originates from a non-linear elastic vibration or the anharmonic phonon interaction. The "piezothermal coupling factor" k^{PT} can be defined from Eqs. (8.19a) and (8.19b) by

$$k^{PT^2} = \frac{(\text{Coupling factor})^2}{(\text{Product of the diagonal parameters})} = \frac{\alpha_L^2}{\left(\frac{c_p}{T}\right)s^E} \qquad (8.22)$$

We can derive the completely "clamped" (strain x free) specific heat capacity and the "adiabatic" elastic compliance in terms of k^{PT^2} theoretically as follows, though these may not be useful physical parameters in practice: Under strain-free condition ($x = 0$), we obtain $X = -\frac{\alpha_L}{s^E}\theta$ from Eq. (8.19b). Since Eq. (8.19a) gives $S = \frac{c_p}{T}\theta + \alpha_L\left(-\frac{\alpha_L}{s^E}\theta\right) = \frac{c_p}{T}(1 - k^{PT^2})\theta$, we finally obtain $c_p^x = c_p^X(1 - k^{PT^2})$. On the other hand, under the adiabatic condition ($S = 0$), we obtain $\theta = -\frac{T}{c_p}\alpha_L X$ from Eq. (8.19a). Since Eq. (8.19b) gives $x = \alpha_L(-\frac{T}{c_p}\alpha_L X) + s^E X = s^E(1 - k^{PT^2})X$, we obtain adiabatic elastic compliance $s_{adia}^{E,S} = s_{iso}^{E,T}(1 - k^{PT^2})$.

8.2.4 *Adiabatic Process 2 — Electrothermal Effect*

When the stress is constant, $X = 0$ in Eqs. (8.8a) and (8.8c), we can obtain the following equations:

$$S = -\left(\frac{\partial^2 G}{\partial T^2}\right)\theta - \left(\frac{\partial^2 G}{\partial T \partial E}\right)E, \qquad (8.23a)$$

$$D = -\left(\frac{\partial^2 G}{\partial T \partial E}\right)\theta - \left(\frac{\partial^2 G}{\partial E^2}\right)E, \qquad (8.23b)$$

or

$$S = \frac{c_p}{T}\theta - pE, \qquad (8.24a)$$

$$D = -p\theta + \varepsilon_0\varepsilon^X E, \qquad (8.24b)$$

where the following notations are used, and denoted as c_p^E specific heat capacity (per unit mass) under $X = 0$ and $E = 0$, $\varepsilon_0\varepsilon^X$ permittivity under constant stress X:

$$\begin{cases} c_p = -T\left(\frac{\partial^2 G}{\partial T^2}\right)_{X,E} \\ \varepsilon_0\varepsilon^X = -\left(\frac{\partial^2 G}{\partial E^2}\right)_{T,X} \\ p = \left(\frac{\partial^2 G}{\partial T \partial E}\right)_X \end{cases} \qquad (8.25)$$

The primary "electrothermal coupling coefficient" p is usually called "pyroelectric coefficient", defined by

$$p = \left(\frac{\partial^2 G}{\partial T \partial E}\right)_X = -\left(\frac{\partial P}{\partial T}\right)_X, \tag{8.26}$$

where we used the relation $\left(\frac{\partial G}{\partial E}\right)_X = D\,(\approx P)$.

8.2.4.1 *Constraint specific heat capacity*

In Eq. (8.25) Top, we introduced c_p specific heat capacity (per unit mass) under $X = 0$ and $E = 0$, that is, under a short-circuit condition of a ferroelectric specimen's surface electrodes. We may consider a different specific heat capacity under an open-circuit condition (i.e., $D = $ constant or zero).[8]

Taking the first derivative of Eq. (8.24a) with respect to T by keeping $X = D = 0$, and $c_p = c_p^E$,

$$\left(\frac{\partial S}{\partial T}\right)_{X,D} = \frac{c_p^E}{T}\left(\frac{\partial \theta}{\partial T}\right)_{X,D} - p\left(\frac{\partial E}{\partial T}\right)_{X,D} \tag{8.27}$$

From Eq. (8.24b) at $D = 0$, we obtain

$$E = \frac{p}{\varepsilon_0 \varepsilon^X}\theta$$

$$\left(\frac{\partial E}{\partial T}\right)_{X,D} = \frac{p}{\varepsilon_0 \varepsilon^X}$$

If we denote $c_p^D = T\left(\frac{\partial S}{\partial T}\right)_{X,D}$ and $\left(\frac{\partial \theta}{\partial T}\right)_{X,D} = 1$, we can obtain

$$c_p^D = c_p^E - \frac{Tp^2}{\varepsilon_0 \varepsilon^X} = c_p^E\left[1 - \frac{p^2}{\left(\frac{c_p^E}{T}\right)\varepsilon_0 \varepsilon^X}\right] = c_p^E(1 - k^{ET^2}) \tag{8.28}$$

Here, we denoted the primary "electrothermal coupling factor" k^{ET} from Eqs. (8.24a) and (8.24b) as

$$k^{ET^2} = \frac{(\text{Coupling factor})^2}{(\text{Product of the diagonal parameters})} = \frac{p^2}{\left(\frac{c_p^E}{T}\right)\varepsilon_0 \varepsilon^X} \tag{8.29}$$

It is important to note that Eq. (8.28) is analogous to Eq. (8.15b) to correlate the D-constant and E-constant parameters in terms of "coupling factors", k^{ET^2} and k^2.

8.2.4.2 *Constraint (Adiabatic) permittivity*

Permittivity was defined isothermally so far. However, we may consider "adiabatic permittivity" theoretically when no heat flow is hypothesized, such as the case where a ferroelectric specimen is suspended in a vacuum chamber.[7]

From Eq. (8.24b), isothermal permittivity ($\theta = 0$) is given by

$$\varepsilon_0 \varepsilon^{X,T} = \left(\frac{\partial D}{\partial E}\right)_{X,T}. \tag{8.30}$$

Under adiabatic condition, by putting $S = 0$ in Eq. (8.24a), we obtain

$$\theta = \frac{pT}{c_p^E}E. \tag{8.31}$$

Inserting Eq. (8.31) into Eq. (8.24b),

$$D = -p\frac{pT}{c_p^E}E + \varepsilon_0\varepsilon^{X,T}E = \varepsilon_0\varepsilon^{X,T}\left[1 - \frac{Tp^2}{c_p^E\varepsilon_0\varepsilon^X}\right]E. \tag{8.32}$$

Thus, "adiabatic (S = constant) permittivity" is related with "isothermal (T = constant) permittivity" again by using primary electrothermal coupling factor k^{ET} as

$$\varepsilon_0\varepsilon^{X,S} = \varepsilon_0\varepsilon^{X,T}(1 - k^{ET^2}). \tag{8.33}$$

8.2.4.3 *Electrocaloric effect*

Equation (8.24a) gives the necessary formula for the electrocaloric effect. When we take "adiabatic" condition; that is, constant entropy $dS = 0$, and Eq. (8.31)

$$\theta = \frac{pT}{c_p^E}E.$$

Here, c_p^E is specific heat capacity (per unit mass) under $E = 0$, and p is pyroelectric coefficient given by $p = -\left(\frac{\partial P}{\partial T}\right)_X$. Temperature decrease is realized by suddenly (adiabatically) applying negative electric field. In practice, increasing the electric field isothermally initially and then sudden short-circuit are utilized in order to escape from the ferroelectric depoling. The material's development strategy should be higher p and lower c_p^E at room temperature (i.e., operation temperature $\sim$300 K).

8.3 Magnetoelectric Devices and Energy Harvesting

8.3.1 *Background of Magnetoelectric Effect*

After entering the 21st century, the ME *Renaissance* started because of a strong demand for it from urgent application developments for society sustainability technologies. Different from high voltage and magnetic field, similar to radio-active radiation, it cannot be sensed by a human body unfortunately. Great anxiety has been growing about brain/human cancer development owing to strong magnetic fields under a high-power transmission cable, by holding a smart phone near the brain or by the leakage of magnetic field from an old microwave oven. We need low-frequency (50 Hz) magnetic field sensors in daily life. More sensitive magnetic sensors are required for surveying submarines, many of which are nowadays equipped with acoustic stealth devices (i.e., military application), discovering land-mines and mineral mines or predicting earthquake occurrence on a weekly level short-term basis (i.e., crisis technologies). Steal and magnetic metals, if they are moving such as a submarine and magma in the earth, in particular, generate a fluctuation of the earth's magnetic field, which can be detected by a sensitive ME device (less than 1 nano Tesla). Further, electric energy harvesting is possible from "stray" magnetic fields under a power transmission cable for operating an unmanned vehicle for checking the cables, or from a magnetic field adjacent to an industrial motor for sending the signal of motor condition remotely to the operation room (i.e., machine safety monitoring). Since single-phase materials cannot satisfy the following requirements from commercialization viewpoints,[9] composite approaches have been highly accelerated these days:

(1) Operation temperature — around room temperature.
(2) High figure-of-merit — higher voltage generation under a unit magnetic field, $\left(\frac{\partial E}{\partial H}\right)$.
(3) Low manufacturing cost — both raw materials and simple fabrication.

In parallel to other research groups, the author's team also chased initially the 0-0 composites by using PZT powders (rather than barium titanate) for enhancing the ME performance.[10] We investigated intensively the effect of the sintering temperature on the sintering behavior, microstructures, piezoelectric and ME properties of this ME particulate composite with Ni-ferrite doped with Co, Cu, Mn particles and PZT matrix.[10] Not only the connectivity of the ferrite phase, but also the sintering temperature is an important parameter for higher ME voltage coefficient (dE/dH). The chemical reaction of the PZT with ferrite and connection of ferrite particles make it difficult to get high ME effects. We obtained the highest ME voltage coefficient from the composite with 20% ferrite added and sintered at 1250°C. A homogeneous and

well-dispersed microstructure, no chemical reaction between the two phases and large grain size of the matrix PZT phase were the most important factors to get a high ME voltage coefficient. Though our value was 45% higher than the previously reported value from Philips Lab., the enhancement was not as significant as our great efforts. Thus, we basically abandoned the "particulate" composite approach, and the target was shifted to the "laminated" composites.

Figure 8.7 shows the paper publication trend in these 50 years. 2907 papers were collected from Scopus data bank under the keywords "magnetoelectric" and "composite". The invention of the *magnetostrictor-piezoelectric laminated composites* (2000) triggered the current research fever, and our papers have been cited more than 2100 times.

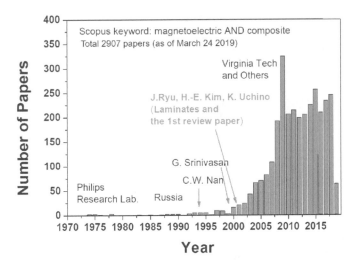

Fig. 8.7: Paper publication trend in these 50 years. 2907 papers collected from Scopus data bank under the keywords "magnetoelectric" and "composite".

8.3.2 *Designing of Magnetoelectric (ME) Laminate Composites*

We describe how to optimize the design of ME laminate composites in this section.

8.3.2.1 *Challenge to the laminate composites*

The particulate ME composites (0-0 type) made of piezoelectric and magnetostrictive ferrite materials showed higher ME properties compared with single-phase ME materials such as Cr_2O_3.[9] However, these composites still need some important issues addressed when fabricating the sintered ME particulate composites to obtain superior ME response. First, no chemical reaction should occur between the piezoelectric and magnetostrictive materials during the sintering process. The chemical reaction may reduce the piezoelectric or magnetostrictive properties of each phase. Second, the resistivity of the magnetostrictive phase should be as high as possible. If the resistivity of the magnetostrictive phase is low, the electric poling of the piezo phase becomes very difficult due to leakage current. Also, the leakage reduces the ME output voltage of the composites. When the ferrite particles make connected chains, the electric resistivity of the composites is reduced significantly because of the low resistivity of the ferrite. Therefore, good dispersion of the ferrite particles in the piezoelectric matrix is required in order to sustain sufficient electric resistivity of the composite. Third, mechanical defects such as pores at the interface between the two phases should not exist in the composite for good mechanical coupling, through which magnetostrictive and piezoelectric materials interact.

These difficulties may be overcome by using a laminar composite (2-2 type), because no chemical reactions and dispersions are involved in the fabrication process. In addition to these advantages, the laminated ME composites have a very simple structure and relatively simple fabrication method, i.e., bonding each disk. The laminated ME composites can be easily applied to practical applications, such as magnetic field sensing devices, leak detectors for microwave ovens and current measurement of high-power electric transmission systems. Thus, we invented the laminated ME composites made by using piezoelectric and magnetostrictive materials.[11-14] Lead Zirconate Titanate (PZT) and Terfenol-D disks were used as the piezoelectric material and magnetostrictive material, respectively, in the ME laminate composites. The composites were manufactured by sandwiching and bonding a PZT disk between two layers of Terfenol-D disks, as shown in Fig. 8.8. When a magnetic field is applied to this composite, the top and bottom Terfenol-D disks shrink or expand. This shrinkage or expansion generates stresses in the sandwiched piezoelectric PZT disk. Hence, electric signals can be obtained when the composite is subjected to a magnetic field. To optimize the PZT and Terfenol-D ME laminate composites, we investigated (1) the effect of the piezoelectric properties of the piezoelectric layer (PZT and PMN-PT single crystals), (2) thickness ratio between the PZT and Terfenol-D disks, (3) the directional dependence of the magnetostriction of the Terfenol-D disks and (4) of the AC magnetic field on the ME response of the PZT/Terfenol-D laminate composites.

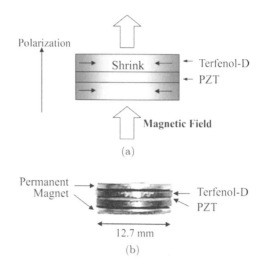

Fig. 8.8: Magnetoelectric laminate composite using Terfenol-D and PZT disks. (a) schematic structure, and (b) photograph of the device.

8.3.2.2 *ME measuring technique*

Prior to the device designing principle, we introduce the ME measuring technique.[11] The ME voltage coefficient was determined by measuring the electric voltage (electric field) generated across the sample when an AC magnetic field with a DC bias was applied to it. The ME property was measured in terms of the variation of the coefficient dE/dH as a function of DC magnetic bias field. An electromagnet (GMW 5403 Magnet, Power and Buckley Inc., New Zealand) was used for the bias magnetic flux density B up to 0.45 T (i.e., magnetic field H = 4.5 kOe). The frequency dependence of the ME voltage coefficient of composites was determined in the range of 10 Hz to 3 kHz under a 1 kOe D.C. bias. The coefficient was measured directly as response of the sample to an AC magnetic input signal at 1 kHz and 2 Oe amplitude (Helmholtz coils were used to give a uniform AC magnetic field in the space between the coils) superimposed on the DC bias field, both parallel to the sample axis (Figs. 8.9(a) and 8.9(b)).

A signal generator (33120A, Hewlett Packard Co., USA) was used to drive the Helmotz coils and generate the AC magnetic field. The voltage generated in the piezoelectric layer was measured under an open circuit condition. A differential amplifier based on the INA121 FET-input Instrumentation Amplifier (Burn-Brown Inc.) was used. This amplifier is specially designed for high impedance transducers, providing differential input impedance in the order of $10^{12} G\Omega/1pF$, which represents almost an ideal open circuit condition. The electric circuit of the amplifier is shown in Fig. 8.9(c). The output signal from the amplifier was measured with an oscilloscope (54645A, Hewlett Packard Co., USA). The output voltage divided by the thickness of the sample and the A.C. magnetic field gives the ME voltage coefficient of the samples.

8.3.2.3 *Effect of Piezoelectrics*

8.3.2.3.1 PZT Case

The effect of the piezoelectric properties of the PZT material on the ME response of the laminate composites was studied first on samples consisting of two Terfenol-D disks and one different kind of PZT material. Note that the permanent magnets in Fig. 8.8(b) were not attached for the basic measurement, but they were used for a practical application to provide the optimized DC bias magnetic field to the ME device. Table 8.1 shows the piezoelectric properties of the PZT materials used for this study. APC 840, PZT-5A and APC 841 were used for their high g_{33}, high d_{33} and high Q_m. respectively.[11] Figure 8.10 shows the ME voltage coefficient variation as a function of the DC magnetic bias with three different PZT types. All PZT disks were machined down to the same thickness (0.5 mm) and diameter (12.7 mm). The ME voltage coefficients of all the composites were increased with increasing DC bias until saturated around 4 kOe. Since the sensitivity is mainly determined by the piezoelectric voltage coefficient (g_{33}); the composite with APC-840 PZT showed the most superior ME property. The maximum ME voltage coefficient for this composite was 4.68 V/cm · Oe under 4.2 kOe D.C. magnetic bias or higher. Note that this 2-2 type laminate composite exhibits more than 30 times higher ME voltage coefficient than the 0-0 type particulate composite introduced in Fig. 8.5.

The dependence of the ME voltage coefficient of composites on the frequency of AC magnetic field was observed under a fixed DC magnetic bias at 1 kOe for all samples, and a maximum ME voltage coefficient at around 150 Hz was found. This frequency dependence seems to occur due to a mechanical resonance of the sample device and sample holder, and is not the property of the composite itself. Since the lowest mechanical resonance frequency of these composite samples can be evaluated to be higher than 100 kHz, a significant frequency dependence cannot be expected in this frequency range.

8.3.2.3.2 PMN-PT Single-Crystal Case

As shown in the previous section, the most important factors needed in order to achieve a high ME voltage coefficient from piezoelectric-magnetostrictive composites are a high piezoelectric voltage coefficient (g_{ij}). Since our discovery, relaxor single crystals such as $Pb(Zn_{1/3}Nb_{2/3})O_3$–$PbTiO_3$ (PZN–PT) and

(a)
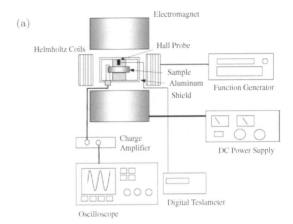

(b)
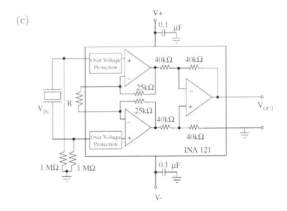

(c)

Fig. 8.9: (a) Schematic diagram of ME measurement system, (b) Photo of the measurement system, (c) Amplifier circuit for ME voltage measurement.

Table 8.1: Piezoelectric properties of PZT materials used in this study.

Material	$\varepsilon_{33}^T/\varepsilon_o$	$\tan\delta(\%)$	d_{33} (pC/N)	g_{33} (mVm/N)	Q_m	k_p
APC840	1250	0.4	320	**25.6**	500	0.59
PZT-5A	1730	1.5	**340**	19.6	68	0.57
APC841	1250	0.35	275	22.0	**1400**	0.60

Pb(Mg$_{1/3}$Nb$_{2/3}$)O$_3$–PbTiO$_3$ (PMN–PT) are very well known to have superior piezoelectric properties,[15–18] and extremely high ME coefficients are expected, when the PZT ceramic layer is replaced by a (001) oriented Pb(Mg$_{1/3}$Nb$_{2/3}$)O$_3$–PbTiO$_3$ (PMN–PT) single crystal, which has a much higher g_{ij} coefficient. Figure 8.11 shows the ME voltage coefficient as a function of applied DC magnetic field. The magnetic field dependence of dE/dH was similar for all three types of piezoelectric specimens. As is evident in the figure, the ME voltage coefficient increased with increasing magnetic bias, saturating at a bias level of ∼4 kOe. The ME laminate composite made using a PMN–PT single crystal had the highest ME voltage coefficient. The value of dE/dH was 10.3 V/cm · Oe, which is ∼80 times higher than that previously reported in the 0-0 particulate composites. This high value of (dE/dH) for PMN–PT single crystals is due to the high piezoelectric voltage constant (g_{33}), as well as high elastic compliance (s_{33}). As introduced later in Eq. (8.9), the output voltage from the composite is directly proportional to the piezoelectric voltage constant g_{31} and the stress on the piezoelectric plate. The stress on the piezoelectric material is dependent on the elastic compliance of the material. The higher the compliance, the higher is the generated stress, as the mechanical coupling is directly correlated with the magnitude of the compliance. Table 8.2 shows the piezoelectric voltage constant and elastic compliance of the three different materials studied in this investigation. It is evident from the table that both the piezoelectric voltage coefficient and elastic compliance are higher for PMN–PT single crystal as compared to PZT ceramic. This difference is responsible for the higher ME properties of the PMN–PT laminate composites. These results indicate that a better ME property can be obtained if the elastic and piezoelectric properties of the materials can be improved.

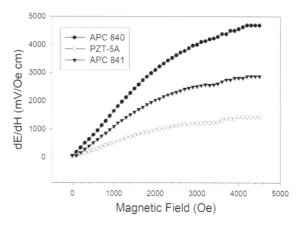

Fig. 8.10: ME voltage coefficient as a function of applied DC magnetic bias for various PZT disks (APC 840, PZT-5A, APC 841) at 1 kHz.[9]

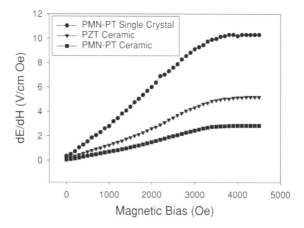

Fig. 8.11: ME voltage coefficient for laminate composites vs. applied DC magnetic bias for PZT, PMN PT ceramic, and PMN–PT single-crystal disks.[12]

8.3.2.4 *Thickness ratio effects on the ME properties*

The ME voltage coefficient was found to increase with decreasing thickness of the PZT layer as depicted in Fig. 8.12(a). This can be explained by the increase in compressive stress in the PZT layer with decreasing

Table 8.2: Material properties of single and polycrystalline piezoelectrics.

Material	$\varepsilon_{33}^T/\varepsilon_o$	d_{33} (pC/N)	g_{33} (mVm/N)	$S_{33}^E(10^{-12}\,\text{m}^2/\text{N})$
PMN-PT single crystal	4344	1710	44.45	56.4
PMN-PT ceramics	5614	570	11.47	9.5
PZT ceramics	1081	250	26.11	17.4

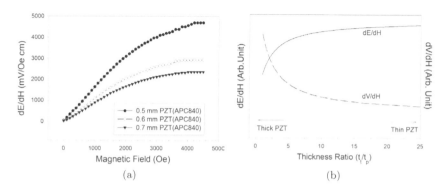

Fig. 8.12: (a) ME voltage coefficient as a function of applied DC magnetic bias with various thickness of PZT layer 1 kHz. (b) Theoretical expectation of the magnetoelectric voltage coefficient as a function of thickness ratio (t_t/t_p) between Terfenol-D and PZT layer.

thickness of PZT. The compressive stress in the PZT layer and the tensile stress in the Terfenol-D layers can be derived from simple beam theory under plane stress conditions, as indicated in Eqs. (8.35) and (8.35):[19]

$$\sigma_{31_t}^E = \frac{E_t E_p t_p \Delta\varepsilon_o}{(1-v)(2E_t t_t + E_p t_p)},\tag{8.34}$$

$$\sigma_{31_p}^E = -\frac{2E_t E_p t_t \Delta\varepsilon_o}{(1-v)(2E_t t_t + E_p t_p)},\tag{8.35}$$

where σ_{31}, $\Delta\varepsilon_o$, E, t and v are the transversal stress normal to the PZT disk (i.e., radial direction), linear strain (in-plane) of the Terfenol-D layer, elastic (Young's) modulus (i.e., stiffness), thickness and Poisson's ratio (Poisson's ratios of Terfenol-D and PZT are assumed to have the same value in these equations), respectively. The subscript "t" or "p" stands for Terfenol-D or PZT, respectively. As shown in these equations, the compressive stress in the PZT layer is increased with the decreasing thickness of the PZT layer or the increasing thickness of the Terfenol-D layer. Since the thickness of the Terfenol-D is fixed at 1 mm (in our design), by decreasing the PZT layer thickness, the compressive stress in the PZT layer is increased. The output voltage from the composite can be expressed by the following equations:

$$V_{out} = 2 \times g_{31} \times t_p \times \sigma_{31p}^E,\tag{8.36}$$

$$\frac{dE}{dH} = \frac{V_{out}}{H_{ac} \times t_p} = \frac{2 \times g_{31} \times \sigma_{31p}^E}{H_{ac}}(V/cm \cdot Oe)\tag{8.37}$$

Therefore, a higher output voltage can be obtained when the compressive stress in the PZT layer is higher, i.e., a thinner PZT layer. From these equations, it can be seen that the output voltage from the composite is also directly proportional to the piezoelectric voltage constant g_{31}. Generally speaking, in PZT ceramics, the piezoelectric voltage constant g_{31} is around 1/3 that of g_{33}. In this regard, the laminate composite made with APC-840, which has the highest g_{33}, exhibits the highest ME voltage coefficient, as shown in Fig. 8.10.

Figure 8.12(b) shows the theoretical expectation for the ME voltage coefficient (dE/dH) and the output voltage (dV/dH) as a function of the thickness ratio (t_t/t_p) between Terfenol-D and PZT. The ME voltage coefficient increases with increasing thickness ratio (t_t/t_p), but output voltage decreases with increasing thickness ratio. Both values saturate above a thickness ratio of 10. The value of the output voltage is more important than the ME voltage coefficient for practical sensor applications. Therefore, a lower thickness ratio (less than 10) is more suitable, even though the ME voltage coefficient is increasing with thickness ratio.

8.3.2.5 *Magnetostriction direction dependence*

Three kinds of laminate composite design were prepared by using two types of Terfenol-D disks to investigate this issue. These are as follows:

(1) Composite with two Terfenol-D disks that have their magnetostriction along the thickness direction (denoted as Comp.T-T).
(2) Composite with one Terfenol-D disk with thickness magnetostriction direction and the other Terfenol-D disk with radial magnetostriction direction (denoted as Comp.T-R).
(3) Composite with two disks that have their magnetostriction along the radial direction (denoted as Comp.R-R).

Figure 8.13(a) shows schematic illustrations of each composite structure. The dielectric polarization direction of the PZT disk and the applied magnetic field direction were in the thickness direction for all the composites. Figure 8.13(b) illustrates the variation of the ME voltage coefficient (dE/dH) as a function of the DC magnetic bias for the three different composites. The ME voltage coefficients of all composites increased with increasing DC bias, and saturated around 4 kOe. The Comp.R-R showed the most superior ME property. Its maximum ME voltage coefficient was 5.90 V/cm · Oe when under a magnetic DC bias equal to or greater than 4.2 kOe. In the Terfenol-D disks, the magnitude of strain in the principal magnetostriction direction is higher than other directions.[20]

8.3.2.6 *Magnetic field direction dependence*

In applications like magnetic field sensing devices, the dependence of the ME response on the magnetic field direction is an important factor. To examine the magnetic field direction dependence, we measured the ME voltage coefficient by changing the applied magnetic field direction. The dependence of the ME voltage coefficient of the composites on the applied AC magnetic field direction is shown in Fig. 8.13(c). The angle of the applied AC magnetic field (1 kHz) indicates the difference in angle between the AC magnetic field and the DC magnetic bias (thickness direction of the sample). The dependence of the AC magnetic field direction exhibited a similar behavior for all the composites, with maxima occurring at 25–45°. Beyond 45°, the ME voltage coefficient decreased with an increase of the AC magnetic field direction. According to

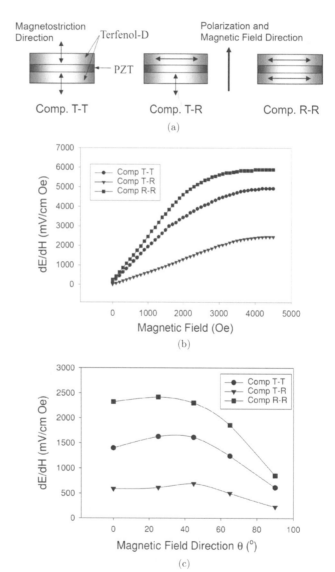

Fig. 8.13: (a) Schematic illustrations for three different PZT/Terfenol-D composites. (b) ME voltage coefficient vs. applied DC magnetic bias with different assembly. (c) ME voltage coefficient as a function of an applied AC magnetic field direction.

the theoretical calculations, these behaviors are basically related to the areal strain changing behavior of Terfenol-D with changing the applied magnetic field direction. The calculated maximum of the areal strain occurs at the orientation angle, $\theta = 51°$.[21] It is expected that the ME coefficient will show a maximum around this angle. This behavior originates from the contribution of the relatively large shear mode strains, i.e., $|d_{15}| > |d_{33}|$ or $|d_{31}|$.

Section 8.3.2 disclosed the ICAT/PSU development strategy on the ME laminate composites. We chased initially the Philips group on the 0-0 type composites with magnetostrictive and piezoelectric powder mixtures, which exhibited 10 times higher ME voltage coefficients than those of most of the single-phase materials. Their idea encouraged us to move into the composites. However, even we used PZT, much better piezoelectric property than barium titanate, our device performance exceeded only 40% that of the Philips. This discouragement instructed us "not to chase the others' idea". Because the author is the inventor of "multilayer" actuators, and also from the "Connectivity" model analysis/simulation, we concluded to use the laminate composite. Once the research direction was fixed, the approaches were not very difficult. Thanks to our former research associates, various composites were prepared by stacking and then bonding piezoelectric materials to Terfenol-D. Laminates were made using several kinds of PZTs, PMN–PT ceramics and a (001)-oriented PMN–PT single crystal. The highest ME voltage coefficient (dE/dH) was found for the PMN–PT single crystal, which was $10.3\,\mathrm{V/cm \cdot Oe}$, under a DC magnetic bias of 0.4 T, another 10 times higher than the 0-0 type powder mixture composites. This unique laminate design concept created our current engineering leading status. You can easily understand the following systematic optimization processes in the above sections: To obtain excellent ME property from the ME laminate composites, a high piezoelectric voltage coefficient (g_{ij}), an optimum thickness ratio between piezoelectric layer and Terfenol-D layers, the direction of magnetostriction in the Terfenol-D disks and higher elastic compliance of piezoelectric material are important factors. Note also that our laminate composite design fits beautifully with so-called MEMS micro-machining technologies to miniaturize the ME devices.

8.3.3 *Magnetoelectric Applications*

8.3.3.1 *Sensor applications*

Similar to nuclear radiation, magnetic irradiation cannot be easily felt by humans. Some reports mentioned that brain cancer may be triggered by a frequent usage of a mobile phone, though there is no strong scientific evidence. Though just a coil can detect the magnetic field theoretically, the problem is the situation that we cannot even purchase a highly sensitive magnetic field detector for a low frequency (50 or 60 Hz). The ME laminate composite developed by the Penn State is for a simple and handy magnetic noise sensor for environmental monitoring, which was the initial motivation for developing the laminate composites. A Japanese real-estate agency sponsored us for the device to monitor the magnetic field below high-power transmission cables, in order to sell "real-estate" there without discounting (Fig. 8.14(a)). Refer to Fig. 8.8(b), where we integrated a pair of permanent magnets to provide a suitable bias magnetic field on the ME laminate sensor, for commercializing portable and handy products. Another application is for the Navy, that is, to survey an enemy submarine which equips the "acoustic stealth system" (that is, difficult to find via SONAR systems). Large Fe (magnetic) mass movement (even slow) modulates the earth magnetic field at around 10 Hz or lower frequency, which can be detected by the laminate ME device down to 1 nT level. The key of this device is high effectiveness for a low-frequency magnetic field modulation. See Fig. 8.14(b) for understanding visually. Similar sensor applications include (a) predicting earthquake occurrence through the earth magnetic field fluctuation via magma movement in a weekly level short-term basis, (b) discovering land-mines and mineral mines, now by sweeping the ME sensor (i.e., crisis and sustainability technologies).

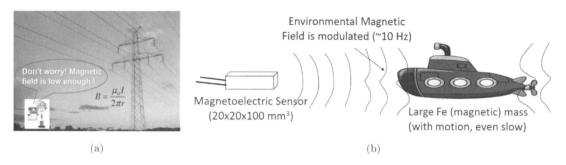

(a) (b)

Fig. 8.14: (a) Stay magnetic field monitoring below high-power transmission cables (50 Hz). (b) Acoustic stealth-submarine monitoring under the sea ($\sim$10 Hz or lower).

8.3.3.2 *Energy harvesting applications*

In addition to just monitoring the "stray" magnetic field, electric energy harvesting is also possible from the magnetic field under the power transmission cable, if the field strength is high enough for obtaining the energy higher than 1 mW. Figure 8.15(a) shows the present human manual inspection procedure of the high-power transmission cables. In order to reduce the required manpower, an unmanned vehicle is being introduced by a Korean electric company for checking

(a) (b)

Fig. 8.15: (a) Human manual inspection of high-power transmission cables. (b) Unmanned inspection robot of transmission cables.

the cables, as shown in Fig. 8.15(b). Though a rechargeable battery is installed at present on this unmanned vehicle, it is much more convenient to equip an energy harvesting device from reasonably high magnetic field very close to the power cable.

8.3.4 *Hybrid Magnetoelectric Energy Harvesting*

Virginia Tech group proposed a hybrid energy harvesting of the magneto-electric type with a piezoelectric-type bimorph element. Figure 8.16 illustrates the cross-section view of the hybrid laminate structure of PZT fiber composite bimorph and Metglas® magnetostrictive foils.[22] Under a magnetic field applied, bending vibration of the laminate composite is excited, then the electric power is generated from the piezo bimorph. Also, under a mechanical vibration, electric voltage is excited from the piezo bimorph simultane-

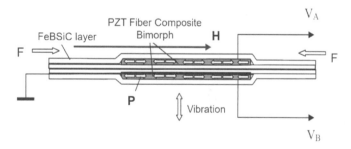

Fig. 8.16: ME devices composed with PZT fiber bimorph composite sandwiched by Metglas® layers.[22]

ously. Zhai *et al.* laminated Metglas (FeBSiC) and demonstrated hybrid energy harvesting. Mechanical vibration generates the electrical energy via a bimorph-type "Push-Pull" PZT MFC, while the magnetic noise can generate electrical energy via an electro-magnetic effect from the Metglas and MFC lamination. They demonstrated 2.5 V output voltage from the pure vibration, 2.5 V output voltage from the pure magnetic noise and 5 V as a sum from a large industrial electromagnetic motor operated at 50 Hz (leading to the same noise modulation frequency).

This sort of mechanical and magnetic hybrid energy harvesting seems to be useful for *health-monitoring* industrial motors in a factory. Since the motor both generates vibration and leaks magnetic field adjacent to the motor at the same 50 Hz range, superposed larger electric energy can be harvested for sending the signal of motor condition remotely to the operation room (i.e., *machine safety monitoring*).

Recently, Finkel *et al.*[23] demonstrated the above idea by using superior piezoelectric single-crystal lead indium niobate–lead magnesium niobate–lead titanate (PIN–PMN–PT) and magnetostrictive single-crystal Galfenol. An AC magnetic field ±250 G under a bias field of 250 G applied to the coupled device causes the magnetostrictive Galfenol element to expand, and the resulting stress forces the phase change in the relaxor ferroelectric PIN–PMN–PT single crystal. They have demonstrated high-energy conversion (2 mW at the matching impedance 1 MΩ) in this ME device by triggering the F_R–F_O transition in the single crystal by a small AC magnetic field in a broad frequency (off-resonance) range that is important for multi-domain hybrid energy harvesting devices.

Chapter Essentials

1. Difference between "sensors" and "energy harvesting" is merely the harvesting energy level, $\langle$ or $\rangle$ 1 mW.
2. Composite Effects:
   a. Sum effect — fishing rod with carbon-fiber-reinforced plastic;
   b. Combination effect — high g piezo constant with 1 (PZT rod) −3 (plastic) composites;
   c. Product effect — ME device with magnetostrictive and piezoelectric laminates.

3. "Functionality matrix" concept is convenient for creating a new phenomenon with product effect.
4. Various Coupling Phenomena: Off-diagonal effects in the material/composite "functionality matrix"

   4.1 Piezoelectric Coupling — piezoelectric effect, converse piezoelectric effect

   $$\begin{cases} x = s^E X + dE \\ D = dX + \varepsilon_0 \varepsilon^X E \end{cases} \quad \text{(piezoelectric constant, elastic compliance, permittivity)}$$

   $$k^2 = \frac{d^2}{\varepsilon^X \varepsilon_0 s^E} \quad \text{(electromechanical coupling factor).}$$

   4.2 Piezothermal Coupling — linear thermal expansion, piezothermal effect

   $$\begin{cases} S = \frac{c_p}{T}\theta + \alpha_L X \\ x = \alpha_L \theta + s^E X \end{cases} \quad \text{(thermal expansion coefficient, specific heat capacity, elastic compliance)}$$

   $$k^{PT^2} = \frac{\alpha_L^2}{\left(\frac{c_p}{T}\right) s^E} \quad \text{(piezothermal coupling factor).}$$

   4.3 Electrothermal Coupling — pyroelectric effect, electrocaloric effect

   $$\begin{cases} S = \frac{c_p^E}{T}\theta - pE \\ D = -p\theta + \varepsilon_0 \varepsilon^X E \end{cases} \quad \text{(pyroelectric coefficient, specific heat capacity, permittivity)}$$

   $$k^{ET^2} = \frac{p^2}{\left(\frac{c_p^E}{T}\right) \varepsilon_0 \varepsilon^X} \quad \text{(electrothermal coupling factor).}$$

5. ME devices can be realized by coupling magnetostrictive and piezoelectric materials.

    5.1 0-0 connectivity ME composites exhibit 10 times higher ME voltage coefficient at room temperature. The 2-2 connectivity, TerFeNOL:PZT laminates exhibit further significant (80 times higher) ME response, which has triggered the current research boom.

    5.2 Designing principle of ME composites — Optimization was studied in terms of (1) crystal orientation, (2) layer thickness, (3) DC bias magnetic field, etc.

    5.3 Sensor applications of ME composites include (1) magnetic noise sensor, (2) geomagnetic sensor, (3) earthquake prediction, (4) anti-stealth submarine technology.

    5.4 ME energy harvesting can realize an unmanned inspection vehicle for checking the high-power transmission cables.

6. A hybrid energy harvesting device which operates under either magnetic and/or mechanical noises was introduced, by coupling magnetostrictive and piezoelectric materials.

7. Other hybrid energy harvesting devices may include the combinations of piezoelectric effect with pyroelectric or photovoltaic effects, or more than two of these effects.

Check Point

1. What is the major difference between "sensors" and "energy harvesting" from the performance viewpoint? Answer it with a simple word.
2. How many are the two-phase composite categorized theoretically with the "connectivity" classes (x-y)?
3. Which type of composite effect do you expect from a steel reinforced concrete composite, sum effect, combination effect or product effect?
4. Which connectivity of PZT-Polymer composites provides larger piezoelectric g constant, parallel connectivity or series connectivity?
5. (T/F) In the 1-3 PZT:Polymer composite, the effective piezoelectric g constant of the composite $g_{33}*$ increases with increasing the volume fraction of the PZT phase ^1V. True or False?
6. (T/F) The converse effect of thermal expansion is the piezothermal effect, that is, temperature change with external stress under adiabatic condition. True or False?
7. Provide an alternative name of the effect to "converse pyroelectric effect".
8. (T/F) In a pyroelectric/ferroelectric material, when we apply positive electric field along the spontaneous polarization direction, we can observe the temperature decrease, which expands refrigeration application. True or False?
9. Nominate two candidate materials for 2 phases to generate the "magnetoelectric" effect.
10. (T/F) A hybrid laminate bimorph structure of a PZT fiber composite and a magnetostrictive foil can generate electric energy for both vibration and magnetic noise signals. True or False?

Chapter Problems

8.1 Taking into account (5×5) input/output components of Table 8.1, we introduce a (5×5) *"functionality matrix"*. If one material has "thermal expansion effect", the functionality matrix of this material can be expressed by

$$
\begin{pmatrix}
0 & 0 & 0 & 0 & 0 \\
0 & 0 & 0 & 0 & 0 \\
0 & 0 & 0 & Piezothermal & 0 \\
0 & 0 & Thermal\ expansion & 0 & 0 \\
0 & 0 & 0 & 0 & 0
\end{pmatrix}.
$$

On the other hand. an "electrocaloric" material has a functionality matrix of the following form:

$$
\begin{pmatrix}
0 & 0 & 0 & Electrocaloric & 0 \\
0 & 0 & 0 & 0 & 0 \\
0 & 0 & 0 & 0 & 0 \\
Pyroelectric & 0 & 0 & 0 & 0 \\
0 & 0 & 0 & 0 & 0
\end{pmatrix}.
$$

When we apply "stress" first. what phenomenon is induced? Derive this from the product of these two functionality matrices.

Hint

Calculate the following products:

$$
\begin{pmatrix}
0 & 0 & 0 & 0 & 0 \\
0 & 0 & 0 & 0 & 0 \\
0 & 0 & 0 & Piezothermal & 0 \\
0 & 0 & Thermal\ expansion & 0 & 0 \\
0 & 0 & 0 & 0 & 0
\end{pmatrix}
\otimes
\begin{pmatrix}
0 & 0 & 0 & Electrocaloric & 0 \\
0 & 0 & 0 & 0 & 0 \\
0 & 0 & 0 & 0 & 0 \\
Pyroelectric & 0 & 0 & 0 & 0 \\
0 & 0 & 0 & 0 & 0
\end{pmatrix}
$$

Only one effect (piezothermal) $\otimes$ (pyroelectric) remains. which is "electric field generation by stress", equivalent to "piezoelectric" effect!

References

1. K. Uchino, "Photostrictive Effect and Its Applications", *Solid State Phys.*, 22(8), 55–60 (1987).
2. R. E. Newnham *et al.*, *Mater. Res. Bull.*, 13, 525 (1978).
3. J. Van Suchetelene, *Philips Res. Rep.*, 27, 28, (1972).
4. K. Uchino, *Solid State Phys.*, 21, 27 (1986).
5. K. A. Klicker, J. V. Biggers and R. E. Newnham, *J. Amer. Ceram. Soc.*, 64, 5 (1981).
6. J. van den Boomgaard, A. M. J. G. Van Run and J. Van Suchetelene, *Ferroelectrics*, 10, 295 (1976).
7. T. Mitsui, T. Tatsuzaki and E. Nakamura, *Ferroelectrics*, Maki Pub. Co., Tokyo (1969).
8. K. Uchino, *Ferroelectric Devices 2nd Edition*, CRC Press, New York (2009).
9. K. Uchino, "Magnetoelectric Composite Materials — A Research & Development Case Study", Chapter 10 of Advanced Lightweight Multifunctional Materials, Edit. Pedro Costa *et al.*, Woodhead Pub. (2020). ISBN: 9780128185018
10. J. Ryu, A. Vazquez Carazo, K. Uchino and H. E. Kim, "Piezoelectric and Magnetostrictive Properties of Lead Zirconate Titanate/Ni-Ferrite Particulate Composites", *J. Electroceramics*, 7, 17–24 (2001).
11. J. Ryu, A. Vazquez Carazo, K. Uchino and H. E. Kim, "Magnetoelectric Properties in Piezoelectric and Magnetostrictive Laminate Composites", *Japan. J. Appl. Phys.*, 40, 4948–4951 (2001).
12. J. Ryu, S. Priya, A. V. Carazo, K. Uchino and H.-E Kim, "Effect of the Magnetostrictive Layer on Magnetoelectric Properties in Pb(Zr,Ti)O₃/Terfenol-D Laminate Composites", *J. Amer. Ceram. Soc.*, 84, 2905–2908 (2001).

13. J. Ryu, S. Priya, K. Uchino and H.-E. Kim, "Magnetoelectric Effect in Composites of Magnetostrictive and Piezoelectric Materials", *J. Electroceramics*, 8, 107–119 (2002).

14. J. Ryu, S. Priya, K. Uchino and H.-E. Kim and D. Viehland, "High Magnetoelectric Properties in $0.68Pb(Mg_{1/3}Nb_{2/3})O_3$-$0.32PbTiO_3$ Single Crystal and Terfenol-D Laminate Composites", *J. Korean. Ceram. Soc.*, 39, 813–817 (2002).

15. J. Kuwata, K. Uchino and S. Nomura, *Ferroelectrics*, 37, 579 (1981).

16. J. Kuwata, K. Uchino, and S. Nomura, *Japan. J. Appl. Phys.*, 21, 1298 (1982).

17. K. Yanagiwara, H. Kanai, and Y. Yamashita, *Japan. J. Appl. Phys.*, 34, 536 (1995).

18. S. E. Park and T. R. Shrout, *Mater. Res. Innovt.*, 1, 20 (1997).

19. A. V. Virkar, J. L. Huang and R. A. Cutler, *J. Amer. Ceram. Soc.*, 70, No. 3, 164 (1987).

20. G. Engdahl, *Handbook of Giant Magnetostrictive Materials*, San Diego, CA: Academic Press, p. 127 (2000).

21. G. Engdahl, *Handbook of Giant Magnetostrictive Materials*, San Diego, CA: Academic Press, p. 175 (2000).

22. Junyi Zhai, Ph. D. Thesis, Virginia Tech (2009).

23. P. Finkel, R. P. Moyet, M. Wun-Fogle, J. Restorff, J. Kosior, M. Staruch, J. Stace and A. Amin, "Non-Resonant Magnetoelectric Energy Harvesting Utilizing Phase Transformation in Relaxor Ferroelectric Single Crystals", *Actuators*, 5, 2; doi:10.3390/act501000 (2016).

Chapter 9

Conclusions and Future Perspectives

9.1 "Don't Read Papers"

This book started from the author's frustration on the recent research papers reporting the piezoelectric energy harvesting systems from four points: (1) Why do they use the "unimorph" design? (2) Why are they reporting the "unrealistic" resonance data on the harvested electric energy? (3) Why are the MEMS researchers reporting resulting energy much lower than 1 mW as energy harvesting? (4) Why don't they report successive energy flow or exact efficiency step by step?

Interestingly, the unanimous answer from these researchers to my question "why" is "because the previous researchers did so!" When the author indicates to research something new, many of my Ph.D. graduate students take the following research steps: (1) searching the recent research papers on the indicated topics, (2) summarizing the results, picking up the unstudied parts by believing the published results correct and (3) setting the research plan for himself/herself. Recently, even the reader's professor may order you to "search the recent published papers on the indicated topics" as your first job task.

My purpose of authoring this textbook is to provide the reader solid theoretical background of piezoelectrics, practical material selection, device design optimization and energy harvesting electric circuits, in order to stop your above "Google Syndrome", and to look forward to the future perspectives in this field. Therefore, I focused on fundamental ideas to understand how to design and develop the piezoelectric energy harvesting devices by intentionally excluding many of "my" frustrating recent 90% papers, in order to keep the consistency of our development philosophy without having a sort of prejudice/biased knowledge.

As the epilogue, the following anecdote may be rather intriguing, contradictory to the current general research strategy (or maybe yours?):

> The author's Ph.D. advisor was Late Prof. Shoichiro Nomura at Tokyo Institute of Technology, who passed away at his age only 50s. He taught me first "Don't read papers", when I joined his laboratory. What is the real meaning of this? I had top academic grades during my undergraduate period fortunately. I read many textbooks and academic journals. Accordingly, whenever Prof. Nomura suggested that I should study a new research topic, I said things such as "that research was done already by Dr. XYZ, and the result was not promising ..." After having a dozen of these sort of negative conversations, partially angrily, partially disappointedly, Prof. Nomura ordered "Hey, Kenji! You are not allowed to read academic papers for a half year. You should concentrate on the following experiment without having any biased knowledge. Having a strong bias, you cannot discover new things. After finishing the experiment and summarizing your results, you are allowed to approach the published papers in order to find whether your result is reasonable, or is explainable by some theory". Initially, I was really fearful of getting totally wrong results. However, I finished it. That led to my first discovery: PMN-PT gigantic electrostrictive materials.

Remember that "knowing too much" suppresses innovative work. A real discovery is usually made by a young less-experienced engineer like the reader. Once he/she becomes an expert professor, unfortunately he/she loses some creativity.

9.2 Summary of Piezoelectric Energy Harvesting

Chapter 1

The 21st Century is called "The Century of Sustainable Society", which requires serious management of the following:[1]

(1) Power and energy (lack of oil, nuclear power plant, new energy harvesting)
(2) Rare material (rare-earth metal, lithium)
(3) Food (rice, corn — bio-fuel)
(4) Toxic material

  - Restriction (heavy metal, dioxin, Pb)
  - Elimination/neutralization (mercury, asbestos)
  - Replacement material

(5) Environmental pollution
(6) Energy efficiency (piezoelectric device).

Though most of the major countries rely on fossil fuel and nuclear power plants at present, renewable energy development becomes important for complementing the energy deficiency (item (1)). Energy recovery from wasted or unused power has been the topic of discussion for a long period. Unused power exists in various electromagnetic and mechanical forms such as ambient electromagnetic noise around high-power cables, noise vibrations, water flow, wind, human motion and shock waves. Industrial and academic research units have focused their attention on harvesting energy from vibrations using piezoelectric transducers. However, the power level of the piezo harvesting device is limited up to 100 W, above which the competitive "electromagnetic generators" have better privilege in efficiency (the typical "break point" is 30 W/30%). The author believes therefore that the target of "piezo energy harvesting" should NOT be aligned on the large energy cultivation line, but is to be set on "elimination of batteries", which are classified as "hazardous wastes", but recycled by less than 0.5% of 10s Billion batteries sold per year in the world. In other words, the research Goal should be related with the above items (4) and (5): elimination of hazardous batteries, rather than mega-power energy harvesting. From this sense, "piezo energy harvesting" is one of the very important technologies in the current "Sustainable Society" from this hazardous waste elimination viewpoint. The efforts so far put have provided the initial research guidelines and have brought light to the problems and limitations of implementing the piezoelectric transducer.

Piezoelectric energy harvesting was initiated from the piezoelectric passive damping historically. Using the mechanical to electrical energy transduction effect in piezoelectrics, we started to develop passive damping by dissipating the converted electric energy via Joule heat in the 1980s, then this energy was accumulated into a rechargeable battery without dissipation in the 1990s.

Chapter 2

Certain materials produce electric charges on their surfaces as a consequence of applying mechanical stress. The induced charges are proportional to the mechanical stress. This is called the *direct piezoelectric effect* and was discovered in quartz by Pierre and Jacques Curie in 1880. Materials showing this phenomenon also conversely exhibit a geometric strain proportional to an applied electric field. This is the *converse piezoelectric effect*, discovered by Gabriel Lippmann in 1881. Origins of the field-induced strains include the following: (a) *Inverse Piezoelectric Effect*: $x = dE$, (b) *Electrostriction*: $x = ME^2$, (c) *Domain reorientation*: strain hysteresis and (d) *Phase transition* (antiferroelectric $\leftrightarrow$ ferro-electric): strain "jump".

- The piezoelectric constitutive equations have two types[2,3]

Intensive parameter description:

$$\begin{pmatrix} x \\ D \end{pmatrix} = \begin{pmatrix} s^E & d \\ d & \varepsilon_0 \varepsilon^X \end{pmatrix} \begin{pmatrix} X \\ E \end{pmatrix},$$

where s^E is the elastic compliance under constant field, ε^X is the dielectric constant under constant stress, d–piezoelectric charge coefficient

Extensive parameter description:

$$\kappa \begin{pmatrix} X \\ E \end{pmatrix} = \begin{pmatrix} c^D & -h \\ -h & \kappa_0 \kappa^x \end{pmatrix} \begin{pmatrix} x \\ D \end{pmatrix} \left[\kappa_0 = \frac{1}{\varepsilon_0} \right],$$

where c^D is the elastic stiffness under constant electric displacement, κ^x is the inverse dielectric constant under constant strain, h is the inverse piezoelectric charge coefficient

Electromechanical coupling factor (k):

$$k^2 = \frac{d^2}{s^E (\varepsilon^T \varepsilon_0)} = \frac{h^2}{c^D (\beta^S / \varepsilon_0)}.$$

Constraint dependence of permittivity and elastic compliance are given by

$$\varepsilon^x / \varepsilon^X = (1 - k^2), \quad s^D / s^E = (1 - k^2), \quad \kappa^X / \kappa^x = (1 - k^2), \quad c^E / c^D = (1 - k^2).$$

- Three losses in piezoelectrics include the following: (a) dielectric loss, (b) mechanical/elastic loss and (c) piezoelectric loss, with loss definitions of

$$\varepsilon^{X*} = \varepsilon^X (1 - j \tan \delta') \qquad \kappa^{x*} = \kappa^x (1 + j \tan \delta),$$
$$s^{E*} = s^E (1 - j \tan \varphi') \qquad c^{D*} = c^D (1 + j \tan \varphi),$$
$$d^* = d (1 - j \tan \theta') \qquad h^* = h (1 + j \tan \theta).$$

These intensive and extensive losses are interrelated by

$$\begin{bmatrix} \tan \delta' \\ \tan \phi' \\ \tan \theta' \end{bmatrix} = [K] \begin{bmatrix} \tan \delta \\ \tan \phi \\ \tan \theta \end{bmatrix} \quad \text{or} \quad \begin{bmatrix} \tan \delta \\ \tan \phi \\ \tan \theta \end{bmatrix} = [K] \begin{bmatrix} \tan \delta' \\ \tan \phi' \\ \tan \theta' \end{bmatrix},$$

where

$$[K] = \frac{1}{1 - k^2} \begin{bmatrix} 1 & k^2 & -2k^2 \\ k^2 & 1 & -2k^2 \\ 1 & 1 & -1 - k^2 \end{bmatrix}.$$

Mechanical quality factors at resonance and antiresonance frequencies are described as

(a) k_{31} mode: intensive elastic loss

$$Q_{A,31} = \frac{1}{\tan \phi'_{11}}, \frac{1}{Q_{B,31}} = \frac{1}{Q_{A,31}} - \frac{2}{1 + \left(\frac{1}{k_{31}} - k_{31} \right)^2 \Omega_{B,31}^2} (2 \tan \theta'_{31} - \tan \delta'_{33} - \tan \phi'_{11})$$

(b) k_t mode: extensive elastic loss

$$Q_{B,t} = \frac{1}{\tan \phi_{33}}, \frac{1}{Q_{A,t}} = \frac{1}{Q_{B,t}} - \frac{2}{k_t^2 - 1 + \Omega_A^2 / k_t^2} (2 \tan \theta_{33} - \tan \delta_{33} - \tan \phi_{33})$$

- Equivalency between mechanical and electrical systems is summarized as

$$M(d^2u/dt^2) + \zeta(du/dt) + cu = F(t), \quad \text{or} \quad M(dv/dt) + \zeta v + c\int_0^t v dt = F(t)$$

$$L(d^2q/dt^2) + R(dq/dt) + (1/C)q = V(t), \quad \text{or} \quad L(dI/dt) + RI + (1/C)\int_0^t I dt = V(t)$$

Mechanical	Electrical (F – V)
Force F(t)	Voltage V(t)
Velocity v / ú	Current I
Displacement u	Charge q
Mass M	Inductance L
Spring Compliance 1/c	Capacitance C
Damping ζ	Resistance R

The solution of *LCR circuit*: $L\left(\frac{d^2q}{dt^2}\right) + R\left(\frac{dq}{dt}\right) + \frac{q}{C} = V(t)$ or $L\left(\frac{dI}{dt}\right) + RI + \frac{1}{C}\int I dt = V$

$$I(t) = \frac{V_0}{Z}\sin(\omega t - \phi); \quad Z = \sqrt{R^2 + \left(L\omega - \frac{1}{C\omega}\right)^2}, \quad \tan\phi = \frac{\left(L\omega - \frac{1}{C\omega}\right)}{R}$$

On the other hand, steady state oscillation for a *mass-spring-dashpot model*: $m\ddot{u} + \xi\dot{u} + cu = f(t)$.

$$\omega_0 = \sqrt{c/m} \quad (\omega_0: \text{resonance angular frequency for zero damping})$$

$$\zeta = \xi/2m\omega_0 \quad (\xi, \zeta: \text{damping factor, ratio})$$

$$u(t) = \frac{f_0\sin(\omega t + \phi)}{\sqrt{(c - m\omega^2)^2 + (2m\zeta\omega_0\omega)^2}}$$

$$u_0 = 7\frac{f_0}{\sqrt{(c - m\omega^2)^2 + (2m\zeta\omega_0\omega)^2}}$$

$$\tan\phi = -\frac{2m\zeta\omega_0\omega}{c - m\omega^2}$$

Bode plot: asymptotic curves $-0\,\text{dB/decade}$, $-40\,\text{dB/decade}$, resonance peak height $= 20\log_{10}\left(\frac{1}{2\zeta}\right)$

Mechanical quality factor: $Q_m = \frac{\omega_0}{\Delta\omega} = 1/2\zeta$

- Admittance of a piezoelectric specimen can be described as
  k_{31} Mode:

$$Y = j\omega C_d\left[1 + \frac{k_{31}^2}{1 - k_{31}^2}\frac{\tan\left(\frac{\omega L}{2v_{11}^E}\right)}{\left(\frac{\omega L}{2v_{11}^E}\right)}\right]$$

k_{33} Mode :

$$Y = \frac{j\omega C_d}{\left[1 - k_{33}^2\frac{\tan\left(\frac{\omega L}{2v_{33}^D}\right)}{\left(\frac{\omega L}{2v_{33}^D}\right)}\right]} = j\omega C_d + \frac{j\omega C_d}{\left[-1 + 1/k_{33}^2\left\{\frac{\tan(\Omega_{33})}{(\Omega_{33})}\right\}\right]}$$

$$= j\omega C_d + \frac{1}{-\frac{1}{j\omega C_d} + \frac{1}{j\tan\left(\frac{\omega L}{2v_{33}^D}\right)\frac{2bwd_{33}^2}{\rho v_{33}^D L^2 s_{33}^{E2}}}}$$

- 2-terminal and 4-terminal equivalent circuits (k_{31} case) are shown below:

2-terminal EC: (a)

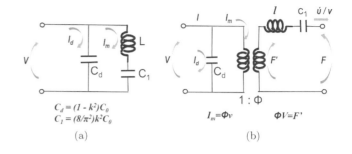

$$L_n = (\rho/8)(Lb/w)(s_{11}^{E2}/d_{31}^2)$$

$$C_n = (8/n^2\pi^2)(Lw/b)(d_{31}^2/s_{11}^{E2})s_{11}^E$$

$$R_n = \sqrt{L_n/C_n}/Q$$

4-terminal EC: (b)

$$\Phi = \frac{2wd_{31}}{s_{11}^E} \quad \text{(force factor)}$$

$$l_n = \Phi^2 L_n = (\rho/2)(Lbw)$$

$$c_n = C_n/\Phi^2 = (2/n^2\pi^2)(L/w\,b)s_{11}^E$$

$$r_n = \Phi^2 R_n = (l_n/c_n)^{1/2}/Q$$

Chapter 3

Piezo energy harvesting research historically originates from the "piezoelectric damper", and the development strategy is basically the same. Piezoelectric passive damping originates from the electric energy dissipation, which is converted from the mechanical noise vibration via the piezoelectric effect. The damping time constant in a piezoelectric passive damper is given by

$$\tau = -\frac{T_0}{\ln(1 - k^2/2)}.$$

Thus, the higher the electromechanical coupling factor k, the better for the damping rate.

The mechanical impedance matching is also an essential factor to increase the system damping factor. The mechanical/acoustic impedance of a material is given by $\sqrt{\rho c}$, where ρ and c are the density and elastic stiffness of the material.

There are three major phases/steps associated with piezoelectric energy harvesting: (i) *mechanical-mechanical energy transfer*, (ii) *mechanical-electrical energy transduction* and (iii) *electrical-electrical energy transfer*.

(i) *Mechanical–mechanical energy transfer* process includes (a) mechanical stability of the piezoelectric transducer under large stresses, and (b) mechanical impedance matching. Even a large acoustic energy exists in water; it does not transmit effectively into a mechanical "hard" PZT ceramics directly.

(ii) *Mechanical–electrical energy transduction* process relates with the electromechanical coupling factor in the composite transducer structure. In order to increase the transduction rate from the initial mechanical energy to the output electric energy, we had better adopt the high k design, escaping from the most inefficient bimorph/unimorph designs.

(iii) *Electrical–electrical energy transfer* process includes electrical impedance matching. A suitable DC/DC converter is required to accumulate the electrical energy from a high impedance piezo device into a rechargeable battery (low impedance).

Chapter 4: Mechanical-to-Mechanical

- Improvement of mechanical-to-mechanical energy transfer between two phases requires the concept of *mechanical impedance matching*. The mechanical/acoustic impedance is expressed by

$$Z = \rho v_p \quad \text{(including liquid, } \rho\text{: mass density, } v_p\text{: sound phase velocity), or}$$

$$Z = \sqrt{\rho c} \quad \text{(solid material, } c\text{: elastic stiffness).}$$

Example of acoustic impedance:

$$\text{Water: } Z = 1.5 \times 10^6 \, \text{kg/m}^2 \cdot \text{s} = 1.5 \, \text{MRayls}$$

$$\text{Polymer: 3.8 MRayls}$$

$$\text{PZT: 20--24 MRayls}$$

Acoustic piezoelectric medical and underwater transducer is mainly composed of three layers: (1) *matching*, (2) *piezoelectric material* and (3) *backing layers*. One or more matching layers are used to increase sound transmissions from the piezo material to the medium, or vice versa. The backing is added to the rear of the transducer in order to damp the acoustic backwave and to reduce the pulse duration. Piezoelectric materials are used to generate and detect ultrasound (20–40 kHz for SONAR, 2–4 MHz for medical diagnosis).

- Compressive DC bias stress effect on piezo ceramic materials:

  (1) Maximize the output electric (or electrical) work under a certain input mechanical (or electric) energy.
  (2) Minimize the fracture probability under tensile stress.
  (3) Increase the elastic stiffness, and mechanical quality factor Q_m (or elastic loss decrease).
  (4) Optimized stress ($\sim$20 MPa) is half of the blocking stress/force ($\sim$40 MPa).

Mechanical fracture strength of PZT ceramics ranges, tensile stress 50–100 MPa; compressive stress $\sim$200 MPa. Bending strength is lower when the force line is perpendicular to the spontaneous polarization direction. When the strain level exceeds 0.1% in piezo ceramics, the lifetime decreases significantly.

Chapter 5: Mechanical-to-Electrical

- Figures of merit for "piezoelectric energy harvesting" can be expressed as

  (a) FOM for stress input — $g \cdot d$ (piezoelectric voltage and strain constants).
  (b) FOM for mechanical energy input — k_{eff}^2 (effective electromechanical coupling factor).

Regarding piezoelectric materials for "energy harvesting"

(a) Soft PZT — High g is preferred.
(b) Piezo single crystal — PMN-PT (high d).
(c) Piezo polymers/composites — PVDF, PZT-polymer composites (acoustic impedance matching).

- There are five definitions of the electromechanical coupling factor:

 (a) *Mason's definition*:

$$k^2 = \text{(Stored mechanical energy/Input electrical energy)}$$

$$k^2 = \text{(Stored electrical energy/Input mechanical energy)}$$

(b) *Material's definition*:

$$U = U_{MM} + 2U_{ME} + U_{EE}$$

$$= (1/2)\sum_{i,j} s^E_{ij}X_jX_i + 2\cdot(1/2)\sum_{m,i} d_{mi}E_mX_i + (1/2)\sum_{k,m} \varepsilon_0\varepsilon^X_{mk}E_kE_m.$$

$$k^2 = U^2_{ME}/U_{MM}U_{EE}$$

(c) *Device definition*:

When the primary constitutive equations are defined in a certain piezo component, as

$$\begin{bmatrix} x \\ D \end{bmatrix} = \begin{bmatrix} s^E & d \\ d & \varepsilon_0\varepsilon^X \end{bmatrix}\begin{bmatrix} X \\ E \end{bmatrix},$$

$$k^2 = \frac{(\text{Coupling factor})^2}{(\text{Product of the diagonal parameters})} = \frac{(d)^2}{(s^E\varepsilon_0\varepsilon^X)}$$

(d) *Constraint condition method*:

Between E-constant, E-constant elastic compliances, s^E, s^D, stiffness c^E, c^D; and stress-free, strain-free permittivity $\varepsilon_0\varepsilon^X$, $\varepsilon_0\varepsilon^x$, inverse permittivity $\kappa_0\kappa^X$, $\kappa_0\kappa^x$:

$$1 - k^2 = \frac{s^D}{s^E} = \frac{c^E}{c^D} = \frac{\varepsilon^x}{\varepsilon^X} = \frac{\kappa^X}{\kappa^x}$$

(e) *Dynamic definition*: 4-terminal equivalent circuit —

Voltage V and current I, mechanical terminal parameters force F and vibration velocity $\dot{u}$:

$$\begin{bmatrix} F \\ I \end{bmatrix} = \begin{bmatrix} Z_1 & -A \\ A & Y_1 \end{bmatrix}\begin{bmatrix} \dot{u} \\ V \end{bmatrix},$$

$$k^2_v = \left|\frac{\left(\frac{A^2}{Z_1Y_1}\right)}{1 + \left(\frac{A^2}{Z_1Y_1}\right)}\right|.$$

The comparison among transverse and longitudinal effect modes is summarized:

	Transverse Effect (k_{31})	Longitudinal Effect (k_{33}, k_t)
Electric condition ($\boldsymbol{k}//x$)	$\frac{\partial E}{\partial x} = 0$	$\frac{\partial D}{\partial x} = 0$
Elastic constant	s^E_{11}	$c^D_{33} = 1/s^D_{33}$
Admittance	$Y = j\omega C_d\left[1 + \frac{k^2_{31}}{1-k^2_{31}}\frac{\tan(\Omega_{11})}{\Omega_{11}}\right]$	$Y = \frac{j\omega C_d}{1 - k^2_{33}\frac{\tan(\Omega_{33})}{\Omega_{33}}}$
Resonance	$\tan(\Omega_{11}) = \infty$	$1 - k^2_{33}\frac{\tan(\Omega_{33})}{\Omega_{33}} = 0$
Half-wave frequency ($\omega_{\lambda/2}$)	ω_R	ω_A
Equivalent circuit		

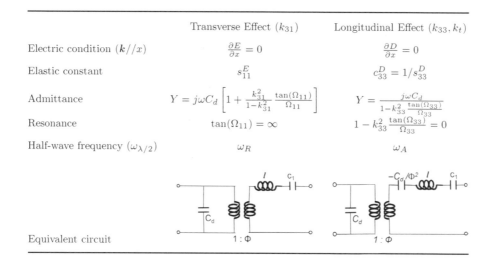

- Various electromechanical coupling factors are numerically compared:

  (a) Transversal effect $k_{31} = \dfrac{d_{31}}{\sqrt{s_{11}^E \varepsilon_0 \varepsilon_{33}{}^X}} - \sim 37\%$

     ➤ Bimorph with d_{31} – less than $\sqrt{\tfrac{3}{4}} k_{31} - \sim 13\%$ (smallest)

  (b) Planar mode $k_p = \sqrt{\dfrac{2}{1-\sigma}} k_{31} = \dfrac{d_{31}}{\sqrt{s_{11}^E \varepsilon_0 \varepsilon_{33}{}^X}} \cdot \sqrt{\dfrac{2}{1-\sigma}} - \sim 62\%$

     ➤ Cymbal transducer based on $k_p - k_{\text{eff}} \sim 30\%$

  (c) Longitudinal effect $k_{33} = \dfrac{d_{33}}{\sqrt{s_{33}^E \varepsilon_0 \varepsilon_{33}{}^X}} - \sim 73\%$

     ➤ Hinge-lever with an ML – $k_{\text{eff}} \sim 60\%$

From the resonance $f_{R,n}$ and antiresonance $f_{A,n}$ frequencies, the dynamic electromechanical factor k_{vn} for the nth harmonic vibration mode can be obtained:

$$
\begin{cases}
k_{vn}^2 = 1 - \dfrac{f_{R,n}^2}{f_{A,n}^2} & \text{(for the mode strain/stress} \perp P_S) \\[2mm]
\dfrac{k_{vn}^2}{1 - k_{vn}^2} = \dfrac{f_{A,n}^2}{f_{R,n}^2} - 1 & \text{(for the mode strain/stress} \mathbin{/\mkern-5mu/} P_S)
\end{cases}
$$

The dynamic k_{vn} is also related with the static k_v as

$$
k_{vn}^2 = P_n k_v^2
$$

where P_n is the *capacitance factor*, which is given by

$$
\begin{cases}
P_n = \dfrac{8}{\pi^2 n^2} & \text{(for the mode strain/stress} \perp P_S) \\[2mm]
P_n = \dfrac{2(1+\sigma)}{\alpha_n^2 - (1-\sigma^2)} & \text{(for the mode strain/stress} \mathbin{/\mkern-5mu/} P_S)
\end{cases}
$$

Output vibration mode differs depending on the input force: (a) impulse mechanical input generates linear displacement change in a piezoelectric component, while it generates sinusoidal reaction in an equivalent circuit (EC), while (b) sinusoidal mechanical input generates sinusoidal displacement change in both piezoelectric component and in an EC. Piezoelectric resonance and antiresonance are both natural mechanical resonance modes. When the piezo component is electrically short-circuited, the resonance mode is realized, while open-circuited, the antiresonance mode is realized.

- In a unimorph design with a piezo-plate and a metal shim, the maximum electromechanical coupling k_{eff} is obtained when the metal thickness is adjusted in a similar level of the piezo-plate. In a multilayer actuator with a hinge-lever mechanism, though the electromechanical coupling k_{eff} is close to k_{33} with $\omega \to 0$, the dynamic k_v (even at an off-resonance frequency) changes significantly according to the hinge-lever structure design. Off-resonance mechanical energy transfer rate e in a hinge-lever mechanism is defined as

$$
e = \frac{\text{Output Energy from the Magnification Mechanism}}{\text{Stored Energy in the Piezoactuator}} = \frac{(\zeta_m^2/2C_m)}{(\zeta_c^2/2C_c)}
$$

where ζ_c is the free displacement (m) of the piezo ceramic device, C_c is the elastic compliance of the ceramic (m/N), ζ_m is the free displacement (m) of the displacement magnification mechanism, and C_m is the elastic compliance of the lever mechanism (m/N).

- There are two types of input forces: one-time impulse force and continuous AC force.

  ➤ Impulse force — For the input mechanical energy U_M, converted/stored electric energy will be $k^2 U_M$. Connecting the matched electrical impedance, $\frac{1}{2}k^2 U_M$ can be harvested every half cycle. Since the input force generates multiple ring-down vibrations on the piezoelectric component, if we can accumulate all cycle energy (if the elastic loss is neglected), the total harvesting energy reaches the initial U_M (one-time energy Joule) in principle.

  ➤ Continuous AC force — For the input mechanical power W_M, converted/stored electrical power will be $k^2 W_M$. Under the matched electrical impedance condition, $\frac{1}{2}k^2 W_M$ electric power (Watts, Joule/s) is harvested continuously.

Chapter 6: Electrical-to-Electrical

- The internal impedance of a piezoelectric component is "capacitive" under an off-resonance operation, $1/j\omega C$, with a value of 100 s kΩ at 100 Hz. For electric impedance matching, external load resistance should be adjusted to the internal impedance of the power supply for obtaining the maximum output power.
- When the input mechanical energy is unlimited (such as environmental water or wind flow), the impedance matching for obtaining the maximum output power is the primary target. Under a continuous AC force X_0 applied, the input mechanical energy into a piezoelectric transducer changes with the shunted external impedance Z:

$$|P|_{in} = \frac{\omega}{2}\left|s^E[1 - \left(\frac{S}{t}\right)\frac{j\omega Z d^2}{s^E(1+j\omega CZ)}]X_0^2\right| \text{ (S: area, t: electrode gap of piezo component)}$$

This originates from the effective complex elastic compliance under Z shunt expressed as

$$s^E_{eff} = \frac{x}{X} = s^E\left[1 - \left(\frac{S}{t}\right)\frac{j\omega Z d^2}{1+j\omega CZ}\right]$$

Recall that s^E or $s^D = s^E(1-k^2)$, when $Z = 0$ or ∞, respectively.

When the input mechanical energy is limited, the energy transmission coefficient is the Figure of Merit:

$$\lambda_{max} = (\text{Output electrical energy/Input mechanical energy})_{max}$$
$$= \left[(1/k) - \sqrt{(1/k^2)-1}\right]^2 = \left[(1/k) + \sqrt{(1/k^2)-1}\right]^{-2}.$$

- Electronic components relevant to the piezo energy harvesting include AC–DC rectification using p–n junction diodes (half-wave and full-wave rectifiers), MOSFET popularly used for an ON–OFF switching regulator. Because the output impedance from the piezo energy harvesting device is too high to accumulate the energy directly into a rechargeable battery, we should adopt a DC–DC converter for the impedance matching purpose. With respect to DC–DC converters, among the Forward converter, Buck converter, Buck-Boost converter and Flyback converter, the Buck converter is best suited for piezo energy harvesting, from the simplest design and high efficiency viewpoints. A piezoelectric transformer may be an alternative device to be used for modifying the impedance matching in the energy harvesting circuit.

Chapter 7

- Energy flow analysis in the Cymbal harvesting process provides

Source Mechanical Energy		Transducer Mechanical Energy		Transducer Mechanical Energy		Circuit-In Electric Energy		Battery Electric Energy
9.48 J	→ ← 87%	8.22 J	→ ← 9%	0.74 J	→ ← 57%	0.42 J	→ ← 81%	0.34 J

In summary, 0.34 J/9.48 J = 3.6% is the energy harvesting rate from the "vibration source energy" to the "storage battery" in the current system. Taking into account the efficiency of popular amorphous silicon solar cells around 5–9%, the piezoelectric energy harvesting system with 3.6% efficiency is promising.

- One of the million-selling piezo energy harvesting devices includes Programmable Air-Burst Munition (PABM), because the competitive products (button batteries) cannot endure in the severe usage condition (high temperature for a long period) in battlefields. Identifying the application target where the competitive products cannot be adopted is a good strategy for piezoelectric product commercialization.

- A "Piezo Tile" is an example of the commercialization failure: technologically neat, but terrible cost/performance for the final application. Consideration of the "human behavior" is essential: because 10 mm sinking on the floor gives much discomfort to a human, they will skip stamping the "tiles" next time.

Chapter 8

- Though the operation principle is the same, difference between "sensors" and "energy harvesting" is merely the harvesting energy level, that is, lower or higher than 1 mW, respectively. In practice, blood soaking syringe and heart pacemaker require 5–10 mW, and MOSFET in harvesting electric circuit spends 3 mW, a blue-tooth transmission device needs 1 mW minimum. From this sense, the current piezo-MEMS "energy harvesting" research reports 1 μW or less power, which is categorized as a sensor, not as energy harvesting devices. An energy "losing" system, in fact. How we can use this small energy is a serious problem.

- Various coupling phenomena are introduced, which are off-diagonal effects in the material/composite "functionality matrix":

(1) Piezoelectric Coupling — piezoelectric effect, converse piezoelectric effect

$$\begin{cases} x = s^E X + dE \\ D = dX + \varepsilon_0 \varepsilon^X E \end{cases} \quad \text{(piezoelectric constant, elastic compliance, permittivity)}$$

$$k^2 = \frac{d^2}{\varepsilon^X \varepsilon_0 s^E} \quad \text{(electromechanical coupling factor)}.$$

(2) Piezothermal Coupling — linear thermal expansion, piezothermal effect

$$\begin{cases} S = \frac{c_p}{T}\theta + \alpha_L X \\ x = \alpha_L \theta + s^E X \end{cases} \quad \text{(thermal expansion coefficient, specific heat capacity, elastic compliance)}$$

$$k^{PT^2} = \frac{\alpha_L^2}{\left(\frac{c_p}{T}\right) s^E} \quad \text{(piezothermal coupling factor)}.$$

(3) Electrothermal Coupling — pyroelectric effect, electrocaloric effect

$$\begin{cases} S = \dfrac{c_p^E}{T}\theta - pE \\ D = -p\theta + \varepsilon_0 \varepsilon^X E \end{cases} \quad \text{(pyroelectric coefficient, specific heat capacity, permittivity).}$$

$$k^{ET^2} = \frac{p^2}{\left(\frac{c_p^E}{T}\right)\varepsilon_0 \varepsilon^X} \quad \text{(electrothermal coupling factor).}$$

- Composite Effects include the following:

  a. Sum effect — fishing rod with carbon-fiber-reinforced plastic
  b. Combination effect — high g piezo constant with 1 (PZT rod)-3 (plastic) composites
  c. Product effect — magnetoelectric device with magnetostrictive and piezoelectric laminates

- "Functionality matrix" concept is convenient for creating a new phenomenon with product effect. Magnetoelectric devices can be realized by coupling magnetostrictive and piezoelectric materials.

  ➤ 0-0 connectivity magnetoelectric (ME) composites exhibit a 10 times higher ME voltage coefficient at room temperature. The 2-2 connectivity, TerFeNOL:PZT laminates exhibit further significant (80 times higher) ME response, which has triggered the current research boom.
  ➤ Designing principle of ME composites – Optimization was studied in terms of (1) crystal orientation, (2) layer thickness, (3) DC bias magnetic field, etc.
  ➤ Sensor applications of ME composites include (1) magnetic noise sensor, (2) geomagnetic sensor, (3) earthquake prediction and (4) anti-stealth submarine technology.
  ➤ ME energy harvesting can realize an unmanned inspection vehicle for checking the high-power transmission cables.

- A hybrid energy harvesting device which operates under either magnetic and/or mechanical noises was introduced, by coupling magnetostrictive and piezoelectric materials. Other hybrid energy harvesting devices may include the combinations of piezoelectric effect with pyroelectric or photovoltaic effects, or more than two of these effects.

9.3 Research Trend and Future Perspectives

9.3.1 *Research Trend in Piezo Energy Harvesting*

Recent research trends of piezoelectric energy harvesting are categorized into the following four:

(a) Mechanical engineers' approach — (1) Machinery vibration (typically sinusoidal 50–100 Hz) from industrial motors and automobiles. (2) Human motion (typically impulse 1 Hz or less) from human action.
(b) Electrical engineers' approach — DC–DC converter development for matching the impedance to rechargeable batteries.
(c) MEMS engineers' approach — Most of the researchers are using 1-μm-thick piezo films, which can generate only nW — μW level energy.
(d) Military application — Programmable Air-Burst Munition is one the "Million-selling" products.

Because the natural vibration sources, such as wind, water, tidal/ocean current and human motion, and constructed objects such as roads and bridges generate usually low-frequency (0.110 Hz) vibration, off-resonance (or impulse) piezo energy harvesting systems are essential. On the other hand, for machinery such as industrial motors (50 Hz) and automobiles (100–200 Hz), rather constant sinusoidal vibration sources, we may use resonance-type piezo energy harvesting systems. In order to achieve this low-frequency resonance, most of the products use a unimorph/bimorph piezo structure, leading to bulky design and low efficiency, which are the current major problems. The current million-selling energy harvesting devices, such as "Programmable Air-Burst Munition" and "Lightning Switch", are based on the impulse operation type for generating 10 mW level.

Though relatively large investments and research efforts are being put on MEMS/NEMS and "nanoharvesting" devices, a positive comment is not provided at the moment. Even for medical applications, obtained/reported energy level pW–nW from one component (this level is called "sensor", not "energy harvester", in practice) is a useless level, which is originated from the inevitable small volume of the used piezoelectric material (i.e., thin films). The reader is requested to take into account the practically required minimum power levels, 30–100 mW for charging electricity into a battery (DC–DC converter itself spends 2–3 mW), 10–20 mW for soaking blood from a human vessel or 1–3 mW for sending an electronic signal, that is to say, minimum 1 mW is necessary. If you recall the maximum handling energy density level $\sim 30\,\mathrm{mW/mm^3}$ for PZT, in order to generate the minimum usable energy level (~ 1 mW), minimum 10-μm-thick PZT films are required (by taking rather large 3×3 mm^3 area). Current studies on 1-μm films are not helpful for the energy harvesting designs (though suitable for "sensor" applications). Without discovering a genius idea of how to combine thousands of these nanodevices in parallel and synchronously in phase, the current efforts will be in vain. "Nano-grid" research to reach to minimum 1 mW level by connecting thousands of nanoharvesting devices is highly encouraged, rather than merely the MEMS fabrication process from an academic viewpoint.

9.3.2 *Uchino's Frustration on Recent Studies*

As one of the pioneers in piezoelectric energy harvesting, the author feels a sort of frustration on 90% of the recent research papers from the industrial application viewpoint, though they are appreciated for their enormous academic and scientific contributions:

(1) Though the electromechanical coupling factor k is the smallest (i.e., the energy conversion rate from the input mechanical to electric energy is the lowest) among various device configurations, the majority of researchers primarily use the "unimorph" design. Why?
(2) Though the typical noise vibration is in a much lower frequency range, the researchers measure the amplified resonance response (even at a frequency higher than 1 kHz) and report the unrealistically harvested electric energy. Why?
(3) Though the harvested energy is lower than 1 mW, which is lower than the required electric energy to operate a typical energy harvesting electric circuit with a DC/DC converter (typically around 2–3 mW), the researchers report the result as an energy "harvesting" system. Does this situation mean actually energy "losing"? Why?
(4) Few papers have reported complete energy flow or exact efficiency from the input mechanical noise energy to the final electric energy in a rechargeable battery via the piezoelectric transducer step by step. Why?

Interestingly, the unanimous answer from these researchers to my questions "why" is "*because the previous researchers did so*"! As explained already, this is the main motivation to my authoring this textbook. I did not include intentionally studies with the above problems. Since some of the current published papers seem to include strategy-misleading contents as described above, the author tried to select only the necessary papers and fundamental knowledge in this "Essentials of Piezoelectric Energy Harvesting". Commenting on the above "frustrations" is actually my "perspectives" for the future required development directions.

9.3.3 *Future Perspectives*

9.3.3.1 *General development directions*

For the future perspectives, there will be two development directions in piezoelectric energy harvesting areas: (1) Remote signal transmission (such as structure health monitoring) in (mW) power level, and (2) Energy accumulation in rechargeable batteries in (W) power level. Less than W energy harvesting does not attract clients in home appliance and automobile manufacturers. The former will use impulse/snap action mechanisms or low-frequency resonance methods, where the efficiency is not a significant issue, but the minimum spec on the energy should be obtained, while the latter requires dramatic research strategy changes. Recall the keys:

- In the case of "impulse force" with the matched external impedance $Z = 1/\omega C$, the harvesting energy is

$$\text{Energy} = \frac{1}{2}k^2 U_M \sum_{n=0}^{\infty} \left[\left(1 - \frac{1}{2}k^2\right) e^{-\frac{\pi}{2Q_M}} \right]^n = \frac{1}{2}k^2 U_M \frac{1}{1 - \left(1 - \frac{1}{2}k^2\right) e^{-\frac{\pi}{2Q_M}}}. \tag{9.1}$$

For reasonable $Q_M > 50$, this approaches U_M (input mechanical impact energy (J) after multiple "vibration ringing", irrelevant to the k value.
- In the case of "continuous AC force" with the matched external impedance $Z = 1/\omega C$, the input mechanical power W_M (W, J/s) is converted to the stored electrical power by the factor of electromechanical coupling factor k, $W_E = k^2 W_M$. Via the matching resistive external impedance, we can harvest the electric power continuously:

$$\text{Power} = \frac{1}{2}k^2 W_M. \tag{9.2}$$

The researchers are requested to forget the current "biased" knowledge, or not to take a strategy, "because the previous researchers did so". Remembering the fundamental principles explained in this book, challenge a totally unique design and/or idea in the future.

9.3.3.2 *High k design*

Limit the usage of unimorph/bimorph designs, which are suitable only for wearable (elastically soft) and wide area harvesting devices. Remember that these "bending" piezo components exhibit the lowest electromechanical coupling factor k_{eff}. When the input source is continuous sinusoidal low-frequency vibration, the bimorph can be commercialized.

Seek much higher k piezo-component designs. Our piezo-Cymbal array usage from the automobile engine vibration has been fully introduced in this textbook, where basically k_p (planar coupling) is used, 70% $\left(\sqrt{\frac{2}{(1-\sigma)}} = 1.7\right)$ higher than k_{31} (base coupling of a bimorph). In the future, the highest k piezo component designs k_{33} and k_t should be challenged. However, these designs are too elastically stiff (mechanical impedance mismatch); we need to consider a so-called *mechanical transformer* which converts the mechanical impedance from PZT high to vibration source low, without losing the energy transfer efficiency via mechanical levers. The mechanical transformer designs with a multilayer (ML) actuator are proposed in Fig. 9.1.

Figure 9.1 summarizes new design concepts:

(a) Unimorph array with wide-frequency-range coverage.
(b) Unimorph topologically long beam design with a very low resonance frequency — See Section 9.3.3.3 for the detailed explanation.
(c) ML & a hinge lever — A similar design to an x-axis stage, used for a reverse application; energy harvesting from in-plane vibration.

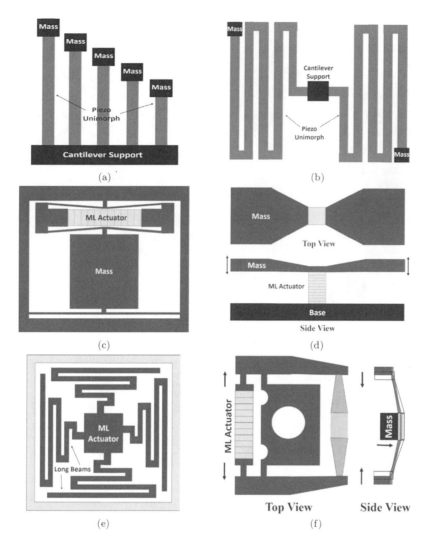

Fig. 9.1: Summary of new design concepts: (a) Unimorph array with wide-frequency-range coverage, (b) Unimorph topologically long-beam design with a very low resonance frequency, (c) ML and a hinge lever for in-plane vibration, (d) ML and wings for out-of-plane vibration, (e) ML and topologically long-wing design with a very low resonance frequency, (f) ML and hinge lever for out-of-plane vibration.

(d) ML and wings — Reverse application of a piezo flying drone, for harvesting from out-of-plane vibration.

(e) ML and topologically long-wing design with a very low resonance frequency.

(f) ML and hinge lever for out-of-plane vibration — Basic mechanical design to change the force and displacement directions by 90°.

As illustrated in Fig. 5.48 already, using an ML-based design, we can expect 10–30 times higher harvesting energy in comparison with the conventional bimorph/unimorph designs.

9.3.3.3 *Resonance or off-resonance usage*

Under off-resonance operation, high k designs are definitely better. However, under resonance operation, such as on a machine or automobile, a bimorph/unimorph design can be adopted if its resonance frequency is adjusted to the low frequency of the vibration source. Figure 9.1 shows some design concepts:

(a) Array of unimorphs with various lengths for various resonance frequencies for wide-frequency-range coverage, which may be suitable for the vibration source with wide frequency deviation.

(b) Unimorph design of topologically long beams with a very low resonance frequency, which is theoretically possible, but the application area is limited practically because of its design fragility (piezo ceramic film may fracture under shaking).

9.3.3.4 *Much higher than 1 mW*

Regarding the thin film MEMS devices, since nW–μW level is useless, two directions should be sought:

(a) Thick film (10–30 μm) — Recall the maximum handling energy density level $\sim$30 mW/mm^3 for PZT. In order to generate the minimum usable energy level ($\sim$1 mW), 10-μm-thick PZT films are required (by taking rather large 3×3 mm^3 area).
(b) Without discovering a genius idea of how to combine thousands of these nanodevices in parallel and synchronously in phase, the current efforts will be in vain. "Nanogrid" research to reach to minimum 1 mW level by connecting thousands of nano harvesting devices is highly encouraged, rather than merely the MEMS fabrication process from an academic viewpoint.

9.3.3.5 *Total energy flow analysis*

The reader learned the energy flow analysis in the Cymbal harvesting process (off-resonance) as follows:

Source Mechanical Energy		Transducer Mechanical Energy		Transducer Mechanical Energy		Circuit-In Electric Energy		Battery Electric Energy
9.48 J	$\rightarrow$ $\leftarrow$ 87%	8.22 J	$\rightarrow$ $\leftarrow$ 9%	0.74 J	$\rightarrow$ $\leftarrow$ 57%	0.42 J	$\rightarrow$ $\leftarrow$ 81%	0.34 J

In summary, 0.34 J/9.48 J = 3.6% is the energy harvesting rate from the "vibration source energy" to the "storage battery" in the current system. Taking into account the efficiency of popular amorphous silicon solar cells around 5–9%, the piezoelectric energy harvesting system with 3.6% efficiency is promising. Note that the most serious bottleneck exists on the electromechanical coupling k_{eff}^2. The Cymbal component with $k_{\text{eff}} = 30\%$ can cultivate the energy 10 times higher than the unimorph/bimorph designs. The total energy flow analysis helps to identify the bottleneck clearly in the energy harvesting system. If we adopt the ML actuator with $k_{\text{eff}} = 70\%$, and supposing no loss via the ideal "mechanical transformer", we can expect 5-time enhancement, leading to 18% efficiency, equivalent to the best solar cell's achievement.

Chapter Essentials

1. Three phases/steps for piezoelectric energy harvesting are (1) mechanical impedance matching, (2) electromechanical transduction and (3) electrical impedance matching.
2. The Cymbal transducer is employed for energy harvesting from a high-power mechanical vibration, while the MFC is suitable for a small flexible energy vibration.
3. Harvesting energy calculation:
   - "Impulse force" (J) with the matched external impedance $Z = 1/\omega C$ — the harvesting energy is

$$\text{Energy} = \frac{1}{2}k^2 U_M \frac{1}{1 - \left(1 - \frac{1}{2}k^2\right)e^{-\frac{\pi}{2Q_M}}}$$

   For reasonable $Q_M > 50$, this approaches U_M (input mechanical impact energy (J)) after multiple "vibration ringing", irrelevant to the k value.
   - "Continuous AC force" with the matched external impedance $Z = 1/\omega C$ — the input mechanical power W_M (W, J/s) and the stored electrical power $W_E = k^2 W_M$. The harvesting electric power

$$\text{Power} = \frac{1}{2}k^2 W_M.$$

4. A Buck-Converter is used as the DC–DC converter for realizing better electrical impedance matching. Further improvement can be realized by using a highly capacitive multilayer design. A piezoelectric transformer is a promising alternative component for the DC–DC converter.

5. The key to dramatic enhancement in the efficiency is to use a high k mode, such as k_{33}, k_t, or k_{15}, rather than flex-tensional modes. Figure 9.1 summarizes promising piezoelectric device designs for energy harvesting applications. The hinge-lever mechanism is an ideal "mechanical transformer" without losing mechanical energy under an off-resonance condition, and it is easily manufactured using micro machining technologies, which may expand the tunability of the mechanical impedance by keeping a high electromechanical coupling factor k in the range of $k_{33} \approx 70\%$.

6. A hybrid energy harvesting device which operates under either magnetic and/or mechanical noises was introduced, by coupling magnetostrictive and piezoelectric materials.

7. Saving energy has become very important in recent years, and the analyses and results in this article can be applied to recover and store the wasted or unused mechanical (and/or magnetic) energy efficiently.

Check Point

1. (T/F) When 1 W mechanical energy is input continuously on a piezoelectric with an electromechanical coupling factor k, we can expect k^2 W electrical energy converted in this piezo material. Thus, this piezoelectric can harvest up to k^2 W electrically to the outside load. True or False?

2. There are three key factors to be considered for developing efficient piezoelectric energy harvesting systems: (1) mechanical impedance matching, (2) electromechanical transduction and (3) electrical impedance matching. Which factor is the primary problem for the inefficiency in the following systems?

   - A piezoelectric energy harvesting bimorph component is connected to a rechargeable battery immediately after a full rectification circuit.
   - A PZT thin film membrane on a Silicon wafer (MEMS) with a unimorph configuration.
   - A PVDF film bonded directly on a thick steel beam.
   - A PZT bulk disk attached directly on a human body, say, for energy harvesting for a pacemaker.

3. (T/F) Elastic compliances are the material's constants in a piezoelectric. Thus, the resonance frequency is determined merely by the sample size, irrelevant to the external electrical load (L, C or R). True or False?

4. (T/F) When the loss of the electronic component is negligibly small, ideally "Input power = Output power" is expected in a *Buck Converter*. If we take the duty ratio (On/Off time period ratio) 2%, the impedance can be reduced by 50 theoretically. True or False?

5. (T/F) When 1 J "impulse" mechanical energy is input on a piezoelectric with an electromechanical coupling factor k, we can expect k^2 J electrical energy converted in this piezo material. Thus, this piezoelectric system can harvest up to k^2 J electrically to the outside load, after accumulating all cycle energy during vibration ringing. True or False?

6. (T/F) In order to develop compact energy harvesting devices, we had better develop the piezo component which can generate the high output electric energy at its resonance frequency (40 kHz). True or False?

7. (T/F) The unimorph piezoelectric structure is most popularly used, because it exhibits the highest energy harvesting rate among various piezo component designs. True or False?

8. What type of piezoelectric component design exhibits the highest electromechanical coupling factor?

9. Calculate the electrical impedance $1/j\omega C$ of a piezoelectric component with capacitance 1 nF at the off-resonance frequency 100 Hz, which is larger than the internal impedance of rechargeable batteries ($\sim$50Ω).

10. (T/F) When we denote the dynamic electromechanical coupling factor as k_{vn} for the nth higher order harmonics, the static electromechanical coupling factor k_v (i.e., off-resonance drive) can be obtained by $k_v^2 = \sum_n k_{vn}^2$. True or False?

Chapter Problems

9.1 Let us consider a piezoelectric component for energy harvesting, the capacitance of which is C with the electrode gap t and area S. We first consider the electrical energy output from the constitutive equations

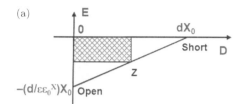

$$\begin{pmatrix} D \\ x \end{pmatrix} = \begin{pmatrix} \varepsilon_0\varepsilon & d \\ d & s^E \end{pmatrix} \begin{pmatrix} E \\ X \end{pmatrix} \quad \text{(P9.1)}$$

The short-circuit condition ($E = 0$) gives $D_0 = dX_0$, while the open-circuit condition ($D = 0$) gives $E_0 = -dX_0/\varepsilon_0\varepsilon$ (i.e., depolarization field). The triangular areas of OD_0E_0 $\left(\frac{1}{2}dX_0 \cdot \left(\frac{dX_0}{\varepsilon_0\varepsilon}\right) = \frac{1}{2}\frac{(dX_0)^2}{\varepsilon_0\varepsilon}\right)$ mean the total electric energy converted from the mechanical vibration (Fig. 9.2). Thus, under an impedance Z shunt condition, we can expect a point (E, D) on the line between the terminals of the above D_0 and E_0:

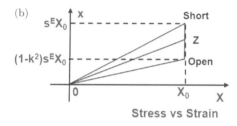

Fig. 9.2: Energy harvesting calculation process: (a) output electrical energy, and (b) input mechanical energy under various external resistances.

$$E = \frac{1}{\varepsilon_0\varepsilon}(D - dX_0). \quad \text{(P9.2)}$$

The output electrical energy can be calculated as

$$U = -DE = -\frac{1}{\varepsilon_0\varepsilon}\left(D - \frac{1}{2}dX_0\right)^2 + \frac{1}{4}\frac{(dX_0)^2}{\varepsilon_0\varepsilon}. \quad (9.3)$$

Note that the rectangular area in the right figure generated by the point (E, D) stands for the output electrical energy on the load Z. It is easy to maximize this output energy by taking the point at the middle of maximum D_0 and E_0; that is the maximum energy (per unit volume) is given by $|U|_{\max} = |\frac{D_0}{2} \cdot \frac{E_0}{2}| = \frac{1}{4}\frac{(dX_0)^2}{\varepsilon_0\varepsilon}$. The same results can be obtained from the electric power calculation. We start from

$$\dot{D} = \varepsilon_0\varepsilon\dot{E} + d\dot{X}. \quad (9.4)$$

Knowing $\dot{D}S = I$, $\dot{E} = j\omega E$, $\dot{X} = j\omega X$, and $E = -V/t$,

$$I = -j\omega\varepsilon_0\varepsilon\left(\frac{S}{t}\right)V + j\omega dXS \quad (9.5)$$

Integrating $I = V/Z$, $C = \varepsilon_0\varepsilon\left(\frac{S}{t}\right)$, and $XS = X_0$, we obtain

$$V = j\omega dX_0/\left(\frac{1}{Z} + j\omega C\right) \quad (9.6)$$

Then, the output power via the load Z is calculated as

$$|P|_{\text{out}} = \text{Re}\left\{ \frac{1}{2} \left| \frac{V\, V^*}{Z^*} \right| \right\} = \frac{1}{2} Z \frac{(\omega d X_0)^2}{(1 + (\omega C Z)^2)} \tag{9.7}$$

which is exactly the same result as Eq. (P9.3). We calculate now the "input mechanical energy" from the second constitutive equation:

$$x = dE + s^E X = -d\left(\frac{V}{t}\right) + s^E X = -\left(\frac{d}{t}\right)\left[\frac{j\omega d X_0}{\frac{1}{Z} + j\omega C}\right] + s^E X \tag{9.8}$$

The last transformation used Eq. (P9.5). We can obtain *effective complex elastic compliance* as

$$s^E_{\text{eff}} = \frac{x}{X} = s^E \left[1 - \left(\frac{S}{t}\right) \frac{j\omega Z d^2}{s^E(1 + j\omega C Z)} \right] \tag{9.9}$$

Now, questions:

(a) Calculate the above "effective elastic compliance" for the following five cases:

- $Z = 0$
- $Z = \infty$
- $Z = 1/\omega C$ (resistive impedance matching)
- $Z = -1/j\omega C$ (conjugate impedance matching)
- $Z = 1/j\omega C$

(b) Calculate the input mechanical energy under the impedance Z.

Hint:

(a) For $Z = 0$ or ∞, s^E or $s^E_{\text{eff}} = s^E[1 - k^2] = s^D$, respectively. Then, for $Z = 1/\omega C$, we obtain $s^E_{\text{eff}} = s^E[1 - k^2 \frac{j}{1+j}]$. Taking the absolute value, $s^E_{\text{eff}} = s^E \sqrt{1 - k^2 + \frac{1}{2}k^4}$. When k is small, $s^E_{\text{eff}} = s^E \left(1 - \frac{1}{2}k^2\right)$.

(b) The input mechanical power is derived as

$$|P|_{\text{in}} = \frac{\omega}{2} \left| s^E [1 - \left(\frac{S}{t}\right) \frac{j\omega Z d^2}{s^E(1 + j\omega C Z)}] X_0^2 \right|. \tag{P9.10}$$

References

1. K. Uchino, *Global Crisis and Sustainability Technologies*, World Scientific Pub., Toh Tuck Link, Singapore (2016), ISBN: 9789813142299.
2. K. Uchino, *Ferroelectric Devices 2nd Edition*, CRC Press, Boca Raton, FL (2010), ISBN: 9781439803752.
3. K. Uchino, *Micromechatronics 2nd Edition*, CRC Press, Boca Raton, FL (2020), ISBN-13: 978-0-367-20231-6.

Appendix

Answers To "Check Point"

Chapter 1

1) False (energy level is too low)
2) False ($< 10\,\text{Hz}$ response is important)
3) False
4) $1.6\,\text{M}\Omega$ (phase $-90°$)
5) False

Chapter 2

1) False (no d_{13}, but d_{31})
2) Electric field direction along 1-axis
3) False
4) False (It is $E = -(\frac{P_S}{\varepsilon_0})$)
5) 70%
6) $1\,\text{kN}$
7) $100\,\text{kHz}$
8) False (max ε)
9) False ($1/2\zeta$)
10) True
11) False ($40\,\text{dB}$)
12) False (counter-)
13) False (combination with mechanical loss)
14) True
15) False
16) False
17) False
18) $Q_M = 1/\tan\phi'$
19) False
20) Q_B
21) False (inversely related)
22) True
23) $Q = \sqrt{L/C}/R$
24) $[K]^{-1} = [K] = \frac{1}{1-k^2} \begin{bmatrix} 1 & k^2 & -2k^2 \\ k^2 & 1 & -2k^2 \\ 1 & 1 & -1-k^2 \end{bmatrix}$

Chapter 3

1) True
2) False (50%)
3) False
4) $1.6\,\text{M}\Omega$ (phase $-90°$)
5) $0.125\,\text{J}$ $((\frac{1}{2})k^2)$
6) No. Resonance is generated without damping.
7) True
8) False (too soft)
9) True
10) U_M

Chapter 4

1) False ($\sqrt{\rho c}$)
2) 20 Mrayls (*use* $\sqrt{\rho/s}$ for evaluation)
3) False
4) Rayl(s)
5) $v_p = \sqrt{c/\rho}$
6) Matching layer
7) False
8) $100\,\text{MPa}$ ($10\,\text{MPa}$ is also okay, if we consider the risk factor)
9) False (Too large strain which makes ceramic crack)
10) False

Chapter 5

1) Small volume of PZT (too thin) limits the total energy less than μW (useless level $< 1\,\text{mW}$).
2) True
3) False
4) False
5) $1.6\,\text{M}\Omega$ (phase $-90°$)
6) $d \cdot g$
7) k
8) False (Triangular displacement)
9) Linearly with 100% overshoot rate
10) (b) (Lower resonance frequency)
11) False (approaches U_M)
12) False
13) Right-side (note the negative capacitance)
14) $k_{31} = 33\%$, $k_p = 56\%$
15) False (Cross section area)
16) False (Bending Res. freq. ratio is not a simple integer.)
17) False ($\sigma = 0.3$)
18) k_{33} ($d_{31}/d_{33} = 1/3$)
19) k_p $\left(\sqrt{\frac{2}{1-\sigma}}k_{31}\right)$

20) False (diameter/depth)
21) True

Chapter 6

1) True
2) 0.5 kJoule
3) $1.6\,M\Omega$ (phase $-90°$)
4) DC-DC converter
5) Buck converter
6) Diode
7) False
8) True
9) False ($50 \rightarrow 2500$)
10) $s^E(1 - \frac{k^2}{2})$

Chapter 7

1) No. Because the total PZT volume is limited in thin films.
2) k_{33} type is better because $k_{33} > |k_{31}|$
3) Small k_{eff} value than other structures
4) True
5) True
6) False (Recall the cost-performance.)
7) True
8) $30\,W/cm^3$ power density in PZT, above which heat generation is significant.

Chapter 8

1) Output electric energy level; lower or higher than $1\,mW$
2) 10
3) Sum effect
4) Parallel connectivity
5) False
6) True
7) Electrocaloric effect
8) False (It should be negative field.)
9) Terfenol-D & PZT
10) True

Chapter 9

1) False (should be $\frac{1}{2}k^2$)
2) (3) elect. imp., (2) elect. mech., (1) mech. imp., (1) mech. imp.
3) False (external load dependent)
4) False (by 2500)
5) False (We can obtain all input 1 J)

6) False (40 kHz is too high frequency)
7) False (lowest k design)
8) k_{33} type piezo-rod
9) $1.6\,M\Omega$ (phase $-90°$)
10) True

Index

305

CPSIA information can be obtained
at www.ICGtesting.com
Printed in the USA
JSRC030402210421
13637JS00005B/7